Lecture Notes in Artificial Intelligence   1930

Subseries of Lecture Notes in Computer Science
Edited by J. G. Carbonell and J. Siekmann

Lecture Notes in Computer Science

Edited by G. Goos, J. Hartmanis and J. van Leeuwen

Springer

Berlin
Heidelberg
New York
Barcelona
Hong Kong
London
Milan
Paris
Singapore
Tokyo

John A. Campbell
Eugenio Roanes-Lozano (Eds.)

Artificial Intelligence and Symbolic Computation

International Conference AISC 2000
Madrid, Spain, July 17-19, 2000
Revised Papers

 Springer

Series Editors

Jaime G. Carbonell, Carnegie Mellon University, Pittsburgh, PA, USA
Jörg Siekmann, University of Saarland, Saabrücken, Germany

Volume Editors

John A. Campbell
University College London
Gower Street, WC1E 6BT London, United Kingdom
E-mail: jac@cs.ucl.ac.uk

Eugenio Roanes-Lozano
Universidad Complutense de Madrid, Dept. Algebra, Fac. Educación
c/ Rector Royo Villanova s/n, 28040 Madrid, Spain
E-mail: eroanes@eucmos.sim.ucm.es

Cataloging-in-Publication Data applied for

Die Deutsche Bibliothek - CIP-Einheitsaufnahme

Artificial intelligence and symbolic computation : revised papers /
International Conference AISC 2000, Madrid, Spain, July 17 - 19, 2000.
John A. Campbell ; Eugenio Roanes-Lozano (ed.). - Berlin ; Heidelberg ;
New York ; Barcelona ; Hong Kong ; London ; Milan ; Paris ; Singapore ;
Tokyo : Springer, 2001
    (Lecture notes in computer science ; Vol. 1930 : Lecture notes in
    artificial intelligence)
    ISBN 3-540-42071-1

CR Subject Classification (1998): I.2.1-4, I.1, G.1-2, F.4.1

ISBN 3-540-42071-1 Springer-Verlag Berlin Heidelberg New York

Springer-Verlag Berlin Heidelberg New York
a member of BertelsmannSpringer Science+Business Media GmbH

http://www.springer.de

© Springer-Verlag Berlin Heidelberg 2001
Printed in Germany

Typesetting: Camera-ready by author, data conversion by Boller Mediendesign
Printed on acid-free paper   SPIN: 10781137   06/3142   5 4 3 2 1 0

Foreword

This volume contains the invited papers, contributed papers, and poster summaries accepted for the Fifth International Conference on Artificial Intelligence and Symbolic Computation (AISC 2000). The conference was held in Spain from 17 to 19 July 2000 at the Hotel NH Zurbano, Madrid, and was organized by the Universidad Complutense de Madrid and the Sociedad Matemática Puig Adam.

One of the reasons for centralizing all activities at one hotel was to avoid both losing time in transportation around Madrid and the distribution of the attendees into disconnected subgroups in diverse locations. In this way a breakfast-to-late-night coexistence (which included some extra-academic events) was ensured, with time for formal and informal conversations. This continued the AISMC/AISC tradition of the creation of a friendly atmosphere, where ideas could be exchanged in a relaxed and effective way.

The conference belongs to a specialized conference series founded by John Campbell and Jacques Calmet with the initial title "Artificial Intelligence and Symbolic Mathematical Computation" (AISMC). AISMC-1 took place in 1992 in Karlsruhe (Germany); AISMC-2 was held in 1994 at King's College (Cambridge, UK), and AISMC-3 in 1996 was located in Steyr (Austria). The proceedings of these conferences were published in Springer's LNCS series as volumes 737, 958, and 1138, respectively.

The Steering Committee then decided to drop the word "Mathematical" from the name of the conference series (and the "M" in the acronym) to emphasize that the conference was not only related to Mathematics but to all aspects of symbolic computation. Therefore, the proceedings from that time onwards have been transferred to Springer's LNAI series from LNCS. Our first conference after that decision, AISC'98, took place in Plattsburgh (NY, USA) during 1998. Its papers appeared as volume 1476 of the LNAI series.

The next conference in the same field, AISC 2002, will be held in Nice (France).

The field includes a wide range of activities, such as Automated Theorem Proving, Logical Reasoning, Mathematical Modeling of Multi-agent Systems, Expert Systems and Machine Learning, Engineering, and Industrial Applications. Despite this breadth of coverage, the program committee has (as in previous conferences) kept the number of accepted papers low, following a strict refereeing process, to avoid any necessity for parallel sessions and to allow longer than usual presentations and periods for questions and discussion. A poster session was included in 2000 for the first time in an AISC conference; short accounts of some of the research covered in posters is included here.

In some of the past AISMC/AISC volumes and in papers in related journals such as the Annals of Mathematics and Artificial Intelligence, forecasts have been made about the areas of likely and promising future research in topics that the conferences cover. As the saying goes, "it is always difficult to predict, especially to predict the future"; so, it is not embarrassing to observe that some of the predictions have still to match what is actually happening. But others, such as the explicit representation (for computer-based use) of mathematical knowledge, are now emerging, as the present volume shows. The fact that this issue of representation has been encouraged indirectly by the expansion of the World-Wide Web and more directly by the existence of HTML and various subsequent ....MLs was not something that was predicted – but such unexpected synergies show that the AISC area is not only alive and well, but is still capable of generating pleasant scientific surprises.

The book begins with the papers from the three invited speakers. The papers that follow have been grouped so that the topics of successive items are as close together as possible.

We acknowledge gratefully the generous sponsors of AISC 2000: the Universidad Complutense de Madrid (through different sources: the "Convocatoria 1999 para la Organización de Reuniones, Congresos y Seminarios", its Department of Algebra and its "Servicios Informáticos"), the Spanish software sales representative "Addlink Software Científico", the companies Texas Instruments and Logic Programming Associates (LPA), the "Real Sociedad Matemática Española", and the "Sociedad Matemática Puig Adam".

We also express our warm thanks to the members of the Steering Committee and Scientific Committee for refereeing contributed papers and their most valuable help in making AISC 2000 a success. We would like to thank additionally the members of the Local Committee, who faced all the "behind the curtains" work, and especially Professor Roanes-Macías for taking care of all the unpleasant and endless economic details. Finally we put on record our thanks to the director of the corresponding BBVA bank office (Mr. Miguel Santos) for his kindness and efficiency; the travel agency "Viajes Eurobusiness" for their good work and for trusting the local organizers of the conference by not asking for the usual imposing financial guarantees in advance, and the management and staff of the Hotel NH Zurbano for the facilities provided for the conference.

January 2001   John Campbell
                                                                   Eugenio Roanes-Lozano

Organization

Steering Committee

Jacques Calmet (Univ. Karlsruhe, Germany)
John Campbell (Univ. College, London, United Kingdom)
Jochen Pfalzgraf (Univ. Salzburg, Austria)
Jan Plaza (SUNY-Plattsburgh, USA)
Eugenio Roanes-Lozano (Univ. Complutense de Madrid, Spain),
Conference Chairman

Program Committee

Luigia C. Aiello (Univ. La Sapienza, Roma, Italy)
José A. Alonso (Univ. de Sevilla, Spain)
Michael Beeson (San Jose State Univ., USA)
Bruno Buchberger (RISC-Linz, Austria)
Alan Bundy (Univ. Edinburgh, UK)
Greg Butler (Univ. Concordia, Montreal, Canada)
Frans P. Coenen (Univ. of Liverpool, UK)
Jim Cunningham (Imperial College London, UK)
Carl van Geem (LAAS-CNRS, Toulouse, France)
Fausto Giunchiglia (Univ. Trento, Italy)
Martin C. Golumbic (Bar-Ilan Univ., Israel)
Leon A. Gonzalez (Univ. de Alcala, Spain)
Anthony J. Guttman (Univ. of Melbourne, Australia)
Reiner Haehnle (Univ. Karlsruhe, Germany)
Deepak Kapur (Univ. New Mexico, USA)
Luis de Ledesma (Univ. Politecnica de Madrid, Spain)
Ursula Martin (Univ. of St. Andrews/SRI, Scottland)
José Mira (UNED, Spain)
Ewa Orlowska (Inst. Telecommunications, Warsaw, Poland)
Angel Pasqual del Pobil (Univ. Jaume I, Spain)
Juan Pazos (Univ. Politecnica de Madrid, Spain)
Zbigniew W. Ras (Univ. North Carolina, Charlotte, USA)
Tomas Recio (Univ. de Santander, Spain)
Joerg Siekmann (Univ. Saarland, Saarbrücken, Germany)
Andrzej Skowron (Warsaw Univ., Poland)
Viorica Sofronie-Stokkermans (Max Planck Institut, Germany)
Stanly Steinberg (Univ. New Mexico, USA)
Karel Stokkermans (Univ. Salzburg, Austria)
Carolyn Talcott (Stanford Univ., USA)
Rich Thomason (Univ. of Pittsburgh, USA)

Peder Thusgaard Ruhoff (Univ. of Southern Denmark, Denmark)
Enric Trillas (Univ. Politecnica de Madrid, Spain)
Dongming Wang (IMAG Grenoble, France)

Local Committee

Francisco J. Blanco-Silva (Purdue Univ., USA)
Martin Garbayo (Univ. Complutense de Madrid, Spain)
Mercedes Hidalgo (Univ. Complutense de Madrid, Spain)
Eugenio Roanes-Macias (Univ. Complutense de Madrid, Spain)
Lola Rodriguez (CPR Leganes)

Table of Contents

George Boole, a Forerunner of Symbolic Computation*

Luis M. Laita[1], Luis de Ledesma[1], Eugenio Roanes-Lozano[2], and
Alberto Brunori[1]

[1] Universidad Politécnica de Madrid, Dept. Artificial Intelligence, Campus de
Montegancedo, Boadilla del Monte, 28660-Madrid, Spain
[2] Universidad Complutense de Madrid, Dept. Algebra, Edificio "Almudena",
c/ Rector Royo Villanova s/n, 28040-Madrid, Spain

Abstract. We examine in this invited presentation Boole's principles of
logic and his method of performing inferences. The principles of Boole's
logic are based on the application of an early symbolic calculus known
in his time as the "method of separation of symbols". His logic's inference procedures are symbolic operations allowed inside this method.
Such inference procedures are reinterpreted and generalized using computer algebra. The lecture also presents a short biography of Boole and
a description of some of the factors that had an influence on the genesis
of his logic.

1 Introduction

George Boole is recognized as one of the precursors of mathematical logic. Nevertheless, some more insight on the genesis of his logic leads one to think that
Boole was also a forerunner of important developments in symbolic computation. He, clearly, could not use computers, but he suggested methods that can be
translated to interesting modern computer algebra results. Section 3 deals with
the influence of Boole's own work in a symbolic method called "the method of
separation of symbols" on the making of his first book on logic, The Mathematical Analysis of Logic [10] (to be hereinafter denoted as MAL). A computer
algebra translation of Boole's inference procedures is provided. This section is
an outline of the more elaborated arguments we have presented in [36].

A relevant part of this lecture (section 2) is dedicated to present a condensed
biography of Boole which includes a short account of two items that deal with influences on the genesis of his logic: the controversy held between the philosopher
William Hamilton and the mathematician Augustus De Morgan and a curious
outlook of Boole that can be extracted from the examination of the writings of
his wife Mary Everest (biographies of Mary Everest are [14] and [15]). These two
items are mentioned to stress how factors, external and objective and internal
and subjective respectively, influence scientific creation. An excellent biography
of Boole is [39]; this and other biographies of Boole, such as [31] and [17], are

---

* Partially supported by project DGES PB96-0098-C04 (Spain).

J.A. Campbell and E. Roanes-Lozano (Eds.): AISC 2000, LNAI 1930, pp. 1–19, 2001.
© Springer-Verlag Berlin Heidelberg 2001

based, mainly but not only, on the one written by Harley shortly after Boole died [27] and on the account of Boole's life given by Mary Everest in "Home Side of a Scientific Mind" (to be cited hereinafter as "HS") [18]. The controversy and Everest's views have been described in detail in [33] and [34] .

2   Aspects of Boole's Life

George Boole was born in Lincoln, England, on the 2nd of November, 1815.

Because of the poverty of his family, his formal education was minimal. His fellow students considered him to be something of a genius ([27], p. 428).

By the age of twelve, George's interests moved from the elementary science taught to him by his father to languages. This early training in languages had its share among the influences which led to the construction of Boole's logic: he built his logic in the same way he felt languages were built.

Several biographers of Boole and Mary Everest tell about his desire when he was about fourteen years old to enter the ministry of the Anglican Church. Curiously, the only one who does not mention this is precisely his first and most reliable biographer, Harley. In a way it seems that Harley tried to avoid all references to Boole's religious attitudes. Nevertheless he quoted in a footnote a letter that Boole wrote to him from London early in 1864. The letter is revealing in regard to Boole's religious feelings, if compared with the accounts of this issue given later by Mary Everest:

> (...) I have just returned from hearing Maurice. To say that I was pleased is to say nothing (...). But I should not express my real feeling if I said less than that I listened to him with a sense of awe. (...) I feel with you that I should not like to leave the Church while Maurice is in it ([27], 460).

Maurice was a preacher of a kind of Christian-socialist theory, who was very much admired by Boole. The last statement of Boole's letter implies that he had doubts about his continuing in the Church, or that at least he questioned some of her teachings.

Whether or not Boole had thought of an ecclesiastical career, it is known that he did not carry out his ideas. Instead he became a teacher, successively in Doncaster, Waddington, and Lincoln.

Boole was a successful teacher. His evenings were spent in the study of mathematics. According to Mary Everest (HS, p. 6) and other biographers, Newton, Lagrange, Laplace, Dirichlet, Jacobi and Cauchy were studied very thoroughly by him without any other help than his own will. They say that the works of these mathematicians were available at the Mechanics Institute of Lincoln, an institution founded by a local squire. MacHale says that Boole had rather started his mathematical studies with Lacroix' book "Differential and Integral Calculus" and he later regretted it. In any case, Boole's first publications in The Cambridge Mathematical Journal (see next section) testify that he had mastered Lagrange

and Laplace while he was very young, no matter whether he had studied these
authors or not in the Mechanical Institute.

In 1835 he gave an address in the Mechanics Institute "On the Genius and
Discoveries of Sir Isaac Newton" [3]. One reads in Boole's address:

> *There was yet another disadvantage attaching to the whole of Newton's
> physical inquiries, (...) the want of an appropriate notation for express-
> ing the conditions of a dynamical problem, and the general principles
> by which its solution must be obtained. By the labours of LaGrange, the
> motions of a disturbed planet are reduced with all their complication and
> variety to a purely mathematical question. It then ceases to be a physical
> problem; the disturbed and disturbing planet are alike vanished; the ideas
> of time and force are at an end; the very elements of the orbit have dis-
> appeared, or only exist as arbitrary characters in a mathematical formula
> ([3], p. 6).*

This quotation shows that Boole, already at the early age of twenty, had
grasped the ideas which were at the base of his whole methodology: first, that
a good symbolism was a necessary tool for the advancement of mathematical
knowledge and secondly, that mathematical manipulation of symbols could be
separated from interpretation at the intermediate steps of proofs.

Boole made a trip in 1839 to Cambridge, where he contacted the Scottish
mathematician Duncan F. Gregory, founder of the Cambridge Mathematical
Journal. In 1841 Gregory published two of Boole's papers (see section 3). In 1844,
his paper "On a General Method in Analysis" was published in the Transactions
of the Royal Society of London, and awarded the Royal Medal. He maintained
periodic contact with mathematicians. Gregory died in 1844, but Boole had met
the mathematician and logician Augustus De Morgan in 1842. From that time
on they had a most cordial relationship.

In 1846 a controversy about a logical issue (the quantification of predicates)
arose between De Morgan and the Scottish philosopher Sir William Hamilton.
Boole became very interested in it and decided to work out a system of his own;
the result was MAL. According to De Morgan, this book appeared in public the
same day that he published his Formal Logic.

William Hamilton (Hamilton's logical ideas can be found in [40]) had an
astonishing erudition referred to many branches of knowledge. Nevertheless, he
had a curious dislike of mathematics. In 1836 he published a paper [26] in re-
sponse to another one written by William Whewell [45], which dealt with the
importance of mathematics in a liberal education.

Hamilton's paper is worthy of careful consideration because some of the ideas
that appeared in it were reflected in the Introduction to MAL. The paper attacks
the opinion that the study of mathematics is important in a liberal education.
Mathematicians themselves, in Hamilton's opinion, were not able to reach an
agreement in regard to the value of mathematics as a gymnastic of the mind.
Some of them held the opinion that analysis does not constitute such a gymnastic
because it transports the student mechanically to the conclusions, whereas the

ancient geometrical constructions led to the end with a clear consciousness of every step in the procedure. Others, on the contrary, held the view that the methods of geometry are tedious. As a result, they recommended the algebraic methods as the most favourable to the powers of generalization.

After having assigned to mathematics a limited place inside logic, Hamilton proceeded to make the following assertions.

(a) It is wholly beyond the domain of mathematics to inquire into the origins and nature of its own principles.
(b) Mathematics does not say anything about necessary truths, but rather about necessary inferences.
(c) The stress on such one-sided disciplines as mathematics produces a disproportionate development of one power at the expense of others.
(d) No other discipline tends to cultivate a smaller number of faculties, in a more partial or feeble manner, than mathematics.

To support these assertions, Hamilton gave a display of erudition, citing author after author from antiquity to his own time, who had held opinions similar to his. One reads in Hamilton's paper that "mathematics are only difficult because they are too easy", so that no pleasure is found when studying mathematics. Hamilton ended his paper with a bitter criticism of the plan of studies followed at Cambridge University at that time.

Boole explicitly mentioned Hamilton's paper in several places of MAL (MAL, pp. 11-14 and p. 81). A footnote in the Introduction to this book shows that Boole had considered Hamilton's arguments very carefully, even to the point of discovering a mistake in one of Hamilton's quotations (MAL, p. 12).

When discussing Hamilton's paper, Boole implied that Hamilton's arguments were incorrect because they were based on the opinion that philosophy deals with causes while sciences deal with the investigation of laws.

Boole's argument was that if the search for causes is a task that does not transcend the limits of the human intellect, and that if the nature of philosophy is that search, then logic forms no part of philosophy. It is at this point that Boole made the statement which lies at the base of his whole system of logic: that logic should not be associated with philosophy but with mathematics.

Regarding Hamilton's statement that to inquire into the origin and nature of its own principles is beyond the domain of mathematics, Boole wrote that if this is so, then the same should be stated of logic. But for him, as for Hamilton, logic "not only constructs a science but also inquires into the nature and origins of its own principles" (MAL, p. 12). Thus his conclusion was that mathematics also has the power and right to inquire into its origins and nature.

Boole had also discussed, in the Introduction of the book, the issue of the relevance of symbols in scientific expositions. His aim was probably to correct Hamilton's opinion that the symbolization of mathematics could lead to the destruction of the reflective powers of students of that discipline. He contended that if symbols are used with full understanding of the laws which render them useful, "an intellectual discipline of high order is provided" (MAL, p. 9).

The reference that Boole made in the Introduction to MAL to the pleasure that the spirit finds in the mathematical study of both nature and mind may have also been intended to correct Hamilton's idea that mathematics is difficult because it is too easy, meaning that no pleasure is afforded by the study of it.

Hamilton's paper had the indirect influence of clarifying (by opposition) Boole's ideas about such points as the relevance of symbols and the relative roles of logic and mathematics. Boole very probably had definite opinions about these issues before he knew of Hamilton's arguments, but it is almost certain that he stated his opinions explicitly because of his desire to correct Hamilton's views.

Some references in MAL to De Morgan's logical ideas (MAL, pp. 41 and 82) show that Boole knew De Morgan's logic well. Such a knowledge being granted, can any kind of influence of De Morgan's ideas upon Boole's be traced to it?

The comparison between MAL and De Morgan's "Syllogism" [16] makes it clear that Boole's logic was a totally different construct from De Morgan's. Nevertheless, deep coincidences regarding methodological principles are noticed. The principles in question are: (a) the crucial importance given to ordinary language as guiding the construction of logic; (b) the possibility of a relevant improvement of logic by means of its mathematization; and (c) the principle of the existence of a universe of discourse embodying terms by pairs, each pair being composed of two opposite elements a and not-a.

As it can be seen, the influences of Hamilton's and De Morgan's logical conceptions on Boole were of an indirect nature, acting mainly as clarifications or confirmations of Boole's already formed opinions. Especially important was the influence on Boole (by contrast) of Hamilton's article on the value of mathematics in education.

In 1849 Boole was appointed professor of mathematics at Queen's College of Cork, Ireland. It seems that De Morgan was instrumental in this appointment. In 1855 Boole married Mary Everest.

George and Mary had five daughters from their marriage, all of whom were later to display special abilities. For instance Lucy, the fourth, became the first female professor of chemistry in England.

Mary Everest, suggested at several places in her writings, collected in [13], that a psychological theory of knowledge with religious implications was at the base of both her husband's logic-mathematical discoveries and his attitudes towards life (HS, p. 40, [19,20,22]). We examine next very shortly Mary Everest's claims; as the reader will see, most of these claims seem at least exaggerated. Nevertheless one gets, after considering them, a sensation that they reflect, thought it is difficult to say to what extent, some of Boole's real feelings and ideology that may have had their influence on the genesis of his logic.

Mary Everest's "Boole's method" reduces to the view that the human mind always faced - in any problem -, pairs of opposite facts, opinions, theories, and so on related to it which had to be weighed and contrasted in order to achieve a synthesis into a superior unity which embodied those opposites. The success of the process of successive unifications was based on the fact that God, being

One, attracts the human mind which in that way feels an instinctive impulse towards Monism. As a matter of curiosity Mary Everest says that this process was discussed by Boole with "a learned Jew" ([20], pp. 951-952).

What and how was Boole's theory of knowledge according Mary Everest?. The best way to describe it is to quote her version of such a theory:

> The mind of a man is encased in a mechanism which, besides receiving impressions through what we call senses, receives information also from some source, invisible and undefinable, access to which opens whenever the mind, after a period of tension on the difference, contrast or conflict between any elements of thought, turns to contemplate the same elements as united or as forming parts of unity ([22], p. 795).

In particular, regarding man's psychology, she says;

> But he seemed to assume, as the first of salutary facts, that there is direct contact between the Divine Magnetism and the nervous system of man ([22], p. 795).

Boole's logic, based on the fundamental equation $x^2 = x$, was then an expression of that philosophy, such an equation being formed by the two opposite elements $(1 - x)$ and x, the sum of which gives 1, the universe of discourse [21].

Regarding Boole's religious beliefs, according to Mary Everest's testimony, Boole was close to Unitarianism. Being convinced that God was the only important matter, he considered particular religious creeds as sources of divisions (HS, p. 3). Thus the true intellectual, and the true religious man, should be impartial (HS, p. 43). One reads in HS that Boole's impartiality made him become, by unanimous acclamation, "a referee in all parties" (HS, p. 24).

No biographer of Boole has taken into consideration the claims that Mary Everest made in regard to the psychological and religious origins and aims of Boolean logic; some have even implied that her judgment was unsound [38,28]; see also ([20], p. 955), where Mary Everest tells about her lack of success in trying to convince scientific people about the existence of a religious message in Boole's logic). Nevertheless there are some arguments that would support the opinion that Mary Everest was basically reflecting the truth, although, as we have speculated above, in a way that was quite exaggerated or distorted.

One argument is inferred from the internal coherence which exists between HS, which has been recognized as reliable by all biographers of Boole, and the rest of Mary Everest's writings, especially in those points which referred to what she called "Boole's method". Some others are provided by the study of several of Boole's own writings, by the consideration of issues such as Boole's community of opinions with other intellectuals he was in contact with, his ideas as a young man expressed in his address on Newton and by the study of the very basic ideas underlying the construction of his logic as presented in MAL. For reasons of space we deal very succinctly with only the last two of these items, referring the reader to [34] for a more elaborated argument.

In his address, Boole said of Newton:

It is generally supposed that his attention was directed to this subject by observing the falling of an apple. If this tradition be correct, it strangly teaches us what effects may arise from trivial or common occurrences, when the latent energies of nature or of mind are thereby roused into action. The falling of an apple was an every-day occurrence, yet its moral consequences have been to all human appearance, greater than the downfall of an empire. It had touched upon some hidden spring, -some sleeping and folded energy: a train of thought was excited, which, though interrupted, was never abandoned, until the foundation was laid of the great science of Physical Astronomy ([3], p. 10).

Thus Boole, already in 1835, believed in "latent energies of the mind" and "folded energies", these concepts resembling the concept of that "source invisible and undefinable" to which "Boole's method" refers. Moreover, the address contains many references to the history of ancient philosophies, by which Boole illustrates the idea that existence of error proved the existence of truth.

But there is more; for Boole cited Zoroaster. Does not this fact imply that he had become interested since the times he was a young man in dualistic philosophies ([3], pp. 21-22)? That Boole knew and had meditated on ancient thoughts is also inferred from a quotation in MAL (MAL, p. 49), and from the testimony of some of his biographers ([27], p. 428 and Taylor, one of Boole's grandsons [44], p. 47).

Regarding MAL, it ought to be recognized that the idea of reaching unities from the contemplation of opposites is repeatedly used by Boole in important parts of his book (MAL: pp. 40, 49-50, 52, 64, 65, 77).

Let us transcribe one of these paragraphs as illustration.

Consider what are those distinct and mutually exclusive cases of which it is implied in the statement of the given Proposition, that some one of them is true, and equate the sum of their elective expressions to unity. This will give the equation of the given Proposition. (MAL, p. 52).

Summarizing, it seems that some psychological and religious ideas contributed to the genesis of Boole's logic. Probably they were not as influential in this genesis as Mary Everest suggested, but there are arguments to suppose they existed.

At one time (1860), the possibility of Boole being nominated Professor of Mathematics at Oxford almost became a reality. But he sent only his name to be included in the list of candidates and the post was assigned to another man. Thus he spent the rest of his short life as a professor in Cork. He was famous for his knowledge, kindness, and total lack of egoism.

As has been noted above, in 1864 Boole made a trip to London. On his return he was almost completely exhausted. One day in November of that year he walked from his house to the College under heavy rain, and lectured in wet clothes. The result was a bad cold which terminated his life a few days later, on December 8, 1864.

3   Boole, Forerunner of Symbolic Computation

3.1   The Method of Separation of Symbols

The philosophy underlying the method of separation of symbols (to be denoted hereinafter as "mss") consisted of separating symbols of operation from their subjects of application and operating with the former as with algebraic entities

The mss had been suggested by several French and British mathematicians working in the second half of the 18th and the first half of the 19th century. It seems to us that Duncan F. Gregory was the one who most clearly stated how the method works. But it was Boole who took the mss to its ultimate consequences: in particular, Boole's logic was one of the branches of mathematics suggested by the mss.

The historical development of the method has been exhaustively studied by Koppelman [30], Knobloch [29], Panteki [41] and Grattan-Guinness [23] as part of the history of symbolic calculi. We have also studied it in [32], [36] and [37]. In this section we refer just to Gregory because this is enough to determine the immediate background of the influence of the mss on the genesis of Boole's logic.

Two of Gregory's papers are examined next to determine how the method works.

The first article under consideration appeared in 1838 [24]. Gregory determines in this paper the symbolic laws used in Newton's binomial scheme. Gregory notes that Euler used only the following laws of combination of symbols in the general application of the binomial development.

- Commutative law: $ab = ba$.
- Distributive law: $c(a + b) = ca + cb$.
- Index law: $a^m a^n = a^{m+n}$.

Gregory states that since it can be proved that the operations of differential calculus and of the calculus of finite differences are subject to those laws, it can be assumed that the Newton's binomial development is valid for such calculi, which means that it is not necessary to repeat the proof for each particular case. Let us consider an example of application taken from Gregory's article: the determination of the nth derivative of a product of functions $u \cdot v$, $\frac{d}{dx}(u \cdot v)$ can be written:

$$\frac{d}{dx}(u \cdot v) = u\frac{dv}{dx} + v\frac{du}{dx} = (\frac{d'}{dx} + \frac{du}{dx})(u \cdot v)$$

where $\frac{d'}{dx}$ is an operation upon v but not upon u, and $\frac{d}{dx}$ is an operation upon u but not upon v.

So, the n-th differential may be considered as a power of just the expression in parentheses with no attention paid to u and v. Gregory says that the result is also valid when n is fractional or negative.

In the article [25], Gregory proposes a characterization of symbolic algebra, a characterization suggested to Gregory by his considerations on the mss.

... it is the science which treats of the combination of operations defined not by their nature, that is, by what they are or what they do, but by the laws of combination to which they are subject. And as many different kinds of operations may be included in a class defined in the manner I have mentioned, whatever can be proved of the class generally, is necessarily true of all the operations included under it ([25], p. 208).

Regarding the reasons for accepting those laws of combination, Gregory says:

It is true that these laws have been in many cases suggested (as Mr. Peacock has aptly termed it) by the laws of the known operations of number; but the step which is taken from arithmetical to symbolical algebra is, that, leaving out the view of the nature of the operations which the symbols we use represent, we suppose the existence of classes of unknown operations subject to the same laws ([25], p. 208).

As Pycior notes [42], Peacock philosophy of mathematics went farther than his own mathematical work. While advocating freedom in algebraic calculation, he was not able to let his work be free because of his need of justifying on the grounds furnished by known mathematics the laws of combination to which the symbols were submitted. Gregory shared Peacock's idea of freedom in algebra, but not his needs for such a limitation. Gregory accepts that even though the different sets of laws which belong to symbolic algebra are suggested by known mathematics, there may be operations subject to the same laws that are not yet well known (but soundly guessed). This allows for the possibility of discovery and construction of different particular algebras each of them an instantiation of a part of symbolic algebra. Then both known mathematical operations and others not yet established (but soundly guessed as said above) may be divided into classes, in such a way that operations that obey formally identical laws belong to the same class. In this context, a theorem is a symbolically expressed result obtained by applying to the operations in a class any mathematical procedure which is permissible inside that class ("permissible" means any procedure valid for the known operations that belong to the class). Nevertheless, it is important to note the remark that Gregory makes: the theorems are true in a particular branch of algebra, "provided always that the resulting combinations are all possible in the particular operation under consideration" ([25], p. 208).

3.2   Boole and the mss

In 1840, Boole sent Gregory two papers for possible publication in The Cambridge Mathematical Journal. Gregory, after suggesting some changes, published them in 1841 [4,5].

It is very likely that Gregory saw an astonishing resemblance with what he himself was doing, in both the underlying philosophy of Boole's papers and particular details. It was then that he must have informed Boole of the specific terms of the mss. This guess is supported by reading Boole's third published paper [6], where he explicitly refers, on page 115, to Gregory's three laws. Moreover, Boole suggests at the end of his paper that the method could be improved if new algebraic processes were found.

Boole improved the method in his longest and most mature paper [7]. The following statement at the beginning of the paper is of interest to what we will go on to say in the next section.

> *Mr Gregory lays down the fundamental principle of the method in these words "there are a number of theorems in ordinary algebra, which, though apparently proved to be true only for symbols representing numbers, admit of a much more extended application". Such theorems depend only on the laws of combination to which the symbols are subject, and are therefore true for all symbols, whatever their nature may be, which are subject to the same laws of combination. The laws of combination which have been hitherto recognized are the following, p and r being symbols of operation and u and v subjects. 1. The commutative law, whose expression is $pru = rpu$ 2. The distributive law, $p(u+v) = pu+pv$ 3. The index law, $p^m p^n u = p^{m+n} u$. Perhaps it may be worth while to consider whether the third law does not rather express a necessity of notation, arising from the use of general indices, than any property of the symbol ([7], p. 225).*

Boole mentions Gregory's three laws, but adds that these are the ones recognized until now, implying that there may be others (see for instance a paper of 1846 [8] and another of 1847 [9] where he presents quite complex symbolic laws to find a solution for Laplace's equation).

Next we shall see, first, that the first principles of the logic as they appear at the beginning of his MAL are an almost direct transcription of laws suggested in the method of separation of symbols, and, second, that the inference procedure consists of applying developments of functions in the equations which transcribe the premises on which such inference is based.

Boole writes at the beginning of his first treatise on logic:

> *Further, let us conceive of a class of symbols x, y, z possessed of the following character. The symbol x, operating upon any subject comprehending individuals or classes, shall be supposed to select from that subject all the Xs which it contains. (...) When no subject is expressed, we shall suppose 1 (the Universe) to be the subject understood, so that we shall have: $x = x(1)$, the meaning of either term being the selection from the Universe of all the Xs which it contains and the result of the operation being in common language, the class X, i.e. the class of which each member is an X. (...)*

> *1st. The result of an act of election is independent of the grouping or classification of the subject...*
> $$x(u + v) = xu + xv$$

2nd. It is indifferent in what order two successive acts of election are performed. (...)

$$xy = yx$$

3rd. The result of a given act of election performed twice, or any number of times in succession, is the result of the same act performed once. (...)

$$x^n = x$$

The third law ($x^n = x$) we shall denominate the index law. It is peculiar to elective symbols and will be found of great importance in enabling us to reduce our results to forms meet for interpretation (MAL, pp. 15-18).

Logic turns out to be a calculus governed by the same laws as some of those in the method of separation of symbols.

From the examination of Gregory's and Boole's work, we may infer that the mss worked as follows. First, symbolic algebra classifies known calculi according to the laws of their combinations. Then one proceeds to examine a new piece of knowledge: if it obeys formally identical laws of combination of symbols such as those of a known class of calculus, it is placed inside that same class. Thereafter, all the mathematical processes that are permissible inside this class lead to theorems in the new theory (provided that the resulting theorems are interpretable).

Boole had the intuition that logic was a piece of knowledge candidate for becoming a part of symbolic algebra. Basing his statement about the laws of combination of the logical symbols on his own study of mental processes, he found that these laws were precisely the distributive, commutative, and index laws.

Choosing algebraic processes, especially equation systems resolution and, curiously, MacLaurin series expansions (MAL, p. 70) as tools for producing proofs was not a mere coincidence, since these processes were allowed inside the class of calculi based on the same three mentioned laws.

3.3   A CA Approach to Boole's Inference Procedures

This subsection reexamines Boole's ideas on inference from today's computer algebra (CA) point of view as follows.

First, Boole's use of MacLaurin series expansions in the translation of logical formulae into polynomials can be justified and extended using a CA System. Second, his method of inference in "hypotheticals" (propositional calculus), can be related directly with a polynomial ideal membership. Third, the final results of his method of inference in "categoricals" (a part of monadic first order logic), also based on MacLaurin series expansions, can be emulated using CA too (if an appropriate interpretation of Boole's logical symbols is made).

A Maple implementation for the first item and a CoCoA one for the second are used. For reasons of space we do not deal with the third item. This choosing of two different languages, Maple and CoCoA, is due only to the fact that each of them has some advantages over the other in the two calculations are needed here. CoCoA is particularlly effective when calculating Gröbner bases.

Logical Statements as Polynomials. Boole presented the following polynomial translation of the basic formulae of logic in chapter 5 (pp. 48-59) of MAL:

- Not X: $(1 - x)$.
- X and Y: xy .
- X or Y (not exclusive): $x + y - xy$.
- X or Y (exclusive): $x + y - 2xy$.
- If X then Y: $x(1 - y) = 0$, that is, $1 - x + xy = 1$, giving the polynomial $1 - x + xy$.

These translations follow directly from a proper use of the equalities (1) and (2) below, which appear in chapter 6 of MAL. Curiously indeed, Boole reached these equalities by basing his argument on MacLaurin series expansion of functions, so far away from the standard mathematical bases of today's logic.

The series expansion of an elective function of just one elective symbol x, gives (MAL, p. 61):

$$f(x) = f(0) + f'(0)x + \frac{f''(0)}{1.2}x^2 + ...$$

that, under the condition $x = x^2$ leads to:

$$f(x) = f(0) + \alpha x \tag{1}$$

and α can be calculated (by hand or using a Maple program) from (1):

$$\alpha = f(1) - f(0) \tag{2}$$

Similarly, the series expansion of a function in two elective symbols gives an expression of the form (MAL, p. 62):

$$f(x, y) = f(0, 0) + \alpha x + \beta y + \delta xy \tag{3}$$

Once the values of α, β, and δ have been found (by hand or using a computer), it is straightforward to check that Boole's translations for "not", "or", "and" and "implies" follow from his MacLaurin's expansions of elective functions as advanced at the beginning of this section.

Boole did not introduce truth values explicitly, but these were implicit in the expressions of elective functions. For instance the polynomial translation $1 - x$ of "Not(X)" follows directly from (1) and (2) if $f(1) = 0$ and $f(0) = 1$. Similarly the polynomial translation $x + y - xy$ of "X or Y" follows from (3) if $f(0, 0) = 0$, $f(0, 1) = f(1, 0) = f(1, 1) = 1$.

Boole's use of MacLaurin series expansions can also be applied to many-valued logic. This is a result that he could not imagine. Let us see how.

¬	
0	2
1	1
2	0

∨	0	1	2
0	0	1	2
1	1	1	2
2	2	2	2

∧	0	1	2
0	0	0	0
1	0	1	1
2	0	1	2

→	0	1	2
0	2	2	2
1	1	1	2
2	0	1	2

Fig. 1. Truth tables for Kleene's three-valued logic

For a function in just one elective symbol, under the condition $x^3 = x$, one obtains the following MacLaurin series expansion:

$$f(x) = f(0) + x \cdot \left(f'(0) + \frac{f'''(0)}{1 \cdot 2 \cdot 3} + ...\right) + x^2 \cdot \left(\frac{f''(0)}{1 \cdot 2} + \frac{f''''(0)}{1 \cdot 2 \cdot 3 \cdot 4} + ...\right)$$

that is, an expression of the form:

$$f(x) = f(0) + \alpha x + \beta x^2.$$

For a function in two elective symbols and under the same condition $x^3 = x$, one obtains the expression:

$$f(x,y) = f(0,0) + \alpha x + \beta y + \delta xy + \varepsilon x^2 + \eta y^2 + \theta x^2 y + \iota xy^2 + \kappa x^2 y^2$$

As above, α, β, δ, ϵ, η, θ, ι, and κ can be calculated by a Maple program. For instance:

$$\alpha = 2f(1) - \frac{3}{2}f(0) - \frac{1}{2}f(2)$$

As an illustration we refer to Kleene's three-valued logic. Its truth tables values can be found in figure 1 (0, 1 and 2 respectively mean "false", "indeterminate" and "true").

Let us denote the function $f(x)$ for negation as $f_\neg(x)$. As $f_\neg(0) = 2$, $f_\neg(1) = 1$, $f_\neg(2) = 0$, by applying the values for $\alpha, \beta, \delta...$ obtained like above, one gets: $f_\neg(x) = 2 - x$ and similarly for $f_\vee(x,y)$, $f_\wedge(x,y)$, and $f_\rightarrow(x,y)$.

Modern Interpretation of Boole's Ideas. Boole made explicit as basic laws of combination of symbols, only the commutative, distributive, and his special index law $x^2 = x$. But he also used products, sums, and implicitly, opposite elements for the sum $(x + (-x)) = x^2 - x = 0$. Thus, he was implicitly working in a polynomial quotient ring, actually the ring $A = \mathbb{Q}[x, y, z, ..., w]/I$, being I the ideal $I = <x^2 - x, y^2 - y, z^2 - z, ..., w^2 - w>$. The ideal I expresses Boole's index law.

$\mathbb{Z}_2$ (which can be extended to $\mathbb{Z}_p$, being p a prime number) plays in our CA approach the role of $\mathbb{Q}$. This is not an essential change from Boole's approach because he, even though allowing 2, 3, etc. as coefficients (for example in his exclusive "or" translated as $x + y - 2xy$) required the final formulae to take only the values 0 and 1. These final values of the polynomial expression of Boole's basic logical statements remain unchanged if performing the following changes:

- Exchange $\mathbb{Q}[x, y, z, ..., w]/I$ by $\mathbb{Z}_2[x, y, z, ..., w]/I$.
- Exclusive "or": exchange $x + y - 2xy$ by $x + y$.
- Inclusive "or": exchange $x + y - xy$ by $x + y + xy$.
- "Implies": exchange $1 - x + xy$ by $1 + x + xy$.

MacLaurin series expansion of functions and Boole's insight justified his translations of statements into polynomials. This can be done by a simple process of determining the coefficients of a polynomial in $\mathbb{Z}_2[x_1, x_2, ..., x_n]/I$, under today's knowledge of truth tables. The process is a little complex, so the reader is referred to [35] for details.

In particular, for Kleene's three-valued logic, the translation into (classes of) polynomials in $A = \mathbb{Z}_3/I$, $I = <x^3 - x, y^3 - y, z^3 - z, ..., w^3 - w>$, is:

- $f_\neg(q) = (2 - q) + I$
- $f_\vee(q, r) = (q^2r^2 + q^2r + qr^2 + 2qr + q + r) + I$
- $f_\wedge(q, r) = (2q^2r^2 + 2q^2r + 2qr^2 + qr) + I$
- $f_\rightarrow(q, r) = (q^2r^2 + q^2r + qr^2 + 2q + 2) + I$

These polynomials are the same as would have resulted by applying Boole's MacLaurin series.

Boole's Inference Methodology. Boole says in MAL, (MAL, p. 55): *"The treatment of every form of hypothetical Syllogism will consist in forming the equations of the premises, and eliminating the symbol or symbols which are found in more than one of them. The result will express the conclusion."*. Let us see an example (MAL p. 56, 5th example):

- If X is true, Y is true: $x(1 - y) = 0$.
- If W is true, Z is true: $w(1 - z) = 0$.
- Either X is true, or W is true: $x + w - xw = 1$.

From these equations, *eliminating w* we have: $x + y - yz = 1$, which expresses the conclusion, in Boole's words, "Either Y is true, or Z is true, the members being non-exclusive".

Boole calls "elimination" the following process; given:

$$ay + b = 0$$
$$a'y + b' = 0$$

multiply the second equation by a and the first by a', and perform the subtraction, obtaining $ab' - a'b = 0$.

Note that *the negation* (Boole negates an expression by making it equal to 0) *of the conclusion results to be an algebraic combination of the negation of the premises*, a fact which will be of utmost importance in the extension of Boole's idea, next.

Such a "Boolean" idea can be extended to the following theorem, which is stated without proof (see again [35]). The theorem refers to any p-valued logic, being p a prime number (and $p - 1$ the truth value corresponding to "true").

Definition 1. *A propositional formula A_0 is a tautological consequence of the propositional formulae $A_1, A_2, ..., A_m$, denoted $\{A_1, A_2, ..., A_m\} \models A_0$, if and only if for any truth-valuation v such that $v(A_1) = v(A_2) = ... = v(A_m) = p-1$, then $v(A_0) = p-1$.*

Theorem 1. *A formula A_0 is a (tautological) consequence of other formulae $A_1, ..., A_m$, if and only if the polynomial that translates the negation of A_0 belongs to the ideal generated by the polynomials that translate the negations of $A_1, ..., A_m$, and the polynomials $x_1^p - x_1, x_2^p - x_2,...,x_n^p - x_n$.*

This theorem has been proved quite recently, independently of Boole's suggestions (see Alonso et al. [2], Chazarain et al. [12], Roanes-Lozano et al. [43] and Laita et al. [35]). What is claimed here is that the theorem both validates Boole's approach to inference in hypotheticals and is an almost natural extension of this approach. The proof in [35] uses a quotient ring with respect to the ideal I, that corresponds to Boole's introduction of the law $x^n = x$).

In computer algebra, the way to check if a polynomial belongs to an ideal is finding if the Normal Form (NF) of the polynomial, modulo the ideal, is 0 (see for instance [1]).

The theorem can be applied to the study of consistency as follows.

A set of propositional formulae $\{A_1, A_2, ..., A_m\}$ is inconsistent if

$$\{A_1, A_2, ..., A_m\} \models A$$

A being *any* formula of the language in which $A_1, A_2, ..., A_m$ are written. This is expressed in terms of ideals by considering the ideal J generated by the negations of $A_1, A_2, ..., A_m$ is the whole ring (i.e. $1 \in J + I$), which means that any formula A is (tautological) consequence of $A_1, ..., A_m$. That 1 belongs to an ideal is expressed in computer algebra by stating that the Gröbner Basis of the ideal is $\{1\}$.

In the remainder of the section the computer algebra language CoCoA is used [11].

We consider now one of Boole's examples for hypotheticals (5th example, MAL, p. 56): this is the example presented as illustration in subsection 3.3.

i) Declare the ring of polynomials and the ideal I (the "elective symbols" x, y, z, w, are respectively denoted as X[1], X[2], X[3], X[4]).

```
A::= Q([x[1..4]]); USE A;
I:=Ideal(x[1]^2-x[1],x[2]^2-x[2],x[3]^2-x[3],x[4]^2-x[4]);
```

ii) Polynomial translation of Boole's basic logical statements (see Subsection 3.3).

```
NEG(M):=NF(1-M, I);
OR1(M,N):=NF(M+N-M*N, I);
OR2(M,N):=NF(M+N-2*M*N, I);
AND1(M,N):=NF(M*N, I);
IMP(M,N):=NF(1-M+M*N, I);
```

iii) As explained above, $\mathbb{Q}$ can be exchanged by $\mathbb{Z}_2$ (then changing -1 by $+1$ and -2 by 1 in NEG, OR1, OR2 and IMP.

```
A::=Z/(2)[x[1],...,x[4]]
```

iv) H1, H2, H3 and C1 respectively denote the hypotheses 1, 2 and 3 and the conclusion

```
H1:=IMP(x[1],x[2]);
H2:=IMP(x[4],x[3]);
H3:=OR1(x[1],x[4]);
C1:=OR1(x[2],x[3]);
```

v) Declare the ideal J generated by the (negations of the) hypotheses:

```
J:=Ideal(NEG(H1),NEG(H2),NEG(H3));
```

vi) Does the negation of the conclusion belong to the ideal J generated by the negations of the premises?. The answer is YES if the following normal form is 0.

```
NF(NEG(NEG(C1), J+I);
```

vii) CoCoA gives 0 as output, as expected.

The same argument can be applied to all other examples in MAL, (MAL, pp. 55-59). As a matter of curiosity, CoCoA finds typographical errors in Boole's examples 6 and 7.

4   Conclusions

Boole's logic was born as a branch of a general symbolic calculus known in his time as the "Method of Separation of Symbols". The laws of logic, according to Boole's intuitions, were no more, no less, that the symbolic expressions of the laws of thought. All mathematical operations allowed inside the class of calculi known to be based on the same laws of symbols he found for logic, regardless their quite differentiated fields of application, were also allowed in logic. In particular, Boole's logic inference procedures were symbolic manipulations based on series expansions and equation systems solving.

Such a conception of logic led Boole to consider propositional logic statements as polynomials and to produce inferences by showing that (the polynomial expression of) consequences were algebraic combinations of (the polynomial expressions of) premises.

In this way, Boole can be considered as a forerunner of both modern symbolic calculus in general and of interesting Computer Algebra approaches lo Artificial Intelligence in particular.

Boole's intuitions that premises and conclusions in logic can be represented as polynomials and that conclusions are found by taking algebraic combinations of premises, advanced to apply Gröbner bases to extraction of consequences in logical systems in general and to verification and extraction of kowledge in rule-based expert systems in AI in particular as follows.

If the Gröbner basis of the ideal generated by the polynomials that translate the (negation of the) rules and facts of a rule-based expert system is $\{1\}$, then the ideal is the whole ring, so theorem 2 says that any formula is a tautological consequence of the information contained in the expert system. It means that the expert system leads to inconsistency, so it has to be corrected. Once inconsistencies have been suppressed, we can ask, by using Normal Forms, whether or not a given formula, formed in the language in which the expert system has been built, is a tautological consequence of the information contained in the system. In [35] we have described the application of this method to verification and extraction of consequences in expert systems based on three-valued logics and containing 50 variables. The verification of this expert system takes two minutes. With just a Pentium-based PC with 128 Mb RAM, expert systems containing 150 variables under three-valued logic can be verified, the computing time being about fifteen minutes.

5 Acknowledgements

We thank Professors Michael J. Crowe and Ivor Grattan-Guinness for their supervision in our studies of Boole. We also thank the anonymous referees of the final version of this paper for their most valuable comments.

References

1. Adams, V., Loustanau, P., An Introduction to Gröbner bases. Graduate Studies in Mathematics 3. Providence, RI, American Mathematical Society Press (1994).
2. Alonso, J.A., Briales, E., Lógicas polivalentes y bases de Gröbner. In: Vide, M. (ed), Proceedings of the V Congress on Natural Languages and Formal Languages (1989), 307-315. Barcelona, Spain, P.P.U. Press.
3. Boole, G, An address on the genius and discoveries of Sir Isaac Newton (Lincoln, 1835).
4. Boole, G., On certain theorems in the calculus of variations. The Cambridge Mathematical Journal 2 (1841), 97-102.
5. Boole, G., Researches on the theory of analytical transformations with a special application to the reduction of the general equation of the second order. The Cambridge Mathematical Journal 2 (1841), 67-73.
6. Boole, G., On the integration of linear differential equations with constant coefficients. The Cambridge Mathematical Journal 2 (1841), 114-119.
7. Boole, G., On a general method in analysis, Philosophical Transactions of the Royal Society of London 134 (1844), 225-286.
8. Boole, G., On the equations of Laplace's function. The Cambridge and Dublin Mathematical Journal 1 (1846), 10-22.
9. Boole, G., On a certain symbolical equation. The Cambridge and Dublin Mathematical Journal 2 (1847), 7-12.
10. Boole, G., The mathematical analysis of logic, being an essay towards a calculus of deductive reasoning, Cambridge and London (1847), 4-5. Reprinted in: Boole, G., Studies in Logic and Probability, Ed. R. Rhees, London (1952).
11. Capani, A., Niesi, G., CoCoA user's manual, Dept. of Mathematics University of Genova (1996).
12. Chazarain, J., Riscos, A., Alonso, J.A., Briales, E.. Multivalued logic and Gröbner bases with applications to modal logic. Journal of Symbolic Computation 11 (1991) 181-194.

13. Cobham ed, E.M., Mary Everest Boole, Collected works (4 vols.), London (1931).
14. Cobham, E.M., Mary Everest Boole, A Memoir with some letters, Ashingdon (1951).
15. Daniel, F., A Teacher of Brain Liberation, London (1923).
16. De Morgan, A., On the structure of the syllogism, Trans. Cambridge Phil. Soc., 8 (1846), 379-408. Reprinted in: On the syllogism (edited with an introduction by Peter Heath), London (1966), 1-17.
17. Diagne, S. B., Boole, 1815-1864; l'oisseau de nuit en plein jour, Belin, Paris (1989)
18. Everest, M., Home side of a scientific mind (1878); Cobham's Mary Everest's Collected Works,, 1-48. Reprinted from The Dublin University Magazine, 91 (1878), 103-114, 173-183, 327-336, 454-460.
19. Everest, M.: Logic taught by love, rhythm in nature and education (printed 1890, publ. 1905); Cobham's Mary Everest's Collected Works, vol. 2, 438.
20. Everest, M.: Indian thought and western science in the nineteenth century (written 1901, publ. 1909), Cobham's Mary Everest's Collected Works, vol. 3, 962.
21. Everest, M., The nineties (publ. or read 1890-1899), Cobham's Mary Everest's Collected Works, vol 2, 565.
22. Everest, M., On education, Cobham's Mary Everest's Collected Works, vol 2, 789.
23. Grattan-Guinness, I., Mathematics and mathematical physics at Cambridge 1815-40: a survey of the achievements and of the French influence. In: Harman, P.M. (ed.), Wranglers and Physicists, Manchester (1985), 84-111.
24. Gregory, D.F., Demonstrations, by the method of separation of symbols of theorems in the differential calculus and calculus of finite differences. The Cambridge Mathematical Journal 1 (1838), 232-244.
25. Gregory, D.F., On the real nature of symbolical algebra. Transactions of the Royal Society of Edinburgh 14 (1840, read 7th May 1838), 208-216.
26. Hamilton, W., On the study of mathematics, as an exercise of the mind. Discussions on Philosophy and Literature, London (1852), 257-313. Reprinted from: Edinburgh Review, 62 (1836), 409-455.
27. Harley, R., George Boole, F.R.S., British Quart. Rev. (1866). Reprinted in: Studies in Logic and Probability. Boole's Collected Logical Works, 425-472.
28. Kneale, W., Boole and the revival of logic, Mind, 58 (1948), 149-175 (155).
29. Kobloch, E., Symbolik und Formalismus im mathematischen Denken des 19. und beginnenden 20. Jahrhunderts. In: Dauben, J.M. (ed.), Mathematical perspectives. Essays on Mathematics and its historical development, New York (1981), 139-165.
30. Koppelman, E., The calculus of operations and the rise of abstract algebra. Archive for the History of Exact Sciences 8 (1971), 155-242.
31. Laita, L. M., A study of the genesis of Boole's logic, Doctoral Dissertation, University of Notre Dame, University Microfilms International (76-4613), An Arbor Michigan, London (1976).
32. Laita, L. M., The influence of Boole's search for a universal method in analysis on the creation of his logic. Annals of science 34 (1977), 163-176.
33. Laita, L. M., Influences on Boole's logic: the controversy between William Hamilton and Augustus De Morgan, Annals of Science 36 (1979), 45-65.
34. Laita, L. M., Boolean algebra and its extralogical sources: The testimony of Mary Everest Boole, History and Philosophy of Logic 1 (1980), 37-60.
35. Laita, L.M., E. Roanes-Lozano, L. de Ledesma, J.A. Alonso: A Computer algebra approach to verification and deduction in many-valued knowledg systems. Soft Computing 3/1 (1999), 7-19.

36. Laita, L. M., de Ledesma, L., Roanes-Lozano, E., Pérez, A., Brunori, A., Boole's logic revisited from computer algebra, Mathematics and Computers in Simulation 51 (2000), 419-439.
37. de Ledesma, L., Pérez, A., Borrajo, D., Laita, L. M., A computational approach to George Boole's discovery of mathematical logic, Artificial Intelligence 91 (1997), 281-307.
38. MacFarlane, A., Lectures on ten British mathematicians of the nineteenth century (New York, 1916) 50-63.
39. MacHale, D., George Boole, his life and work, Boole Press, Dublin (1985).
40. Mansel H.L., Veitch, J. (Eds.), Lectures on metaphysics and logic (4 vols.), vols 3 and 4, Edinburgh and London (1859-1861).
41. Panteki, M., Relationships between algebra, differential equations, and logic in England 1808-1860. Ph.D. Thesis, Council for National Academic Awards, UK (1991)(chapter 4).
42. Pycior, H.M., Peacock and the British origins of symbolical algebra, Historia Mathematica 8 (1981), 23-45.
43. Roanes-Lozano, E., Laita, L.M., Roanes-Macías, E., A polynomial model for multivalued logics with a touch of algebraic geometry and computer algebra. Mathematics and Computers in Simulation 45/1-2 (1998), 83-99.
44. Taylor, G., George Boole F.R.S., 1815-1864, Notes and records of the Royal Society of London, 12 (1956), 44-52 (47). Also in: Harley's biography, 428.
45. Whewell, W., Thoughts on the study of mathematics as part of a liberal education, Cambridge (1835).

Artificial Intelligence as a Decision Tool for Efficient Strategic and Operational Management

Marc Knoppe

EuWiM AG
Siemensstraße 4
D-61449 Steinbach (Frankfurt/Main)

knoppe@euwim.de

Aspects of short-range planning dominate the strategic decision-making process in management. The capability of managers to carry out their own functions effectively tends to be reduced by the increasing complexity and pressure of this job. Intelligent management systems (IMS) are needed to improve the decision-making process in management. Artificial intelligence and neural networks are very well matched to this need. The decision-making process is described in some detail, to illustrate what kinds of IMS functionalities are required, and thus to present the problem to specialists in artificial intelligence.

The Challenges of a Modern Management Process

Enterprises are more and more confronted with the consequences of dynamic change caused by the globalization of commerce and by the information flood that is due to the Internet. The radical change of the environment in which companies operate reduces the ability of managers to act on the strategic aspects of their task, and to conduct effective management of operations, in a timely way. Catching and structuring the complexity of the area that is being managed therefore tends to happen less, or less effectively.

Existing management systems are not in a position to give optimal treatment to the daily data streams, or (therefore) to prepare the best management decisions. The evaluation of chances and risks becomes more difficult day by day. Thus, in practice, the percentage of risky or emotional decisions will be increased, and the strategic decision-making process will be dominated by the pressing considerations of short-range planning.

The current trend of mergers and acquisitions shows one reason for the pressure of short-term considerations: the beliefs of shareholders about what constitutes value. In this respect, the idea of being the biggest bank in the world, and actual success as measured on stock exchanges, was a primary factor for the recent attempted merger between Deutsche Bank and Dresdner Bank. The key facts and milestones for any long procedure for achieving a functional merger between businesses received relatively little attention. A precise analysis showing the fitting of the different

J.A. Campbell and E. Roanes-Lozano (Eds.): AISC 2000, LNAI 1930, pp. 20-31, 2001.

cultures of the two banks was missing. There was no detailed implementation planning. There was no worst-case scenario and no emergency plan. In consequence, the deal failed.

This example underlines the actual situation of complex value chains, insufficient information, deficient evaluation of key facts, deficient research, inadequate tests, and decisions that were at best merely incorrect. If intelligent managemennt systems (IMS) had been available for identifying relevant data, checking the available relevant data, and providing a best available evaluation in time, some of these difficulties could have been reduced, or at least pointed out so that managers could give them proper consideration. Moreover, to be useful in a given management environment, IMS must include components for learning from the data and cases available. Producing good IMS is still a challenge, despite past achievements in some particular areas.

In order for the challenge to be appreciated in detail, it is necessary to explain the managerial decision-making process (Knoppe, M., "Strategische Allianzen in der Kreditwirtschaft", Oldenbourg Verlag, p. 18-23, München 1997).

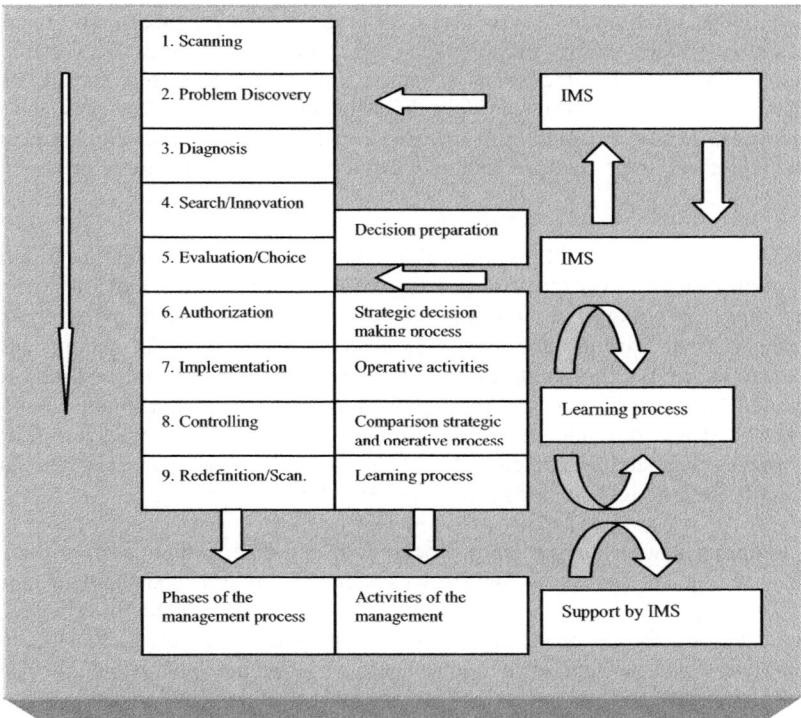

Fig. 1. Phases and activities of a modern decision making process

The Phases of a Modern Decision Making Process

Figure 1 shows the different phases of modern management. This management is an ongoing problem-solving process. To show the complexity of the activity, it helps to split the management process into different phases. Each phase has its own demands with respect to IMS, and consideration of the different phases side by side allows one to start forming impressions about what an appropriate computer-based architecture for IMS should be.

The modern management process begins with a scanning phase. Scanning detects a possible opportunity, threat, variation from a norm, or disturbance. Problem discovery then defines the problems that have been uncovered by the scanning. Diagnosis calls for more detailed information about the problems. Discovery and diagnosis determine the direction and location of search. Search and innovation produce redefinitions of the problems, changes in level of aspiration, and reinterpretations of what consititutes an ideal solution. Search and innovation provide what is to be evaluated and chosen. Evaluation and choices narrow the range of possibilities for what will be sought. Search is conducted to justify what has already been chosen tentatively as a solution. Evaluation and choice must be authorized before being implemented. Rejected authorization or failed implementation forces re-evaluation, redesign or redefinition. Problem diagnosis determines evaluation and choice. Search is then eventually eliminated. The solutions to the problem are given by the diagnosis. The results of evaluation and choice modify the diagnosis and raise new problems. Implementation experience changes the focus of scanning and thereby improves the whole management process. "Controlling" checks whether or not the decisions and their consequences are still acceptable, fit the time constraints, and represent best practice. It is evident from their purpose that redefinition and scanning are a never-ending story.

Phases 8 and 9 especially - controlling and redefinition - set up a modern management process. This process is characterized by permanent learning activities, directed towards strategic issues and also towards tactics and management of operations. Good learning activities guarantee a management's quality and its competence to survive in a complex business world. Learning is not confined to the management; it also refers to any management systems that are in place. A modern IMS must be expected to integrate whatever learning processes are needed to fulfil the management's demands (Kirsch, W., "Die Handhabung von Entscheidungs-problemen", Barbara Kirsch Verlag, p. 180–191, München 1988).

By comparison with management processes in the past (which were without the presence of the Internet and high-tech systems), modern management and its decision-making are dominated by a habitual lack of time. The element of time is the first consideration for the profit of a management system. Therefore, an IMS has to combine relevant time-dependent components and components that deal with the learning process(es).

What Is a Management System?

A management system is in essence an additional organization (see Figure 2). These additional organizations overlay the actual business systems and management structures (basic organization). The staff has to undertake tasks derived from the basic organization and the management systems. Each task demands a special management system such as information systens, planning systems and "controlling" systems. According to its function, any single management system is part of either a strategic

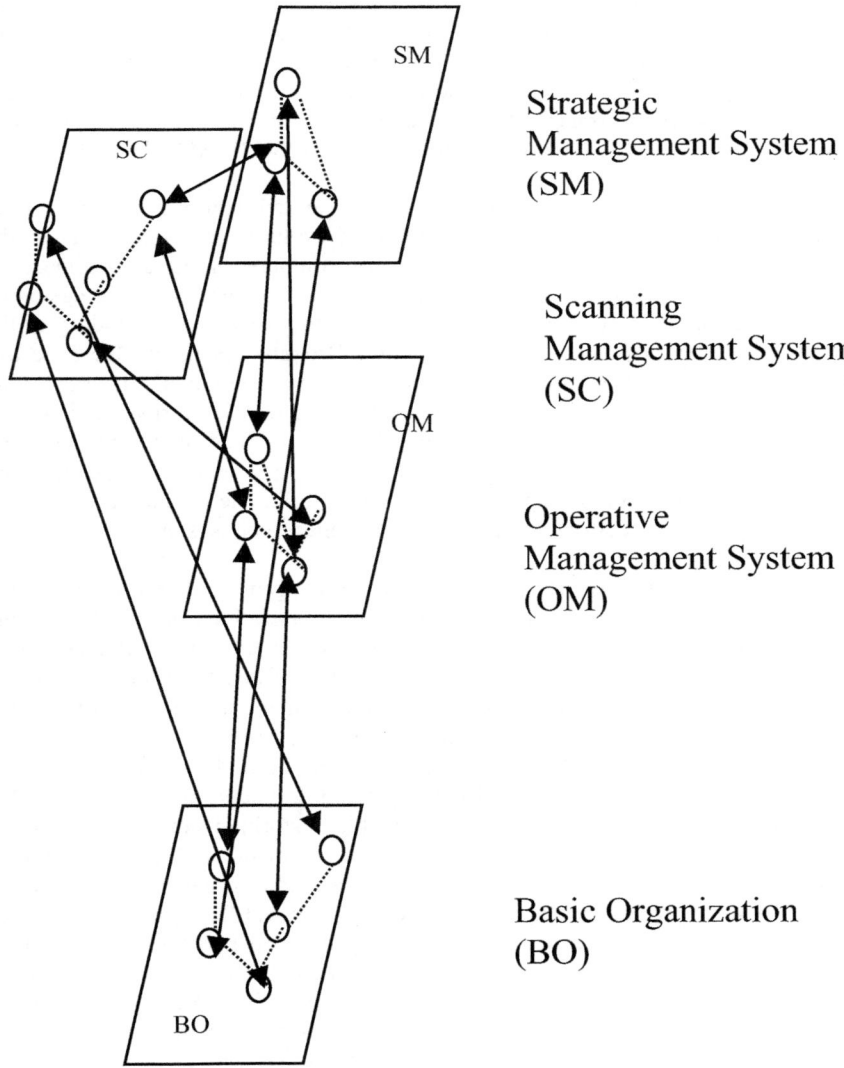

Strategic
Management System
(SM)

Scanning
Management System
(SC)

Operative
Management System
(OM)

Basic Organization
(BO)

Fig. 2. Management systems as additional organizations

or operational architecture of management systems (see Figure 3). Strategic management systems provide information and support for the long-range planning process. "Operative" management systems are defined as those that support the short-term decision-making process. The differentiation between steategic and operative management systems shows that different features are required in the design and implementation of the two types of system.

The point of a management system is to reduce the apparent complexity of an enterprise. To reach this target, the overall system of a company has to be split into a number of different subsystems (modules). Each module has a place in the architecture shown in Figure 3. The modules there have different tasks, and not every company needs to use all those modules. The modules are exchangeable, and should be adaptable to special needs. Among the modules there are different relationships which represent personal, organizational, technical and social connections (Kirsch, W., "Managementsysteme", Barbara Kirsch Verlag, p. 128-139, München 1989)

In the past, management systems had no integrated learning process. They therefore did not evolve internally; all aspects of evolution remained outside the systems. Today management systems are routinely computer-based, and to meet the challenges of modern management processes, internal evolution is highly desirable. Something like this view turned up in the 1960s, accompanying the building of the first computer-supported management systems. Broadly speaking, these failed, because appropriate hardware and software technology were not available at the time. During that period, Ackhoff (Ackhoff, R.L., "Management Misinformation System", in: Management Science, H. 4, p. 147–156, 1967) even talked about "management misinformation systems". But during the last 10-15 years, hardware and software technology have developed so rapidly that they now offer the chance to realise the performance spectrum necessary for an intelligent management system.

To survive in the new world of globalization, management needs qualified and detailed information for planning and controlling an organization's daily activities and for arriving at well-checked decisions quickly. To handle the challenges of a company's environment and compensate for the effects of rapidly-changing markets, an IMS must collect, filter, store and evaluate the information, must categorize data, and take advantage of any opportunity to improve any aspect of the management process. Furthermore, an IMS should be able to formulate scenarios, show alternatives, propose solutions, and deliver arguments to underpin the decisions that they propose. In order to do this, it should be capable of storing its experiences, and of adapting past experiences and past decisions to new situations. This means that it must be able to learn. In this respect it can provide a kind of organizational memory, as a way of making up for the fact that most companies feel the lack of well-trained staff and/or staff with long experience of the organization and the reasons for its particular procedures and decisions. We may never be able to say that an IMS is an intelligent as a managerial brain, but if we can achieve similar levels of performance in some knowledge-intensive areas of management, then the human managers will be helped by the IMS to check, evaluate and commit in a timely way to decisions that are good even from the strategic point of view.

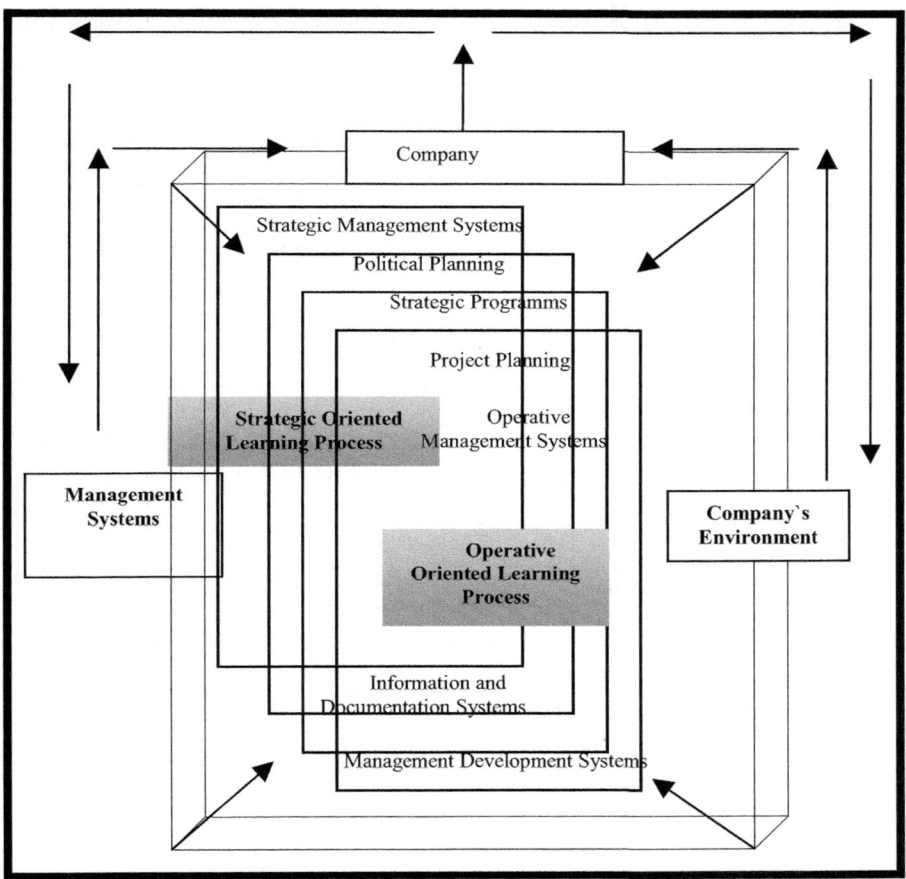

Fig. 3. An architecture of management systems

Dynamic change offers no time for human management to evaluate all the details of decision parameters carefully. While an IMS should not be expected to take over the job of a manager, who should continue to make the final evaluation and the selection of a best decision, the IMS can deal with the initial steps, generation of alternatives etc., and can also reduce the risk of reaching incorrect decisions.

Figure 4 shows the interfaces between the different phases of a decision process, IMSs, computer systems and managers (Alex Björn, „Künstliche neuronale Netze in Management-Informationssystemen", Grundlagen und Einsatzmöglichkeiten, Gabler Verlag, p. 74-76, Wiesbaden 1998).

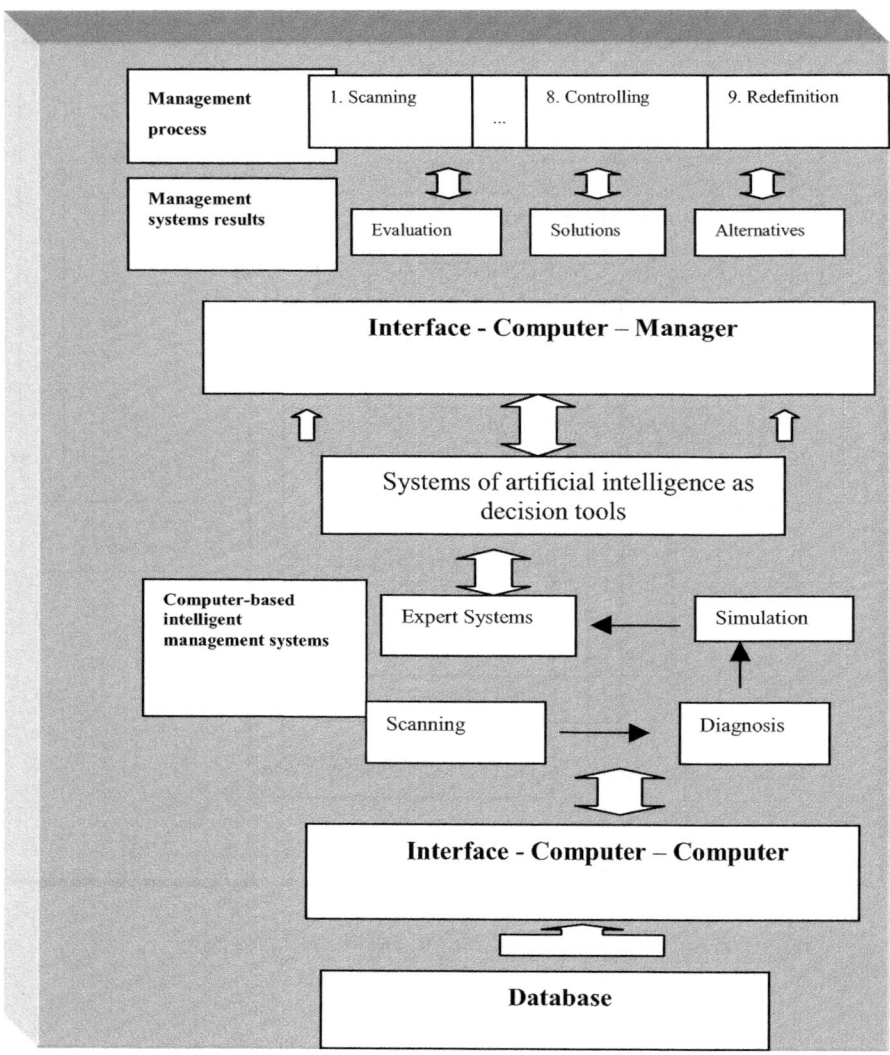

Fig. 4. The interfaces between the phases of a decision process, intelligent management systems, computer systems and managers

Applicability of Artificial Intelligence as a Management Tool

The human brain has the competence to interpret complex signals within milliseconds, and to make sense of new situations of a given general type by reference to knowledge about old situations of the same general type. This kind of activity is what a good manager does. Where artificial intelligence (AI) has analogues of this behaviour, its techniques are likely to be of the greatest value within IMS implementations.

The closest analogue in AI is in connectionism, and in the use of neural nets in particular. Compared to a traditional managemnent system, a neurally-based IMS has the advantage that not much knowledge about the internal structure of a problem class is needed, but just knowledge about what it takes to specify problems in the class. Samples (problem examples, with solutions) are adequate to train the neural networks, so that they can be said to have learned about the associations between problems and solutions, and can exploit this knowledge in use on problems that arrive subsequently. As stated above, learning is a key to the future acceptability of IMS. Classical (symbolic) learning techniques in AI are not irrelevant, but it is likely that neural and other subsymbolic methods will match better the actual needs of IMS users. For example, the users (who are not AI specialists) will want to treat the learning components of an IMS as black boxes, while symbolic learning methods appear to demand that the user should know something about the details of what goes on inside the box.

The most immediately evident applications of neural networks are in the three phases of diagnosis, evaluation and controlling. In fact, applications of this kind are quite numerous already in commerce, particularly in the financial sector (Goonatilake S. and Treleaven P. (eds.), Intelligent Systems for Finance and Business, John Wiley & Sons, Chichester,1995). For example, banks use neural networks for assessing the creditworthiness of companies and individuals (the contribution by D. Leigh, "Neural Networks for Credit Scoring" at p. 61-69 in the book quoted above). Traders on stock exchanges use neural networks to classify shares according to their potentialities and risks (contribution by Refenes A.N., Zapranis A.D., Connor J.T. and Bunn D.W., "Neural Networks in Investment Management", p. 177-208). Neural networks also support market analysis in various industries, e.g. direct-mailing, (Furness P., "Neural Networks for Data-Driven Marketing", p. 73-96) food products (coffee), toiletries (shampoo). Other applications include assessing the development of market shares, and optimizing solutions to problems in production processes. These examples cover all of the three phases of managerial activity mentioned above. While they are not specifically "managerial", each activity is of direct value for some managerial process, and strong analogues of the kinds of computation involved are relevant for IMS applications. For instance, an analogue of the creditworthiness check would be a check of the credibility of arguments/data in favour of a business merger. The aim of a diagnosis phase here should be to filter similar features of past mergers. By comparing different features, and matching problem situations and risks and outcomes of similar mergers, neural networks may be able to make interesting suggestions about the problem of a merger currently under consideration. Because of the heterogeneity of the data about this problem, and the likelihood that the volume or quality of data will not permit safe conclusions to be drawn from statistical methods, neural networks should be able to come up with better results.

It can be said that the examples like those above have validated the use of neural networks for IMS applications in the future.

In principle there are no phases in Figure 1 that subsymbolic AI cannot handle (with neural networks, though more recent techniques such as genetic algorithms are also demonstrating their worth where the underlying problem-solving activity is search rather than the classification or association that occurs when neural networks are used (J. Koza, "Genetic Programming for Economic Modeling", p. 251-269, same book). Genetic programming is also useful more widely, e.g. for operations involving design and/or uncertainty (Smith A.E. and Norman B.A., "Evolutionary Design of Facilities Considering Production Uncertainty", in I. Parmee (ed.), Adaptive Computing in Design and Manufacture 2000, p. 175-186, Springer-Verlag, Berlin, 2000). And there are methods located somewhere between subsymbolic and traditional symbolic AI which need to be understood better and applied more in managerial areas. Bayesian belief networks are a good example; one of several existing applications with direct interest to management (assessing dependability of "systems") has been reported by Neil. Littlewood and Fenton (Neil M., Littlewood B. and Fenton N., "Applying Bayesian Belief Networks to Systems Dependability Assessment", Proceedings of Safety Critical Systems Club Symposium, Leeds, p. 71-93. Springer-Verlag, Berlin, 1996) Moreover, there are situations where symbolic AI still has some relevance. Even if methods based explicitly on mathematical logic happen to be too slow or unwieldy for the time-limited needs of an IMS, more flexible symbolic methods such as case-based reasoning (Kolodner, J., "Case-Based Reasoning", Morgan Kaufman, San Mateo, 1993) can be considered; they are intended to deal with just the kinds of activity mentioned near the end of the previous section. Furthermore, it is possible that some applications, particularly quantitative ones (as in production optimisation, and treatment of business-oriented econometric data), some hybrid treatment involving a mixture of symbolic and subsymbolic methods will be advisable. There are many open problems and open areas of applicability, with plenty of data; it is a challenge for an AI audience to produce new results and new tools to assist in IMS development and application.

With respect to neural networks at least, Figure 5 shows some conceivable applications within an IMS.

Some of these applications are not purely activities that can be reduced to making associations or classifications with respect to past and current examples of problems. It may therefore appear that traditional neural-network schemes are not adequate to express all the contents of the applications. But there are interesting and relatively new extensions of the traditional approach, which could benefit from exposure to such applications - and vice versa. In the present book, there are examples of interesting extensions in the contributions by J. Pfalzgraf and by A. Iglesias and A. Galvez.

A further challenge for suppliers of neural networks - and, indeed, any tool derived from AI - is that the usefulness of a technique to management, and especially top management, depends positively on how little the users are expected to know about its theoretical side (or, in other words, about the inside of the black box). The approach to educating users about a technique has to be easy, and the time needed for the approach has to be short. Neural networks respect these considerations, which is why they are very attractive as IMS components.

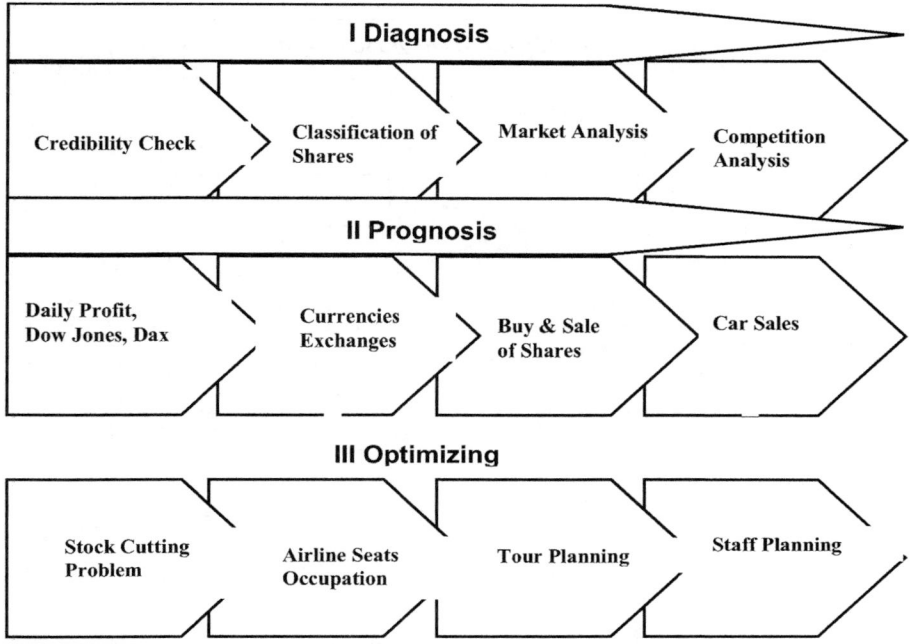

I Diagnosis

| Credibility Check | Classification of Shares | Market Analysis | Competition Analysis |

II Prognosis

| Daily Profit, Dow Jones, Dax | Currencies Exchanges | Buy & Sale of Shares | Car Sales |

III Optimizing

| Stock Cutting Problem | Airline Seats Occupation | Tour Planning | Staff Planning |

Fig. 5. The field of neural networks as a tool of a management system

Knowledge about methods of neural networks needs to be kept separate from the knowledge of users about the decision-making and the commercial process that is being managed. The user should not be required to know anything substantial about the former. Figure 6 illustrates a design for the functional behaviour of an IMS, involving neural components, which respects these considerations.

This has been a look at the area of IMS from a user's perspective, identifying needs and some possible ways for AI specialists to fill those needs. It has not considered explicitly mathematical knowledge or reasoning, though there are many areas of management practice where dealing with large amounts of quantitative information implies the effective use of mathematical knowledge. Integrating the special perspective and skills of scientists working on "artificial intelligence and symbolic computation" with the problems of the IMS area is a general challenge which contains many particular problems. The AI community can support practice by developing theoretical models and methods adapted to the methods and demands of economically-based management decision-making. To end with just one example, general methods for the plausible explanation and justification of the outputs from neural networks would obviously have substantial immediate value, but we are still in need of them (Alex Björn, "Künstliche neuronale Netze in Management-Informationssystemen, Grundlagen und Einsatzmöglichkeiten", Gabler Verlag, Wiesbaden 1998).

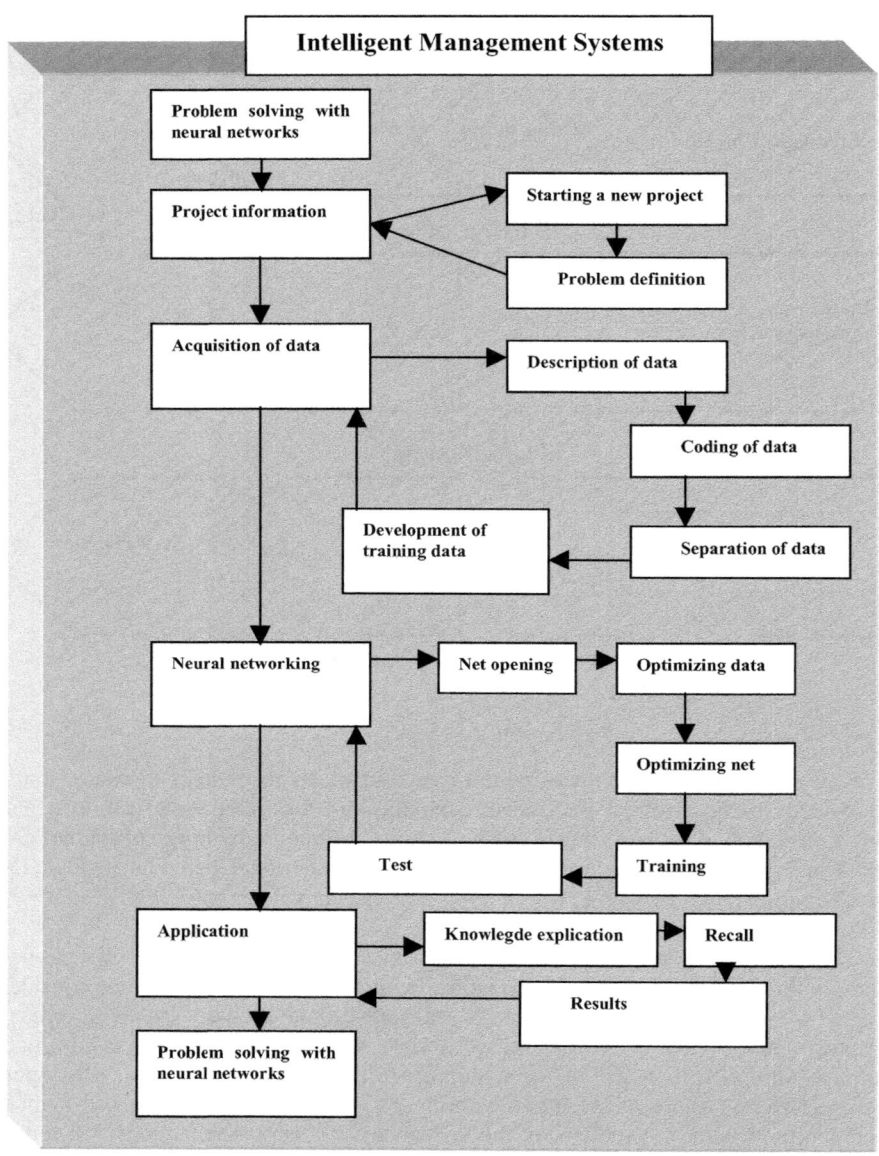

Fig. 6. A general design of a management process software based on neural networks

References

Ackhoff, R. L., A concept of corporate planning, New York 1970

Ackhoff, R. L., Management Misinformation System, in: Management Science,   H. 4, P. 147–156, 1967

Alex Björn, Künstliche neuronale Netze in Management-Informationssystemen, Grundlagen und Einsatzmöglichkeiten, Gabler Verlag, Wiesbaden 1998

Bass, B., Organizational Decision Making, Homewood (III) 1983

Beltratti A. et al., Neural Networks for Economic and Financial Modelling, London 1996

Furness P., "Neural Networks for Data-Driven Marketing", p. 73-96

Goonatilake S. and Treleaven P. (eds.), Intelligent Systems for Finance and Business, John Wiley & Sons, Chichester,1995

Kirsch, W., Die Handhabung von Entscheidungsproblemen, Barbara Kirsch Verlag, München 1988

Kirsch, W., Managementsysteme, Barbara Kirsch Verlag, München 1989

Knoppe, M., Strategische Allianzen in der Kreditwirtschaft, Oldenbourg Verlag, München 1997

Kolodner, J., Case-Based Reasoning, Morgan Kaufman, San Mateo, 1993

Neil M., Littlewood B. and Fenton N., "Applying Bayesian Belief Networks to Systems Dependability Assessment", Proceedings of Safety Critical Systems Club Symposium, Leeds, p. 71-93. Springer-Verlag, Berlin, 1996

Refenes A.N., Zapranis A.D., Connor J.T. and Bunn D.W., "Neural Networks in Investment Management", p. 177-208

Smith A.E. and Norman B.A., "Evolutionary Design of Facilities Considering Production Uncertainty", in I. Parmee (ed.), Adaptive Computing in Design and Manufacture 2000, p. 175-186, Springer-Verlag, Berlin, 2000

OMDoc: Towards an Internet Standard for the Administration, Distribution, and Teaching of Mathematical Knowledge

Michael Kohlhase

FB Informatik, Universität des Saarlandes, Saarbrücken
http://www.ags.uni-sb.de/~kohlhase

Abstract. In this paper we present an extension OMDoc to the OPEN-MATH standard that allows the representation of the semantics and structure of various kinds of mathematical documents, including articles, textbooks, interactive books, courses. It can serve as the content language for agent communication of mathematical services on a mathematical software bus.

1 Introduction

It is plausible to expect that the way we do (conceive, develop, communicate about, and publish) mathematics will change considerably in the next ten years. The Internet plays an ever-increasing role in our everyday life, and most of the mathematical activities will be supported by mathematical software systems (we will call them *mathematical services*) connected by a commonly accepted distribution architecture, which we will call the *mathematical software bus*. We have argued for the need of such an architecture in [SHS98, FHJ+99], and we have in the meantime gained experiences with the MATHWEB system that provides a general distribution architecture (see [FK99b]); other groups have conducted similar experiments [DCN+00, AZ00] based on other implementation technologies, but with the same vision of creating a world wide web of cooperating mathematical services. In order to avoid fragmentation, double inventions and to foster ease of access it is necessary to define interface standards for MATHWEB[1]. In [FHJ+99], we have already proposed a protocol based on the agent communication language KQML [FF94] and the emerging Internet standard OPENMATH [AvLS96, CC98] as a content language (see Fig. 1). This layered architecture which refines the unspecific "application layer" of the OSI protocol stack is inspired by the results from agent-oriented programming [Sho90], and is based on the intuition, that all agents (not only mathematical services) should understand the agent communication language, even if they do not understand the content language, which is used to transport the actual mathematical

---

[1] We will for the purposes of this paper subsume all of the implementations by the term MATHWEB, since the communication protocols presented in this paper will make the constructions of bridges between the particular implementations simple, so that that the combined systems appear to the outside as one homogenous web.

J.A. Campbell and E. Roanes-Lozano (Eds.): AISC 2000, LNAI 1930, pp. 32–52, 2001.
© Springer-Verlag Berlin Heidelberg 2001

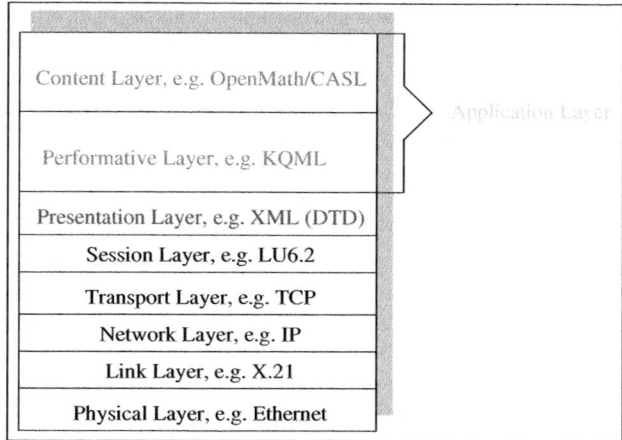

Fig. 1. Artificial Communication: KQML and the OSI Reference Model

content. The agent communication language is used to establish agent identity, reference and – in general – model the communication protocols (see [AK00] for details in the case of mathematical services). Thus we can concentrate on the content language in this paper.

The experience with MathWeb in general, and with the Ωmega system – a mathematical assistant system based on several MathWeb services (see [BCF+97]) – in particular have shown that it is not sufficient to be able to communicate *mathematical objects*, but also *mathematical knowledge* in general. Support for the communication of mathematical objects is already provided by OpenMath, which is

> [...] a standard for representing mathematical objects, allowing them to be exchanged between computer programs, stored in databases, or published on the worldwide web. [...]   [CC98]

This is sufficient for symbolic computation services like computer algebra systems, which manipulate (simplify) or compute objects like equations or groups. Even though the logical formulae constructed or manipulated by reasoning systems like the Ωmega system can be expressed as OpenMath objects, mathematical services like reasoners or presentation systems need more information e.g.:

1. is this formula an axiom, a definition, or a theorem to be proven?
2. what is a good strategy to proceed with the proof in this domain?
3. is this constant basic, or defined (so that it can be expanded to a formula involving simpler concepts)?
4. what is the common name of this concept (and its grammatical category)?

Unfortunately, OpenMath fulfills this goal only partially, since it deals exclusively with the representation of the mathematical objects proper. Of course it would be possible to characterize an axiom by applying a predicate "axiom" to a formula or using a special variant of the equality relation for definitions, but this would only solve item 1 above.

This paper is concerned with the question of a communication standard for *mathematical knowledge*. We propose an extension OMDOC of the OPENMATH standard to alleviate this perceived limitation. We will use mathematical documents as a guiding intuition for mathematical knowledge, since almost all of mathematics is currently communicated in this form (publications, letters, e-mails, talks,...). To ensure widespread applicability, we will use the term document in an inclusive, rather than exclusive way (including papers, letters, interactive books, e-mails, talks, communication between mathematical services (see for instance [FK99b, FHJ+99]) on the Internet,...), claiming that all of these can be fitted into a common representation. Since such documents normallly have a complex structure of their own, the specific task to be achieved in the extension to OPENMATH is to provide a standardized infrastructure for this as well. As we will use the Internet standard XML [BPSM97] (see section 2) as a basis for this, we can consider the syntax problem for communication in MATH-WEB as solved by the imminent wider acceptance of XML (OPENMATH is based on XML and we have defined an XML representation for KQML in [FK99a]).

Another piece of infrastructure which will play a role for understanding OM-DOC is the MBASE system [FK00, KF00], a MATHWEB service that acts as a distributed mathematical knowledge base system that can answer questions such as the ones shown above. OMDOC serves as a *input output language* for MBASE, so that MBASE can be used as a and as document preparation language. Thus the system offers a service that allows storage and (flexibly) reproduction of (parts of) OMDOC documents. As OMDOC can be transformed directly to e.g. LATEX, external input to MBASE can be published directly.

To evaluate the scope of OMDOC, let us look at a few possible applications. OMDOC can serve as

- a *communication standard* between mechanized reasoning systems, e.g. the CLAM-HOL interaction [BSBG98], or the ΩMEGA-TPS [BBS99] integration.
- a data format that supports the *controlled refinement* from informal presentation to formal specification of mathematical objects and theories. Basically, an informal textual presentation can first be marked up, by making its discourse structure[2] explicit, and then formalizing the textually given mathematical knowledge in logical formulae (by adding FMP elements; see sections 5 and 2).
- a basis for *individualized (interactive) books*. OMDOC documents can be generated from MBASE making use of the discourse structure information encoded in MBASE.
- an interface for *proof presentation* [HF97, Fie99]: since the proof part of OMDOC allows small-grained interleaving of formal (FMP) and textual (CMP) presentations.

---

[2] classifying text fragments as definitions, theorems, proofs, linking text, and their relations; we follow the terminology from computational linguistics here.

These and similar applications are pursued in the ΩMEGA project at the Saarland University, Saarbrücken (see `http://www.ags.uni-sb.de/` omega) in cooperation with the RIACA project at Eindhoven.

In the next section we will review the Internet standards and their architecture that are the basis before we come to the definition of OMDoc proper.

2   Markup, XML, OpenMath, MathML, and OMDoc

Mathematical (and other) texts are often written on text processors (which are often WYSIWYG type). Many authors consistently confuse information and document structure with presentation by associating formatting characteristics with various textual document components. Even in LaTeX, one can mix structural markup like `\chapter{Title}` or

$$\begin{Definition}[Title]\dots\end{Definition}$$

with presentation markup, such as font size information, or using

$${\bf proof}:\dots\hfill\Box$$

to indicate the extent of a proof.

The problem with presentation markup is that it is specified for human consumption, and although it is *machine-readable*, the data presented in the document is not *machine-understandable*. Generally, it is very hard to automate anything for documents, when their structure is specified by presentation markup.

With the advent of the Internet, which is quickly becoming the world's fastest growing repository of mathematical documents, it is not possible to manage all the available knowledge manually, because of the volume of information distributed over the Web.

The generally accepted solution is to use *logical* or *generic* markup, i.e. to describe the structure of the data contained in the documents. In this markup scheme, the logical function of all document elements – title, section, paragraphs, figures, tables, bibliographic references, or mathematical equations or definitions – must be clearly defined in a machine-understandable way.

This motivation has led to the development of the "Simple Generalized Markup Language" SGML, and more recently to the "eXtensible Markup Language" XML [BPSM97] family of markup languages. XML was designed as a simplified subset of SGML that can serve as a rational reconstruction of the "Hypertext Markup Language" HTML [RHJ98], which carries most of the markup on the Internet today. From SGML, XML inherits the concept of a "document type definition" (DTD), i.e. a grammar that defines the set of well-formed documents in a given XML language and in particular, allows documents to be validated by generic tools (parsers). Moreover, presentation markup for the data specified in an XML document can be flexibly generated by using the XSL style sheet mechanism [Dea99]. In particular, it is possible to use more than one XSL style sheet for a given document to generate specialized presentations (e.g. personalized to the tastes of a specific reader) of contained data using the content markup in the document.

Thus the "content markup" paradigm gives improved presentation (for human consumption) and improved machine readability at the same time. This has led to considerable activity in developing specialized markup schemes for specific application areas. (This paper is an instance of this activity).

OPENMATH is a content markup language for communicating mathematical objects realized as an XML language. Its syntax (given by a DTD) and semantics are specified in the evolving OPENMATH standard [CC98]. The central construct of OPENMATH is that of an OPENMATH object (OMOBJ), which has a tree-like representation made up of applications (OMA), binding structures (OMBIND using OMBVAR to tag the bound variables), variables (OMV) and symbols (OMS).

Fig. 2 shows an OPENMATH representation of the law of commutativity for addition on the reals (the logical formula $\forall a, b . a \in \mathbf{R} \land b \in \mathbf{R} \to a + b = b + a$). The mathematical meaning of a symbols (that of applications and bindings is

```
<OMOBJ id="commutativity-formula">
  <OMBIND>
    <OMS cd="quant1" name="forall"/>
    <OMBVAR>
      <OMV name="a"/>
      <OMV name="b"/>
    </OMBVAR>
    <OMA><OMS cd="logic1" name="implies"/>
      <OMA><OMS cd="logic1" name="and"/>
        <OMA><OMS cd="set1" name="in"/><OMV name="a"/><OMS cd="barshe" name="real"/></OMA>
        <OMA><OMS cd="set1" name="in"/><OMV name="b"/><OMS cd="barshe" name="real"/></OMA>
      </OMA>
      <OMA><OMS cd="relation" name="eq"/>
        <OMA><OMS cd="barshe" name="plus-real"/><OMV name="a"/><OMV name="b"/></OMA>
        <OMA><OMS cd="barshe" name="plus-real"/><OMV name="b"/><OMV name="a"/></OMA>
      </OMA>
    </OMA>
  </OMBIND>
</OMOBJ>
```

Fig. 2. An OPENMATH representation of $\forall a, b . a + b = b + a$.

known from the folklore) is specified in a so-called content dictionary, which contain formal (FMP "formal mathematical property") or informal (CMP "commented mathematical property") specifications of the mathematical properties of the symbols. For instance, the specification

```
<CDDefinition>
  <Name>plus</Name>
  <Description>Addition on real numbers</Description>
  <CMP>Addition is commutative</CMP>
  <FMP><OMOBJ xref="commutativity-formula"/></FMP>
</CDDefinition>
```

could be part of the content dictionary[3] `barshe.cd` for elementary properties of real numbers (cf. section 4.2 for the relation of content dictionaries with OMDOC documents).

MATHML [IM98] is another XML-based markup scheme for mathematics. In contrast to OPENMATH, it is more concerned with presentation markup (trying to reach LATEX quality on the web) than with logical markup. Moreover, it is mainly concerned with the K-12 fragment of mathematics (Kindergarten to 12th grade). OPENMATH is well-integrated with MATHML:

- the basic content dictionaries of OPENMATH mirror the MATHML constructs, and there are converters between the two formats.
- MATHML supports the `semantics` element that can be used to annotate MATHML presentations of mathematical objects with their OPENMATH encoding, and OPENMATH supports the `presentation` attribute that can be used for annotating with MATHML presentation.
- OPENMATH is the designated extension mechanism for MATHML beyond K-12 mathematics.

Therefore, it is not a limitation of the presentational capabilities to use OPENMATH for marking up mathematical objects. As MATHML can be viewed by the WEBEQ plug-in and is going to be natively supported by the primary browsers MS INTERNET EXPLORER and NETSCAPE NAVIGATOR in version 6 (see `http://www.mozilla.org` for MOZILLA, the open source version), MATHML will be the primary presentation language for OMDOC.

Since OMDOC is an extension of OPENMATH, it inherits its connections to XML and MATHML. The structure of OMDOC documents is defined in the OMDOC document type definition DTD (cf. [Koh00b] or `http://www.mathweb.org/ilo/omdoc`, where you can also find worked examples (including part of a mathematical textbook [BS82] and an interactive book [CCS99] (IDA))).

An OMDOC document is bracketed by the XML tags `<omdoc>` and `</omdoc>`, and consists of a sequence of OMDOC elements, which contain specialized representations for text, assertions, theories, definitions,... (see below). In contrast to markup languages like LATEX, OMDOC does not partition the documents into specific units like chapters, sections, paragraphs, by tags and nesting information, but makes these document relations explicit with `omgroup` elements (see section 7.3). This choice is motivated by the generality of the document classes and the fact that the relative position of OPENMATH documents can be determined in the presentation phase. In particular, since OPENMATH documents can be hypertext documents, or generated from a database, it can be impossible to determine the structure of a document in advance, therefore we consider

---

[3] In fact the reference `<OMOBJ xref="commutativity-formula"/>` pointing to the OMOBJ with the `id` attribute `commutativity-formula` uses an extension of OMDOC to OPENMATH that allows us to represent formulae as directed acyclic graphs preventing exponential blowup. It is licensed by the OPENMATH standard, since pure OPENMATH trees can be generated automatically from it.

document structure information as presentation information and describe it in section 7.3.

The general pattern "definition, theorem, proof" has long been considered paradigmatic of mathematical documents like textbooks and papers. To support this structure, OMDOC provides elements for mathematical items and theory items which we will describe in sections 4 and 5. Since proofs have a more complex internal structure, we will defer them to section 6. Before we come to these, we will describe the structure of intermediate explanatory text (section 3). Finally, we will reserve section 7 for auxiliary items like exercises, applets, etc.

3   Text Elements

The OMDOC text elements are XML elements that can be used to accommodate and classify the explanatory text parts in mathematical documents. We have two kinds of them:

CMP These text elements are used for comments and describing mathematical properties inside other OMDOC elements. They have an xml:lang attribute that specifies the language they are written in; thus using groups of CMPs with different languages can promote OMDOC internationalization. Conforming with the XML recommendation, we use the ISO 639 two-letter country codes (en $\widehat{=}$ English, de $\widehat{=}$ German, fr $\widehat{=}$ French, nl $\widehat{=}$ Dutch...).
CMPs may contain arbitrary text interspersed with OPENMATH objects (OMOBJ elements) (see the OPENMATH standard [CC98] for details), omlets (see section 7) and hyperlinks (see below). No other elements are allowed. In particular, presentation elements like paragraphs, emphases, itemizes,... are forbidden, since OMDOC is concerned with *content markup*. Generating presentation markup from this is the duty of specialized presentation components, e.g. XSL style sheets, which can base their decisions on presentation information (see section 7.3) and the rsrelation information described in this section.

ref elements are used to specify hyperlinks via the XLINK/XPOINTER specification (see http://www.w3c.orgTR/{xlink/xptr}). If the reference object is defined in the same document, then it is sufficient to specify its id attribute in the xlink:href attribute, otherwise, it must include the relevant URL or xpointer material.

omtext OMDOC text elements can appear on the top level (inside omdoc elements). They have an id attribute, so that they can be cross-referenced, an (optional) rsrelation attributes specifying the rhetorical structure relation of the text to other OMDOC elements and contain
1. an (optional) metadata declaration (we use the well-known Dublin Core schema, cf. http://purl.org/dc/ or see [Koh00b])
2. a non-empty set of CMP elements that contain the text proper.

The rsrelation attributes allow us to markup the discourse structure of a document in form of so-called discourse relations following the the well-known

"Rhetorical Structure Theory" RST [MT83, Hor98] content model, which models a text as a tree whose leaves are the sentences (or phrases) and whose internal nodes model the relations between their daughters. This generalizes markup schemes of text fragments offered e.g. by LaTeX into categories like "Introduction", "Remark", or "Conclusion". This is sufficient for simple markup of existing mathematical texts and to replay them verbatim in a browser, but is insufficient e.g. for generating individualized, presentations at multiple levels of abstractions from the representation. The OMDoc text model – if taken to its extreme – can be used to pinpoint the respective role and contributions of smaller text units, even down to the sub-sentence level, and can make the structure of mathematical texts "machine understandable".

Concretely, the `rsrelation` attributes specifies the relation type in a `type` attribute and the RST tree daughters in attributes `for` (for the head daughter) and `from` for the others. At the moment OMDoc uses a variant of the RST [MT83] content model that supports the relation types `introduction`, `conclusion`, `thesis`, `antithesis`, `elaboration`, `motivation`, `evidence`, `linkage` with the obvious meanings, motivated by the application to mathematical argumentative texts (see also [Hor98]). The relation type also determines the default presentation.

4 Theory Elements

Traditionally, mathematical knowledge has been partitioned into so-called **theories**, often centered about certain mathematical objects like groups, fields, or vector spaces. Theories have been formalized as collections of

- signature declarations (the symbols used in a particular theory, together with optional typing information).
- axioms (the logical laws of the theory).
- theorems; these are in fact logically redundant, since they are entailed by the axioms.

In software engineering a closely related concept is known under the label of an (algebraic) specification, which is used to specify the intended behavior of programs. There, the concept of a theory (specification) is much more elaborated to support the structured development of specifications. Without this structure, real world specifications become unwieldy and unmanageable.

In OMDoc, we support this structured specification of theories; we build upon the technical notion of a development graph [Hut99], since this supplies a simple set of primitives for structured specifications and also supports management of theory change. Furthermore, it is logically equivalent to a large fragment of the emerging CASL standard [CoF98] for algebraic specification (see [AHMS00]).

Theories are specified by the `theory` element in OMDoc. Since signature and axiom information is particular to a given theory, the `symbol`, `definition`, `axiom` elements must be contained in a theory as sub-elements.

```
<theory id="monoid-thy">...
  <symbol id="monoid">
    <commonname xml:lang="en">monoid</commonname>
    <commonname xml:lang="de">Monoid</commonname>
    <commonname xml:lang="it">monoide</commonname>
    <type system="simply-typed">
       set[any] -> (any -> any -> any) -> any -> bool
    </type>
  </symbol>...
</theory>
```

Fig. 3. An OMDoc symbol declaration

symbol This element specifies the symbols for mathematical concepts, such as 1 for the natural number "one", + for addition, = for equality, or group for the property of being a group. The symbol element has an id attribute which uniquely identifies it. This information is sufficient to allow referring back to this symbol as an OPENMATH symbol. For instance the symbol declaration in Fig. 3 gives rise to an OPENMATH symbol that can be referenced as <OMS cd="monoid" name="monoid"/>. If the document containing this symbol element were stored in a data base system, the OPENMATH symbol could be looked up by its common name. The type information specified in the signature element characterizes a monoid as a three-place predicate (taking as arguments the base set, the operation and a neutral element).

definition Definitions give meanings to (groups of) symbols (declared in a symbol element elsewhere) in terms of already defined ones. For example the number 1 can be defined as the successor of 0 (specified by the Peano axioms). Addition is usually defined recursively, etc.

The OMDoc definition element supports several kinds of definition mechanisms specified in the type attribute currently:

The FMP (see section 5) contains an OPENMATH representation of a logical formula that can be substituted for the symbol specified in the for attribute of the definition.

The formal part is given by a set of recursive equations whose left and right hand sides are specified by the pattern and value elements in requation elements. The termination proof necessary for the well-definedness of the definition can be specified in the just-by attribute of the definition.

Here, the FMP elements contain a set of logical formulae that uniquely determines the value of the symbols that are specified in the for slot of the definition. Again, the necessary proof of unique existence can be specified in the just-by attribute.

This can be used to directly give the concept defined here as an OPENMATH object, e.g. as a group representation generated by a computer algebra system.

Fig. 4 gives an example a (simple) definition of a monoid.

For a description of abstract data types see [Koh00b]

```
<definition id="mon.d1" for="monoid" type="simple">
  <CMP>
    A structure (M,*,e), in which (M,*) is a semi-group
    with unit e is called a monoid.
  </CMP>
</definition>
```

Fig. 4. A Definition of a monoid

4.1 Complex Theories and Inheritance

Not all definitions and axioms need to be explicitly stated in a theory; they can be inherited from other theories, possibly transported by signature morphism. The inheritance information is stated in an `imports` element.

imports This element has a `from` attribute, which specifies the theory which exports the formulae.

For instance, given a theory of monoids using the symbols `set`, `op`, `neut` (and `axiom` elements stating the associativity, closure, and neutral-element axioms of monoids), a theory of groups can be given by the theory definition using `import` in Fig. 5.

```
<theory id="group">
  <imports id="group.import" from="monoid" type="global"/>
  <axiom><CMP> Every object in
    <OMOBJ><OMS cd="monoid" name="set"/></OMOBJ> has an inverse.
  </CMP></axiom>
</theory>
```

Fig. 5. A theory of groups based on that of monoids

morphism The morphism is a recursively defined function (it is given as a set of recursive equations using the `requation` element, described above). It allows to import specifications modulo a certain renaming. With this, we can e.g. define a theory of rings, where a ring is given as a tuple $(R, +, 0, -, *, 1)$ by importing from a group $(M, \circ, e, i)$ via the morphism $\{M \mapsto R, \circ \mapsto +, e \mapsto 0, i \mapsto -\}$ and from a monoid $(M, \circ, e)$ via the $\{M \mapsto R^*, \circ \mapsto *, e \mapsto 1\}$, where R^* is R without 0 (as defined in the theory of monoids).

inclusion This element can be used to specify applicability conditions on the import construction. Consider for instance the situation given in Fig. 6, where the theory of lists of natural numbers is built up by importing from the theories of natural numbers and lists (of arbitrary elements). The latter imports the element specification from the parameter theory of elements, thus to make the actualization of lists to lists of natural numbers, all the symbols and axioms of the parameter theory must be respected by the natural numbers. For instance if the parameter theory specifies an ordering relation on elements, this must also be present in theory Nat, and have the same properties there. These requirements can be specified in the inclusion element of OMDOC. Due to lack of space, we will not elaborate this and refer the reader to [Hut99, Koh00b].

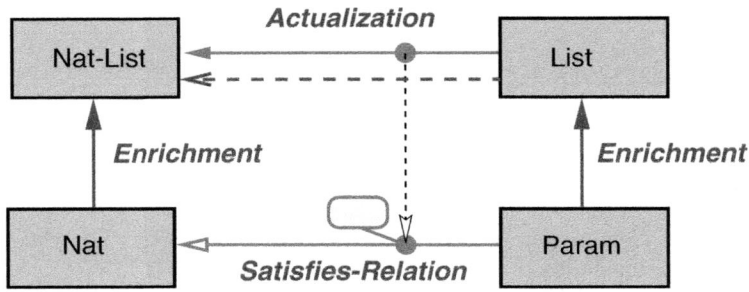

Fig. 6. A Structured Specification of Lists

4.2   OMDOC **Theories and** OPENMATH **Content Dictionaries**

In the examples we have already seen that OMDOC documents contain definitions of mathematical concepts, which need to be referred to using OPEN-MATH symbols. In particular, documents describing theories like barshe.omdoc or ida.omdoc even reference OPENMATH symbols they define themselves. Thus it is necessary to generate OPENMATH content dictionaries from OMDOC documents, or develop an alternative mechanism to establish symbol identity in OMS. The generation of content dictionaries is already supported in the MBASE system, but can also be achieved by writing specialized XSL style sheets. For the purposes of this paper, we will only assume that one of these measures has been taken.

5   Mathematical Elements

We will now present the mathematical elements that are not integral parts of a theory, since they are optional (they can be derived from the material specified in the theory). We have the following elements:

FMP This is the general element for representing mathematical formulae as OpenMath objects, for instance the formula in Fig. 2. As logical formulae often come as sequents, i.e. a conclusion is drawn from a set of assumptions, OMDoc also allows the content of an FMP to be a (possibly empty) set of assumption elements followed by a conclusion. The intended meaning is that the FMP asserts that the conclusion is entailed by the assumptions in the current context. As a consequence, `<FMP>A</FMP>` is equivalent to `<FMP><conclusion>A</conclusion></FMP>`. The assumption and conclusion elements allow to specify the content by an OpenMath object (OMOBJ) or in natural language (using CMPs).

assertion This is the element for all statements (proven or not) about mathematical objects (see Fig. 7). Traditional mathematical documents discern various kinds of these: theorems, lemmata, corollaries, conjectures, problems, etc. These all have the same structure (formally, a closed logical formula). Their differences are largely *pragmatic* (theorems are normally more important in some theory than lemmata) or proof-theoretic (conjectures become theorems once there is a proof). Therefore, we represent them in the general assertion element and leave the type distinction to a type attribute. These type specifications in OMDoc documents should only be regarded as defaults, since e.g. reusing a mathematical paper as a chapter in a larger monograph, may make it necessary to downgrade a theorem (e.g. the main theorem of the paper) and give it the status of a lemma in the overall work.

```
<assertion id="ida.c6s1p4.11" type="lemma">
  <CMP> A semi-group has at most one unit.</CMP>
</assertion>
```

Fig. 7. An assertion about semigroups

alternative-def Since there there can be more than one definition per symbol, OMDoc supplies the alternative-def. It not only contains the new definition, but also points to two assertions that state the equivalence with definitions of the concepts that are already known.

example In mathematical practice, examples play an equally great role as proofs, e.g. in concept formation (as witnesses for definitions, or as either supporting evidence or as counterexamples for conjectures). Therefore, examples are given status as primary objects in OMDoc. Conceptually, we model an example for a mathematical concept $\mathbf{C}$ as a triple $(\mathcal{W}, \mathbf{A}, \mathcal{P})$, where $\mathcal{W} = (\mathcal{W}_1, \dots, \mathcal{W}_n)$ is an n-tuple of mathematical objects, $\mathbf{A}$ is an assertion of the form $\mathbf{A} = \exists W_1 \dots W_n . \mathbf{B}$, and $\mathcal{P}$ is a proof that shows $\mathbf{A}$ by exhibiting the witnesses $\mathcal{W}_i$ for W_i. The example $(\mathcal{W}, \exists W_1 \dots W_n . \neg \mathbf{B}, \mathcal{P})$ is a counterexample to a conjecture $\mathbf{T} := \forall W_1 \dots W_n . \mathbf{B}$, and $(\mathcal{W}, \mathbf{A}, \mathcal{P}')$ a supporting example for $\mathbf{T}$.

OMDoc specifies this intuition in an element example that contains a set of OpenMath objects (the witnesses), and has the attributes

- `for` (for what concept or assertion is it an example),
- `type` (one of the keywords or for the function)
- `assertion` (a reference to the assertion **A** mentioned above)
- `proof` (a reference to the constructive proof $\mathcal{P}$)

Consider for instance the structure $\mathcal{W} := (A^*, \circ)$ of the set of words over an alphabet A together with word concatenation $\circ$. Then $(\mathcal{W}, \exists W.\mathtt{monoid}(W), \mathcal{P}_1)$ is an example for the concept of a monoid (with the empty word as the neutral element), if e.g. $\mathcal{P}_1$ uses $\mathcal{W}$ to show the existence of W. The example $(\mathcal{W}, \exists V_{\mathtt{monoid}}.\neg\mathbf{group}(V), \mathcal{P}_2)$ uses $\mathcal{W}$ as a counterexample to the conjecture $\mathbf{C} := \forall V_{\mathtt{monoid}}.\mathbf{group}(V)$, since $\mathbf{Q} \to \neg\mathbf{C}$ ($\mathcal{P}_2$ uses $\mathcal{W}$ as a witness for V). Fig. 8 gives the OMDOC representation of this example of an example.

```
<example id="mon.ex1" for="monoid" type="for"
         assertion="strings-are-monoids" proof="sam-pf">
  <CMP>The set of strings with concatenation</CMP>
  <OMOBJ><OMS cd="simple-monoids" name="strings"/></OMOBJ>
</example>
<example id="mon.ex2" for="monoid" type="against"
         assertion="monoids-are-groups" proof="mag-pf">
  <CMP>The set of strings with concatenation is not a group</CMP>
  <OMOBJ><OMS cd="simple-monoids" name="strings"/></OMOBJ>
</example>
```

Fig. 8. An OMDOC representation of an example

Finally, there are OMDOC elements that support structuring the knowledge in theories. We have already seen the possibility to define (parts of) theories by so-called theory morphism specified in `imports` and `include` elements in section 4.1. Following Hutter's development graph [Hut99], we can use the knowledge about theories to establish so-called inclusion morphisms that establish the source theory as included (modulo renaming by a morphism) in the target theory. This information can be used to add further structure to the theory graph and help maintain the knowledge base with respect to changes of individual theories.

An `axiom-inclusion` element contains a `morphism` (see section 4.1), and the attributes `from` and `to` specify the source and target theories. For any axiom in the source theory there must be an assertion in the target theory (whose FMP is just the image of the FMP of the axiom under the morphism) with a proof. These are represented by an empty `by` element, which has the attributes `axiom`, `assertion`, and `proof` with the obvious meanings.

A `theory-inclusion` is a global variant of `axiom-inclusion` that can be obtained as a path of `axiom-inclusions` (or other `theory-inclusion`) which are specified in the `by` attribute.

6   Proofs

Proofs are representations of evidence for the truth of *assertion*. As in the case of definitions, there can in general be more than one proof for a given assertion. Furthermore, it will be initially infeasible to formalize totally all mathematical proofs needed for the correctness management of the knowledge base in one universal proof format, therefore OMDOC supports a proof format whose structural and formal elements are derived from the $\mathcal{PDS}^4$ structure developed for the ΩMEGA system, but also allows natural language representations at every level. In the future, it may be necessary and advantageous to allow various other proof representations there like proof scripts (ΩMEGA replay files, ISABELLE proof scripts,...), references to published proofs, resolution proofs, etc, to enhance the coverage.

This mixed representation enhances multi-modal proof presentation [Fie97], and the accumulation of proof information in one structure. Informal proofs can be formalized [Bau99]; formal proofs can be transformed to natural language [HF96].

The OMDOC `proof` environment contains a list of proof steps. Such `derive` steps have the attributes `id` (so it can be referred to) and the optional `type` attribute. It can contain the following child elements (in this order)

CMP This gives the natural language representation of the proof step.
The rest of the children form the formal content of the derive step. Together, they represent the information present e.g. in a $\mathcal{PDS}$ node.

FMP A formal representation of the assertion made by this proof step, they contain CMP and FMP elements. Local assumptions from the FMP should not be referenced to outside the derive step they were made in. Thus the derive step serves as a grouping device for local assumptions.

method is an OPENMATH symbol representing a proof method or inference rule that justifies the assertion made in the FMP element.

premise These are empty elements whose `xref` attribute is used to refer to the proof- or local assumption nodes that the `method` was applied to to yield this result. These attributes specify the DAG structure of the proof.

proof If a derive step is a logically (or even mathematically) complex step that can be expanded into sub-steps, then the embedded `proof` element can be used to specify the sub-derivation (which can have similar expansions in embedded `proof` environments again).
This embedded `proof` allows us to specify generic markup for the hierarchic structure of proofs.

---

[4] The **P**roof plan **D**ata **S**tructure ($\mathcal{PDS}$) was introduced in the ΩMEGA [BCF+97] system to facilitate hierarchical proof planning and proof presentation at more than one level of abstraction. In a $\mathcal{PDS}$, expansions of nodes justified by tactic applications are carried out, but the information about the tactic itself is not discarded in the process as in tactical theorem provers like ISABELLE or NUPRL. Thus proof nodes may have justifications at multiple levels of abstraction in a hierarchical proof data structure.

```
<derive id="barshe.2.1.2.proof.a.proof.D2.1">
  <CMP>By <OMOBJ><OMS cd="barshe" name="alg-prop-reals.A2"/></OMOBJ>
     we have z + (a + (−a)) = a + (−a)
  </CMP>
  <conclusion>(z + a) + (−a) = z + (a + (−a))</conclusion>
  <method><OMS cd="omega-base-calc" name="foralli*"/>c
     <parameter><OMOBJ><OMV name="z"/></OMOBJ></parameter>
     <parameter><OMOBJ><OMV name="a"/></OMOBJ></parameter>
     <parameter>−a</parameter>
  </method>
  <premise xref="alg-prop-reals.A2"/>
</derive>
```

Fig. 9. A derive proof step

7 Auxiliary Elements

In this section we will present OMDoc elements that are not strictly mathematical content, but have useful functions in mathematical documents or knowledge bases. For the OMDoc representations of things like exercises we refer the reader to [Koh00b] and concentrate on the representation of applets and presentation information instead.

7.1 Non-Xml Data and Program Code in OMDoc

Sometimes mathematical services have to be able to communicate (e.g. to the MBase system for storage) data in non-Xml syntax, or whose format is not sufficiently fixed to warrant for a general Xml encoding. Examples of this are pieces of program code, like tactics of tactical theorem provers, linguistic data of proof presentation system, etc. One characteristic of such data seems to be that it is private to certain applications, but may be relevant to more than one user. For this, OMDoc provides the private element, which contains the usual CMPs and a data element described below. It has the attributes

pto specifies the system to which the data are private.
pto-version is its version; Specifying this may be necessary, if the data or even their format change with versions.
format/type the type of the data and the format the data are in, the meaning of these fields is determined by the system itself.
requires specifies the identifiers of the elements that the data depend upon, which will often be code elements.
theory allows the specification of the mathematical theory (see section 4) that the data is associated with.

The data element contains the data of a in a CDATA section (this is the Xml way of allowing data that cannot be parsed by the Xml parser). If the content of this field is too large to store directly in the OMDOC or often changes, then it can be substituted by a link, specified in the xref attribute.

The code element is for embedding pieces of code into an OMDOC document. This element has the same attributes as the private element, like it, it can contain CMP, and data elements. Furthermore, it can contain documentation elements input, output and effect that specify the behavior of the procedure defined by the code fragment.

7.2   Applets in OMDOC

omlet elements contain OMDOC specifications of applets (program code that can in some way executed during document manipulation). omlets generalize the well-known *applet* concept in two ways: The computational engine is not restricted to *plug-ins* of the browser (current servlet technology can be used and specified using code and omlet elements in OMDOCs) and the program code can be specified and distributed more easily, making document-centered computation easier to manage.

```
<code id="callmint">
  <input>None</input>
  <output>The result</output>
  <effect>None</input>
  <data><![CDATA[...  the call-mint code goes here ...]]></data>
</code>
<derive id="monp_1">
  <CMP> <omlet type="js" function="callMint">Intros.</omlet></CMP>
  <method><OMS name="Intros" cd="COQ"/></method>
</derive>
```

Fig. 10. An omlet

Like the HTML applet tag, the omlet element can be used to wrap any (set of) well-formed elements. It has the following attributes.

type This specifies the computation engine that should execute the code. Depending on the application, this can be a programming language, such as javascript (js) or Oz, or a process that is running (in our case the $\mathcal{L\Omega UI}$ or ΩMEGA services).

function The code that should be executed by the omlet is specified in the function attribute. This points to an OMDOC code element that is accessible in some way (e.g. in the same OMDOC). This indirection allows us to reuse the machinery for storing code in OMDOCs. For a simple example see Fig. 10.

argstr allows specification of an (optional) argument string for the function. A call to the $\mathcal{L\Omega UI}$ interface would then have the form in Fig. 11. Here, the code in the `code` element `sendtoloui` (which we have not shown) would be java code that simply sends the argstr to $\mathcal{L}\Omega\mathcal{U}\mathcal{I}$'s remote control port.

The expected behavior of the `omlet` can be implemented in the XSL style sheet, which in the case of e.g. translation to MOZILLA will put the `callmint` code directly into the generated `html`.

```
<CMP> Let's prove it
  <omlet id="bla" type="java" function="sendtoloui"
       argstr="load(problem='monoid_uniq')">
    interactively
  </omlet>
</CMP>
```

Fig. 11. An `omlet` calling an external process

7.3 Presentation

In the introduction we have stated that one of the design intentions behind OM-Doc is to separate content from presentation, and leave the latter to the user. In this section, we will briefly touch upon presentation issues. The technical side of this is simple: OMDoc documents are regular XML documents that can be processed by XSL [Dea99] style sheet to produce conventional presentations from OMDoc representations of mathematical documents. At the moment, we have XSL style sheets to convert OMDoc to HTML (one each specialized to the respective browsers), LaTeX, and to the input languages of the ΩMEGA, INKA, and λ*Clam* systems (they can be found at http://www.mathweb.org/ilo/omdoc). At the moment, these hard-code certain presentation decisions for the overall appearance of the documents, but we are working on style sheet generators that make these user-adaptive.

The mathematical concepts and symbols introduced in an OMDoc document (`symbol` elements) often carry typographic conventions, which cannot be determined by general principles alone. Therefore, they need to be specified in the document itself, so that typographically good representations can be generated from this (and subsequent) documents. The `presentation` element in Fig. 12 allows the addition of XSL style sheet information to symbols, where they are defined. In this case, the style sheet information will cause an OPENMATH expression

```
<OMA>
  <OMS cd="ida" name="monoid"/><OMV name="M"><OMV name="o"><OMV name="e">
</OMA>
```

to be rendered as $(M, o, e) \in \mathbf{MOD}$ in a TeX or LaTeX document derived from `ida.xml` via a suitable XSL style sheet. Of course, this information will need to

```
<presentation format="TeX">
  <xsl:template match="OMA[OMS[position()=1 and
                               @name='monoid' and
                               @cd='ida.monoid']]">
    (<xsl:apply-templates select="*[2]"/>,
     <xsl:apply-templates select="*[3]"/>,
     <xsl:apply-templates select="*[4]"/>)\in{\bf MON}
  </xsl:template>
</presentation>
```

Fig. 12. XSL Presentation for the symbol in Fig. 3

be included into the respective style sheets. This is easily realized by a two-stage style sheet process: in the first pass, a general (higher-order) style sheet extracts the presentation information from the relevant OMDOC documents, and in the second stage, this is used to present the OMOBJs in the source OMDOC.

The presentation elements discussed up to now, allow specification of the presentation of OPENMATH elements. To specify the overall structure of mathematical texts, such as books, chapters, sections, or paragraphs, but also enumerations, itemizes, lists, we use the omgroup element. We use a general construct that specifies the presentation in the type attribute, since the presentation component (style sheet) may need to decide on that. omgroup elements contain an optional metadata element and then a sequence of omgroup and ref elements. The first allow the definition of a recursive document structure, and elements of the second kind are used to refer to other OMDOC elements by the use of xlink attributes (most notably xlink:href for hyperlinks).

Note that this representation, which relies on explicit (hyper)-references instead of nesting information allows the specification of more than one document using the mathematical material specified in the other OMDOC elements. In particular, it becomes possible to specify and store more than one linearization of the material in a document, or generate linearization or "guided tours" (see [SBC+00] for details).

8   Conclusion

We have proposed an extension to the OPENMATH standard that allows the representation of the semantics and structure various kinds of mathematical documents, including articles, textbooks, interactive books, courses. We have motivated and described the language and presented an XML document type definition for it.

We are currently testing this in the development of a user-adaptive interactive book including proof explanation based on IDA [CCS99] in close collaboration with the authors. This case study unites several of the application areas discussed in the introduction. The re-representation of IDA in the OMDOC format makes it possible to machine-understand the structure of the document, read it into the MBASE [FK00, KF00] knowledge base system without loss of information, preserving the structure, and generate personalized sub-documents or linearizations of the structured data based on a simple user model. Furthermore, the OMDOC representation supports the formalization of (parts of) the mathematical knowledge in IDA and makes it accessible to the ΩMEGA mathematical assistant system [BCF+97], which can find proofs that solve some of the problems either fully automatically (by proof planning) or in interaction with the authors. This newly developed stock of formal data (it is not present in IDA now) will enable the reader to read and experiment with the proofs behind the mathematical theory, much as she can in the present version with the integrated computer algebra system GAP [S+95]. Finally, OMDOC will serve as the input format for the LIMA system (see [Bau99]), an experimental natural language understanding system specialized to mathematical texts (this can be used to develop formalization in FMPs from the text in the respective CMPs).

In the context of this project, we have developed first authoring tools for OMDOC that try to simplify generating OMDOC documents for the working mathematician. There is a simple OMDOC mode for emacs, and a LaTeX style [Koh00a] that can be used to generate OMDOC representations from LaTeX sources and thus help with the migration of existing mathematical documents. A second step will be to integrate the LaTeX to OPENMATH conversion tools. Michel Vollebregt has built a program that traverses an OMDOC and substitutes various representations for formulae (including the MATHEMATICA, GAP, and MAPLE representations) by the corresponding OPENMATH representations.

Acknowledgments

The work presented in this report was supported by the "Deutsche Forschungsgemeinschaft" in the special research action "Resource-adaptive cognitive processes" (SFB 378), Project ΩMEGA.

The author would like to thank Armin Fiedler, Andreas Franke, Martin Pollet, and Julian Richardson for productive discussions, and the RIACA group (specifically Arjeh Cohen, Olga Caprotti, Michel Vollebregt, and Manfred Riem) for valuable input from the IDA case study.

References

[AHMS00]   Serge Autexier, Dieter Hutter, Heiko Mantel, and Axel Schairer. Towards an evolutionary formal software-development using CASL. In C. Choppy and D. Bert, editors, *Proceedings Workshop on Algebraic Development Techniques, WADT-99*. Springer, LNCS 1827, 2000.

[AK00]   Alessandro Armando and Michael Kohlhase. Communication protocols for mathematical services based on KQML and OMRS. In Manfred Kerber and Michael Kohlhase, editors, *CALCULEMUS-2000, Systems for Integrated Computation and Deduction*, 2000, AKPeters.

[AvLS96]   J. Abbot, A. van Leeuwen, and A. Strotmann. Objectives of Open-Math. Technical report 12, RIACA, Technische Universiteit Eindhoven, The Netherlands, June 1996.

[AZ00]   Alessandro Armando and Daniele Zini. Towards Interoperable Mechanized Reasoning Systems: the Logic Broker Architecture. In A. Poggi, editor, *AI*IA-TABOO Workshop 'From Objects to Agents: Evolutionary Trends of Software Systems'*, 2000.

[Bau99]   Judith Baur. Syntax und semantik mathematisher texte — ein prototyp. Master Thesis, Saarland University, 1999.

[BBS99]   Christoph Benzmüller, Matthew Bishop, and Volker Sorge. Integrating Tps and ΩMEGA. *Journal of Universal Computer Science*, 5(2), 1999.

[BCF$^+$97] The ΩMEGA group. ΩMEGA: Towards a mathematical assistant. In William McCune, editor, *CADE-14*, LNAI 1249, pages 252–255, 1997. Springer Verlag.

[BPSM97]   Tim Bray, Jean Paoli, and C. M. Sperberg-McQueen. Extensible Markup Language (XML). W3C Recommendation TR-XML, December 1997. http://www.w3.org/TR/PR-xml.html.

[BS82]   Robert G. Bartle and Donald Sherbert. *Introduction to Real Analysis*. Wiley, 2 edition, 1982.

[BSBG98]   R. Boulton, K. Slind, A. Bundy, and M. Gordon. An interface between CLAM and HOL. In Jim Grundy and Malcolm Newey, editors, *TPHOLS-98*, pages 87–104, 1998.

[CC98]   Olga Caprotti and Arjeh M. Cohen. Draft of the Open Math standard. The Open Math Society, http://www.nag.co.uk/projects/OpenMath/omstd/, 1998.

[CCS99]   Arjeh Cohen, Hans Cuypers, and Hans Sterk. *Algebra Interactive!* Springer Verlag, 1999. Interactive Book on CD.

[CoF98]   Language Design Task Group CoFI. Casl - the CoFI algebraic specification language - summary, version 1.0. Technical report, http://www.brics.dk/Projects/CoFI, 1998.

[DCN$^+$00] Louise A. Dennis, Graham Collins, Michael Norrish, Richard Boulton, Konrad Slind, Graham Robinson, Mike Gordon, and Tom Melham. The Prosper toolkit. In *Proc. TACAS-2000*, 2000.

[Dea99]   Stephen Deach. Extensible stylesheet language (xsl) specification. W3c working draft, W3C, 1999. http://www.w3.org/TR/WD-xsl.

[FF94]   T. Finin and R. Fritzson. KQML — a language and protocol for knowledge and information exchange. In *Proceedings of the 13th Intl. Distributed Artificial Intelligence Workshop*, pages 127–136, 1994.

[FHJ$^+$99] Andreas Franke, Stephan M. Hess, Christoph G. Jung, Michael Kohlhase, and Volker Sorge. Agent-oriented integration of distributed mathematical services. *Journal of Universal Computer Science*, 5:156–187, 1999.

[Fie97]   Armin Fiedler. Towards a proof explainer. In Siekmann et al. [SPH97], pages 53–54.

[Fie99]   Armin Fiedler. Using a cognitive architecture to plan dialogs for the adaptive explanation of proofs. In Thomas Dean, editor, *Proceedings IJCAI-99*, pages 358–363, 1999. Morgan Kaufmann.

[FK99a]   Andreas Franke and Michael Kohlhase. Communicating with MBASE in KQML. Internet Draft http://www.mathweb.org/mbase, 1999.

[FK99b]   Andreas Franke and Michael Kohlhase. System description: MATHWEB, an agent-based communication layer for distributed automated theorem proving. In Harald Ganzinger, editor, *Proceedings CADE-16*, LNAI 1632, pages 217–221. Springer Verlag, 1999.

[FK00]   Andreas Franke and Michael Kohlhase. System description: MBASE, an open mathematical knowledge base. In David McAllester, editor, *CADE-17*, LNAI 1831, pages 455–459. Springer Verlag, 2000.

[HF96]   Xiaorong Huang and Armin Fiedler. Presenting machine-found proofs. In M.A. McRobbie and J.K. Slaney, editors, *Proceedings CADE-13*, LNAI 1104, pages 221–225, 1996. Springer Verlag.

[HF97]   Xiaorong Huang and Armin Fiedler. Proof verbalization in *PROVERB*. In Siekmann et al. [SPH97], pages 35–36.

[Hor98]   Helmut Horacek. Generating inference-rich discourse through revisions of RST-trees. In *Proceedings AAAI-98*, pages 814–820. MIT Press, 1998.

[Hut99]   Dieter Hutter. Reasoning about theories. Technical report, DFKI, 1999.

[IM98]   Patrick Ion and Robert Miner. Mathematical Markup Language (MathML) 1.0 specification. W3C Recommendation 1998. http://www.w3.org/TR/REC-MathML/.

[KF00]   Michael Kohlhase and Andreas Franke. Mbase: Representing knowledge and context for the integration of mathematical software systems. *Journal of Symbolic Comutation*, 2000. forthcoming.

[Koh00a]   Michael Kohlhase. Creating OMDOC representations from LaTeX. Internet Draft http://www.mathweb.org/omdoc, 2000.

[Koh00b]   Michael Kohlhase. OMDOC: Towards an OPENMATH representation of mathematical documents. Seki Report SR-00-02, Fachbereich Informatik, Universität des Saarlandes, 2000.

[MT83]   William Mann and Sandra Thompson. Rhethorical structure theory: A theory of text organization. Technical Report ISI/RR-83-115, ISI at University of Southern California, 1983.

[RHJ98]   Dave Raggett, Arnaud Le Hors, and Ian Jacobs. HTML 4.0 Specification. W3C Recommendation 1998. http://www.w3.org/TR/PR-xml.html.

[S+95]   Martin Schönert et al. *GAP – Groups, Algorithms, and Programming*. RWTH Aachen, Germany, 1995.

[SBC+00]   The ΩMEGA group. Adaptive course generation and presentation. In P. Brusilovski, editor, *Proceedings of ITS-2000 workshop on Adaptive and Intelligent Web-Based Education Systems*, Montreal, 2000.

[Sho90]   Y. Shoham. Agent-Oriented Programming. Technical report, Stanford University, 1990.

[SHS98]   M. Kohlhase S. Hess, Ch. Jung and V. Sorge. An implementation of distributed mathematical services. In Arjeh Cohen and Henk Barendregt, editors, *CALCULEMUS and TYPES*, 1998.

[SPH97]   J. Siekmann, F. Pfenning, and X. Huang, editors. *Proceedings of the First International Workshop on Proof Transformation and Presentation*, Schloss Dagstuhl , Germany, 1997.

On Communicating Proofs in Interactive Mathematical Documents

Olga Caprotti and Martijn Oostdijk

Eindhoven University of Technology
The Netherlands
{olga,martijno}@win.tue.nl

Abstract. There is a wealth of interactive mathematics available on the web. Examples range from animated geometry to computing the nth digit in the expansion of π. However, proofs seem to remain static and at most they provide interaction in the form of links to definitions and other proofs. In this paper, we want to show how interactivity can be included in proofs themselves by making them executable, human-readable, and yet formal. The basic ingredients are formal proof-objects, OpenMath-related languages, and the latest eXtensible Markup Language (XML) technology. We exhibit, by an example taken from a formal development in number theory, the final product of which we believe to be a truly interactive mathematical document.

Keywords: Interactive mathematical documents, Formal mathematical proofs, Type theory, Markup languages.

1 Introduction

One may broadly classify the many examples of online mathematical documents promising interactivity into two categories: textual documents that are hyperlinked and allow readers to consult several pieces of information by comfortably clicking through the links, and documents that are interfaces to software agents performing computations (for instance, to visualize graphically a surface, to solve a system of equations or to find references of papers in which some integer sequence occurred). Both kinds of documents are considered interactive in that the user is actively involved in producing the final reading material. Unfortunately each document has its own notation and if mathematical objects are used, then they can hardly be directly utilized in a different setting. In this form, the mathematical knowledge cannot be shared.

Although JAVA applets and cgi-scripts have provided the support for embedding graphical and computational facilities into an interactive mathematical document, little interactivity is currently available in proofs. One area in which applets have actually been used in "proofs" is planar geometry [25] where they can be applied naturally. If we adhere to the idea that the essence of doing mathematics is proving, then we must conclude that mathematical knowledge is still

J.A. Campbell and E. Roanes-Lozano (Eds.): AISC 2000, LNAI 1930, pp. 53–64, 2001.
© Springer-Verlag Berlin Heidelberg 2001

poorly communicated interactively. On the other hand, symbolic computation systems, like computer algebra packages and automated proof environments, can play a key role in supporting the development of interactive proofs, to generate the content, validate it, and act as back engines during online presentation.

In this paper we present a possible approach. We consider tactics-based proof assistants, in the tradition of the AUTOMATH project [23], thus excluding resolution-based theorem provers. We focus on type-theoretical proof assistants such as Coq, Lego, and NuPRL [11, 22, 10], yet some of the technologies involved can be applied to proof plans in the general sense.

The main appeal of using proofs developed in type-theoretical systems is that proofs are terms in the formal language of the system. Proving occurs by interaction with the system through specialized user-friendly interfaces [20] designed to produce formal proofs. These formal proofs are often too detailed and become hard to read. This explains the many efforts made in producing more natural descriptions of the resulting proofs [12, 2, 21, 17]. Still, these efforts produce only "flat text" akin to conventional informal proofs and not directly suited for supporting interaction. In fact, such text is meant to be read by humans and not by programs. For interactive use, a published proof also needs to access the original formal proof that produced the text; the flat text version of the proof alone is not enough.

In our approach, presentation and content are kept distinct. The formal proof is used as content, and multiple views, i.e. multiple presentations of the same content, are generated. In order to do so, the formal proof is encoded using OpenMath [9], a standard markup language for mathematical content. The overall mathematical document is generated automatically in the OpenMath Document [19] format. This allows the inclusion of OpenMath objects such as formal proofs, flat text containing informal mathematics and tactic scripts. In this way, sharing of mathematical knowledge and transparent interaction with computational tools are achieved. In the paper we present some of the technologies that we have developed and that are currently being used for the next version of "Algebra Interactive!" [8, 6].

The paper is structured as follows. Section 2 discusses the different options for including proofs in interactive mathematical documents. In Section 3, some of the basic knowledge needed to understand type-theoretical theorem proving is briefly recalled using an example. The technologies involved in bringing proofs online are described in Section 4. The concluding remarks are found in Section 5.

2   Proofs in Interactive Mathematical Documents

In this section we study the options for embedding proofs within an interactive mathematical document.

The first and most obvious is to include a textual version of the proof in natural language, similar to proofs in traditional (paper) mathematical texts. Indeed, one of the first tools used to publish mathematics on the web was a converter from TEX to HTML. Nowadays, most word processing software is able

	NL Text	Tactics Scripts	Proof Objects
OM encodable	-	-	+
mathematical object	-	-	+
readable	+	+/-	-
machine checkable	-	+/-	+

Table 1. Pros and cons of different proof embedding options.

to output HTML and in some cases the mathematics is not replaced by an image (a gif file) but by its presentation in the MathML language [3].

Although this form of presentation does contain structure, for example references to lemmata or definitions, the format is usually plain text. Because it consists of mathematical vernacular, it is very easy to read but hard to reconstruct a fully formal proof from it. Having the mathematical content semantically marked-up is desirable for two reasons. First, it can be communicated to computational software such as a computer algebra system or a proof checker. Second, the structure can be used to open different views on the proof. Allowing the user to change views is one form of interactivity we achieve in our approach.

The second way to embed proofs is to include high-level tactics scripts which serve as input for theorem provers. This approach is opposite to the former one in that it takes the notion of proof from a theorem prover, i.e. a formal system, instead of from informal mathematics. Such scripts are system dependent and, in general, it is easy to communicate the proof to the specific system and have it checked. Users of the system usually regard tactics scripts as real proofs and do not find it difficult to understand them even though they show only one half of a dialogue. In recent work, high-level tactic scripts are used to produce natural language verbalization of the proof [17].

The third way to embed proofs is to include a formal proof, for instance a derivation tree. In a type-theoretical setting, the corresponding lambda term is a good candidate for the formal representation of the proof. Because it is a mathematical term, it can be encoded using a markup language for mathematics and it or parts of it can be communicated to computational software in a standard way. A major drawback to this option is that formal proofs contain many details and hence are not very readable, moreover their marked-up encoding becomes even larger. However, this drawback can be overcome by creating suitable views on the formal tree leaving out the undesired details and adding information to clarify the formal structure, see for instance our natural language view in Figure 1.

Table 1 sums up the pros and cons of the above three options.

In this paper we propose to mix the different options, thus retaining the readability of textual proofs, the ability to step through the proof interactively, and the rigor of formal proofs. Type theoretical systems are based on the "propositions as types" paradigm which is recalled in the next section through examples.

3   Example of a Formalization in Type Theory

In this section we briefly sketch the process of formally developing a mathematical theory in a proof assistant. The examples we use arise during the formalization, done in Coq, of Pocklington's Criterion [24, 7, 4] for finding whether a positive number is prime.

In the type theory of Coq, mathematical concepts are encoded as typed expressions. For instance, one can introduce the notions of division and primality of natural numbers by the following definitions. Divides is defined in the usual way as a predicate on two natural numbers n and m (usually denoted as $n|m$), and Prime is defined as a predicate on a natural number n.

```
Definition Divides: nat -> nat -> Prop :=
   [n,m:nat](EX q:nat | m=(mult n q)).

Definition Prime: nat -> Prop :=
   [n:nat](gt n (1)) /\ (q:nat)(Divides q n) -> q=(1)   q=n.
```

Besides concepts from the mathematical object language, also concepts from the meta language such as theorems and proofs are encoded as typed expressions. This principle is called the *Curry-Howard correspondence*, also known as *propositions as types*.

A proof is encoded as an object which has, as its type, the encoding of the statement of the theorem. A typed expression which represents a proof is called a *proof-object*. If an expression of type Prop has more than one inhabitant, those inhabitants represent different proof-objects of the proposition. If it has no inhabitants, then it cannot be proved.

Consider as example the observation that if all prime divisors of a positive number n are greater than $\sqrt{n}$, then the number n is prime. This can be stated in Coq as a new theorem (primepropdiv) that uses library functions (mult and gt) and newly defined predicates (Divides and Prime):

```
Lemma primepropdiv:   (n:nat)(gt n  (1))            ->
                      ((q:nat)(Prime q)             ->
                            (Divides q n)      ->
                            (gt (mult q q) n)) ->
                      (Prime n).
```

To assist the user in constructing proof-objects, Coq uses a high-level language of *tactics*. A user of Coq, trying to prove primepropdiv, would be presented with the initial goal and interactively arrive at the following script.

```
Intros. Elim (primedec n). Intro. Assumption.
Intros. Elim (nonprime_primewitness n).
Intros. Elim H2. Intros. Elim H4. Intros.
Elim H6. Intros. Elim (le_not_lt (mult x x) n).
Assumption. Unfold gt in HO. Apply HO.
Assumption. Assumption. Assumption. Assumption.
```

The script uses lemmata from the Coq `arith` library and previously proven lemmata, like `primedec` which proves the decidability of the `prime` predicate and `nonprime_primewitness` which proves that non-prime numbers greater than 1 have prime-divisors. The proof-object `primepropdiv` is given below.

```
[n:nat;
H:(gt n (1));
H0:((q:nat)(Prime q)->(Divides q n)->(gt (mult q q) n))]
(or_ind (Prime n) ~(Prime n) (Prime n) [H1:(Prime n)]H1
[H1:(~(Prime n))]
(ex_ind nat
[d:nat](lt (1) d) /\ (le (mult d d) n) /\ (Divides d n) /\ (Prime d)
(Prime n)
[x:nat;
H2:((lt (1) x) /\ (le (mult x x) n) /\ (Divides x n) /\ (Prime x))]
(and_ind (lt (1) x) (le (mult x x) n) /\ (Divides x n) /\ (Prime x)
(Prime n)
[_:(lt (1) x);
H4:((le (mult x x) n) /\ (Divides x n) /\ (Prime x))]
(and_ind (le (mult x x) n) (Divides x n) /\ (Prime x)
(Prime n)
[H5:(le (mult x x) n); H6:((Divides x n) /\ (Prime x))]
(and_ind (Divides x n) (Prime x) (Prime n)
[H7:(Divides x n); H8:(Prime x)]
(False_ind (Prime n)
(le_not_lt (mult x x) n H5 (H0 x H8 H7))) H6) H4) H2)
(nonprime_primewitness n H H1)) (primedec n))
```

If available, such a proof-object can be used in an interactive document describing Pocklington's Criterion in several ways. First of all, it provides evidence of the truth of the assertion it proves: such a term can be easily checked by Coq to be of type `primepropdiv`. If it is encoded in a system-independent standard language such as OpenMath, then it can be shared. Moreover, it can also be used to produce a natural language view of the proof it represents since in its current form it is not readable. Part of our technology includes a tool that, when given a context and a proof object, produces a natural language view of the associated proof. The tool can interactively adapt the level at which the proof is displayed by collapsing or expanding certain sentences. The major technologies involved are described in Section 4.

4   Enabling Technologies

This section introduces the core technologies developed to support our approach to proofs in interactive mathematical documents. The natural language view is strongly connected to formal proof objects based on type theory. OpenMath and related technologies enable the representation of structured mathematical information such as a proof term, its context, a tactics script and natural language explanations.

4.1   The Natural Language View

The natural language viewer, described in [?], extends the standard algorithm for translating Coq proof-objects presented in [13]. The input is a Coq context and a proof-object. The output obtained by the standard algorithm is a detailed natural language proof. Our implementation takes into account requirements for interactivity and improves upon it in two ways.

First, instead of producing flat text, the final presentation is an adjustable view. It is generated from an object which is tightly connected to the original formal proof-object. This intermediate object contains both natural language parts and formal parts. When rendering such an object on the screen, the formal parts of a sentence may be treated differently. For example, whenever the name of a definition occurs, the renderer produces a hyperlink to the place in the context where the definition is introduced.

Second, recursive calls of the translation algorithm result in a sentence which can be expanded or collapsed by the reader, similar to the display of a directory structure in file-browser in modern GUI systems. When the sentence is collapsed, instead of recursively translating the corresponding subproof, the renderer displays the type of the subproof. Figure 1 shows a combination of the natural language view with Fitch-style natural deduction notation.

The translation algorithm is rule based, and is driven by the structure of the proof-object. Some of the rules are given below. Note that the translations uses type inference to guide the translation process. Here $\boxed{M}_\tau$ on the left hand side means "M is of type τ" and ▣ (M) on the right hand side means "recursively translate M when the folder is open, display the type of M when it is closed".

$$\boxed{h}_\tau \mapsto \left\{ \text{▯ "By" } h \text{ "we have" } \tau \right.$$

$$\boxed{\boxed{M}_{(\forall x:A.B)} \boxed{N}_A}_\tau \mapsto \left\{ \begin{array}{l} \text{▣ } (M) \\ \text{▯ "By taking" } N \text{ "for" } x \text{ "we get" } \tau \end{array} \right.$$

$$\boxed{\boxed{M}_{(A \to B)} \boxed{N}_A}_\tau \mapsto \left\{ \begin{array}{l} \text{▣ } (N) \\ \text{▣ } (M) \\ \text{▯ "We deduce" } \tau \end{array} \right.$$

$$\boxed{\lambda h : \boxed{A}_{\mathsf{Prop}}. \boxed{M}_B}_\tau \mapsto \left\{ \begin{array}{l} \text{▯ Assume } A \ (h) \\ \text{▣ } (M) \\ \text{▯ "We have proved" } \tau \end{array} \right.$$

$$\boxed{\lambda x : \boxed{A}_{\mathsf{Set}}. \boxed{M}_B}_\tau \mapsto \left\{ \begin{array}{l} \text{▯ "Consider an arbitrary" } x \text{ "∈" } A \\ \text{▣ } (M) \\ \text{▯ "We have proved" } \tau \text{ "since" } x \\ \text{"is arbitrary"} \end{array} \right.$$

The sentences are kept as abstract as possible by not unfolding definitions unless needed, using expected types instead of derived types as done in [12] and by repeated introduction of variables in one sentence.

Many of the usual symbols from mathematics and logic can be defined in type theory. The fact that notions such as the natural numbers and their induction principle are not primitive but definable, is regarded as a strength of type-theoretical systems. In fact, even connectives like ¬, ∧, ∨, and ∃. and relations like = (general Leibniz equality) and < (on ℕ) are all definable in Coq. They are defined in the standard library together with their introduction and elimination rules. In theory, it is sufficient to add to the above rules translation rules for inductive types and their inhabitants. However, the translation algorithm treats some defined objects as primitives in order to choose a translation as close as possible to the language of informal mathematics.

The example from Section 3 uses many lemmata from the standard Coq library **Arith**. Some of these should also be treated as primitive by the natural language viewer since they would be considered trivial in informal mathematics.

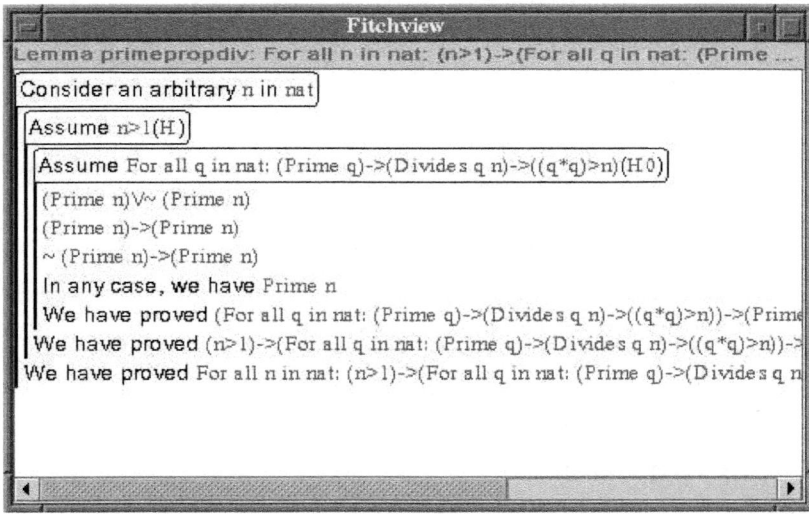

Fig. 1. A Natural Language View of `primepropdiv`

4.2   OpenMath, OpenMath Documents, and MathML

The eXtensible Markup Language (XML) is becoming an increasingly popular choice as source language in which to represent semantically rich information that can be searched, stored and presented in different formats. The World Wide Web Consortium is currently recommending several technologies related to XML and the next generation of browsers will be XML-enabled, namely will be able to directly display XML documents. This section discusses how some of these technologies can be used to provide interactivity to "online" proofs.

OpenMath The predominant XML languages for mathematics are MathML and OpenMath. Ideally these two languages complement each other: MathML-Presentation can be used for presenting mathematical content written in Open-Math. A detailed description of OpenMath is given in [9]. In this paper we assume a certain level of familiarity with the general OpenMath ideas and describe it by examples.

OpenMath is a language for the representation of mathematical content. The symbols used in the OpenMath objects are defined in XML documents called Content Dictionaries (CDs). Official CDs are available for public use from the OpenMath Society [28] but users may also write private CDs of symbols used in own applications. As example, consider the definition of the OpenMath symbol `<OMS cd="pock" name="Divides"/>`, denoted in short by `pock:Divides`, for representing the predicate `Divides` on natural numbers used in the example in Section 3. The formal definition of the example is represented as an OpenMath "defining mathematical property", `DefMP`:

```
<DefMP name="Divides">
<OMOBJ><OMBIND><OMS cd="lc" name="Lambda"/>
      <OMBVAR><!-- n:N, m:N -->
      <OMATTR><OMATP><OMS cd="icc" name="type"/>
                     <OMS cd="setname" name="N"/>
      </OMATP>      <OMV name="n"/>     </OMATTR>
      <OMATTR><OMATP><OMS cd="icc" name="type"/>
                     <OMS cd="setname" name="N"/>
      </OMATP>      <OMV name="m"/>     </OMATTR>
      </OMBVAR>
      <OMBIND><OMS cd="quant1" name="exists"/>
      <OMBVAR><!-- q:N -->
      <OMATTR><OMATP><OMS cd="icc" name="type"/>
                     <OMS cd="setname" name="N"/>
      </OMATP>      <OMV name="q"/>     </OMATTR>
      </OMBVAR><OMA><OMS cd="relation1" name="eq"/>
                    <OMV name="m"/>
                    <OMA><OMS cd="arith1" name="times"/>
                       <OMV name="n"/>
                       <OMV name="q"/>
                    </OMA></OMA>
        </OMBIND></OMBIND>
  </OMOBJ>
  </DefMP>
```

The formal signature can be given similarly in terms of an OpenMath object.

As mentioned before, a proof-object is a term and as such it can be represented in OpenMath provided some primitive symbols are available in some CD. Symbols for constructing and eliminating inductive types in the Inductive Calculus of Constructions used by Coq are given in the CD called `icc`. Additionally, we have a private CD for symbols that Coq uses in proof-objects and refer to introduction and elimination rules of inference. For example, the Coq symbol `and_ind` in the proof-object in Section 3 is represented by the OpenMath symbol `coq:and_ind` in the private CD `coq` and can be exactly defined in terms of the primitive inductive constructors. There are two reasons for representing the proof-object at a higher level than necessary. Firstly, the extra information conveyed by the specific inference rule used in the proof can be directly used for tuning the natural language presentation of the term. The second reason is compactness and readability. The proof-object can be easily transformed into one which uses only the primitive inductive constructors.

OpenMath is not designed to express hierarchies of mathematical knowledge since it lacks the mechanisms to relate definitions, theories and theorems. Moreover, the command language used during an interactive session with a computational tool is hard to express formally in OpenMath. Variable declarations and tactic scripts are examples of this limitation and help motivate the introduction of OpenMath Documents.

OpenMath Documents The OpenMath Document Specification (OMDOC) [19], currently under development by Kohlhase and ourselves, is an XML document type definition that can be used to represent general mathematical knowledge the way it is written in lecture notes and in scientific articles, but also in mathematical software like algebraic specification modules or library files of a proof checker. It is being used as source format for the next release of the *Algebra Interactive!* book [8], an interactive textbook used in teaching first year university algebra. OpenMath Documents are intended to be the input format for a knowledge base of mathematics, Mbase [18].

The mathematical objects within an OpenMath Document are expressed using an extension of the XML encoding[1] of OpenMath. Most important, the usage of OpenMath, conveying the semantical content and not the presentational content of the mathematics, offers two major advantages:

- It allows the mathematical knowledge base to use techniques such as pattern matching or unification to implement search, e.g. modulo and equational theory.
- It equips OpenMath Documents with a standard language for the communication among mathematical services, thus making them suitable to be exchanged between systems for symbolic computation and reasoning.

In this paper we focus on using OMDOC for representing and publishing proofs in interactive mathematical documents.

The format of an OpenMath Document provides an interface for proof presentation by allowing fine-grained interleaving between the formally specified part of the proof and the informal, vernacular text. There are several options for writing a proof to an assertion. All of the views discussed in Section 2 may coexist in the same document and, depending on the viewer, presented upon request.

As the techniques for producing natural language descriptions from proof objects or from tactics/proof plans improve, we may well envision that simple conventional "informal" proofs will become reproducible by automatic machinery. For now, it is possible to simply include informal proofs mixing formal OpenMath statements with flat text.

Interactive proofs described by a proof or a tactic script are put directly in an OpenMath Document and become executable once the document is loaded

---

[1] It contains extra attributes for linking OpenMath subtrees

in a browser. Most of the interactivity available to the reader clicking through a proof is supported by a combination of Javascript and JAVA that is invoked on the correct data by relying on the Document Object Model [14]. Each step in the proof is a request of a performative that is sent to the appropriate software package server or to a broker in a network like [15, 27]. More interesting is the possibility of choosing different mathematical servers depending on the nature of the requested computation. For an example, see [16]. It is well known that, for instance, type-theoretical proof checkers are not very good at equational reasoning [1]. Trying to mimic equational reasoning in such systems produces long and unintelligible proof-objects, not to mention the fact that it requires deep knowledge of the proof assistant for understanding them. This also means that in such cases, an interactive proof gives little insight to the reader and misses the point of the proof.

The OpenMath encoding of the proof-object is also stored in the document directly and can trigger the natural language viewer. Moreover, because it is encoded in standard OpenMath, the proof-object can be exchanged easily among proof checkers implementing similar type theories.

MathML As we said, OpenMath is about conveying content and not about presentation. The same can be said of OpenMath documents. Both are not meant to be read in their XML encoding but transformed to more convenient presentation formats for online browsing.

Two technologies under development that produce customized output formats from an XML input source are based on Cascading StyleSheet (CSS) language and the eXtensible Stylesheet Language (XSL) [26]. Using these, it is possible to convert OpenMath Documents to various flavors of dynamic HTML using MathML-Presentation for the OpenMath objects [5]. Similarly, it is possible to generate LaTeX documents for producing printed version of the material.

5   Conclusions and Future Work

We have presented some of the latest technologies we are developing in order to support true interaction in mathematical documents and in particular in mathematical proofs. The key issue is being able to distinguish content from presentation. The content of a mathematical proof is a formal proof-object encoded in OpenMath, whereas the various presentations are given in an OpenMath document as conventional informal descriptions, or are automatically generated from the proof-object. Computational content of a proof is conveyed by including tactic scripts which become executable by the presentation in browser.

Future work lies in further developing the natural language viewer by including more libraries for primitives, and adding more authoring capabilities allowing user-customized translations. More work needs to be done on the OpenMath document and associated translation tools. In the end, we want to be able to create large interactive mathematical documents semi-automatically by using, for instance, the knowledge contained in a formal development done in a proof assistant.

Acknowledgements The authors would like to thank Arjeh M. Cohen, Herman Geuvers, Michael Kohlhase, and Manfred N. Riem for suggestions, discussions and inspiration.

References

[1] P. Bertoli, J. Calmet, F. Giunchiglia, and K. Homann. Specification and Integration of Theorem Provers and Computer Algebra Systems. In J. Calmet and J. Plaza, editors, *Artificial Intelligence and Symbolic Computation: International Conference AISC'98*, volume 1476 of *Lecture Notes in Artificial Intelligence*, Plattsburgh, New York, USA, September 1998.

[2] Yves Bertot and Laurent Théry. A Generic Approach to Building User Interfaces for Theorem Provers. *Journal of Symbolic Computation*, 25:161–194, 1998.

[3] Stephen Buswell, Stan Devitt, Angel Diaz, Bruce Smith, Neil Soiffer, Robert Sutor, Stephen Watt, Stéphane Dalmas, David Carlisle, Roger Hunter, and Ron Ausbrooks. Mathematical Markup Language (MathML) Version 2.0. W3C Working Draft 11 February 2000, February 2000. Available at http://www.w3.org/TR/2000/WD-MathML2-20000211/.

[4] O. Caprotti and M. Oostdijk. How to formally and efficiently prove prime(2999). In *Proceedings of Calculemus 2000, St. Andrews*, 2000.

[5] O. Caprotti and M. N. Riem. Server-side Presentation of OpenMath Documents using MathML. Submitted.

[6] Olga Caprotti and Arjeh Cohen. Connecting proof checkers and computer algebra using OpenMath. In *The 12th International Conference on Theorem Proving in Higher Order Logics*, Nice, France, September 1999.

[7] Olga Caprotti and Arjeh Cohen. Integrating Computational and Deduction Systems Using OpenMath. In *Proceedings of Calculemus 99*, Trento, July 1999.

[8] A. M. Cohen, H. Cuypers, and H. Sterk. *Algebra Interactive, interactive course material.* Number ISBN 3-540-65368-6. SV, 1999.

[9] The OpenMath Consortium. The OpenMath Standard. OpenMath Deliverable 1.3.3a, OpenMath Esprit Consortium, http://www.nag.co.uk/projects/OpenMath.html, August 1999. O. Caprotti, D. P. Carlisle and A. M. Cohen Eds.

[10] R. L. Constable, S. F. Allen, H. M. Bromley, W. R. Cleaveland, J. F. Cremer, R. W. Harper, D. J. Howe, T. B. Knoblock, N. P. Mendler, P. Panangaden, J. T. Sasaki, and S. F. Smith. *Implementing Mathematics with the Nuprl Proof Development System.* Prentice Hall, 1986. URL http://simon.cs.cornell.edu/Info/Projects/NuPrl/nuprl.html.

[11] Projet Coq. *The Coq Proof Assistant: The standard library*, version 6.3-1 edition. Available at http://www.ens-lyon.fr/LIP/groupes/coq.

[12] Y. Coscoy. A natural language explanation for formal proofs. In C. Retoré, editor, *Proceedings of Int. Conf. on Logical Aspects of Computational Liguistics (LACL), Nancy*, volume 1328. Springer-Verlag LNCS/LNAI, September 1996.

[13] Y. Coscoy, G. Kahn, and L. Théry. Extracting text from proof. In M. Dezani and G. Plotkin, editors, *Proceedings of Int. Conf. on Typed Lambda-Calculus and Applications (TLCA), Edinburgh*, volume 902. Springer-Verlag LNCS, April 1995.

[14] Document Object Model (DOM) Level 2 Specification version 1.0. W3C Candidate Recommendation 07 March, 2000, March 2000.
http://www.w3.org/TR/2000/CR-DOM-Level-2-20000307/.

[15] A. Franke, S. Hess, Ch. Jung, M. Kohlhase, and V. Sorge. Agent-Oriented Integration of Distributed Mathematical Services. *Journal of Universal Computer Science*, 5(3):156–187, March 1999. Special issue on Integration of Deduction System.

[16] Online Pocklington Test generating OMDOC.
http://crystal.win.tue.nl/~olga/openmath/pocklington/omdoc/.

[17] Amanda M. Holland-Minkley, Regina Barzilay, and Robert Constable. Verbalization of high-level formal proofs. In *Sixteenth National Conference on Artificial Intelligence*, 1999.

[18] M. Kohlhase. MBASE: Representing mathematical knowledge in a relational data base. In *Calculemus '99 Workshop*, Trento, Italy, July 1999.

[19] M. Kohlhase. OMDOC: Towards an Openmath Representation of Mathematical Documents. Technical report, DFKI, Saarbrücken, 1999.

[20] Pascal Lequang, Yves Bertot, Laurence Rideau, and Loïc Pottier. Pcoq: A graphical user-interface for Coq. http://www-sop.inria.fr/lemme/pcoq/, February 2000.

[21] Z. Luo and P. Callaghan. Mathematical Vernacular and Conceptual Wellformedness in Mathematical Language . In *Proceedings of the 2nd International Conference on Logical Aspects of Computational Linguistics 97*, volume 1582 of *Lecture Notes in Computer Science*, Nancy, 1997.

[22] Z. Luo and R. Pollack. *LEGO Proof Development System: User's Manual.* Department of Compter Science, University of Edinburgh, 1992.

[23] R. P. Nederpelt, J. H. Geuvers, and R. C. de Vrijer, editors. *Selected Papers on Automath*, volume 133 of *Studies in logic and the Foundations of Mathematics.* Elsevier, 1994.

[24] Paulo Ribenboim. *The New Book of Prime Number Records.* Number ISBN 0-387-94457-5. SV, 1996.

[25] Jurgen Richter-Gebert and Ulrich H. Kortenkamp. *The Interactive Geometry Software Cinderella (Interactive Geometry on Computers).* Number ISBN 3540147195. SV, July 1999.

[26] W3C. Extensible Stylesheet Language (XSL) Specification. W3C Working Draft, 21 Apr 1999. http://www.w3.org/TR/WD-xsl/.

[27] Paul Wang. Design and Protocol for Internet Accessible Mathematical Computation. Technical Report ICM-199901-001, ICM/Kent State University, January 1999.

[28] OpenMath Society Webs

Composite Distributive Lattices as Annotation Domains for Mediators

Jacques Calmet, Peter Kullmann, and Morio Taneda

IAKS, University of Karlsruhe, 76128 Karlsruhe, Germany
calmet@ira.uka.de, kullmann@ira.uka.de, taneda@ira.uka.de

Abstract. In a mediator system based on annotated logics it is a suitable requirement to allow annotations from different lattices in one program on a per-predicate basis. These lattices however may be related through common sublattices, hence demanding predicates which are able to carry combinations of annotations, or access to components of annotations.
We show both demands to be satisifiable by using various composition operations on the domain of complete bounded distributive lattices or bilattices, most importantly the free distributive product.
An implementation of the presented concepts, based on the KOMET implementation of SLG-AL with constraints, is briefly introduced.

Keywords: Annotated Logic, Distributive Lattices, Dual Transform, Free Distributive Lattice Product, Mediator, SLG Resolution.

1 Introduction

Complete distributive lattices and bilattices have been generally recognized as classes of lattices which due to their properties are highly suitable for use in annotated logic (AL). Recently, larger lattices than the well-known ones with a small finite number of elements, such as $\mathcal{FOUR}$, have been investigated. One problem is the high computational complexity associated with larger lattices when used in general deduction procedures over annotated logic. Therefore AL did not seem very well suited to solving complex problems where deep searches over large search spaces are required.

Various mechanized reasoning paradigms may be regarded as spread over a spectrum from highly specialized methods which are very efficient in their problem domain, to more general methods which suffer complexity disadvantages.

Computer Algebra Systems (CAS) are one class of specialized systems, containing large libraries of algorithms for different fields of mathematics. Another class at this end of the spectrum are natural language processing systems.

Moving away from specialized single-domain systems, automated theorem provers (ATP) are suited to limited mediatory roles, at the same time remaining specialized to deep searches, as shown in [1] where a hierarchical relationship between a CAS and a strategic theorem prover is outlined. A limitation of ATPs in this situation is their lack of paraconsistent reasoning capabilities.

J.A. Campbell and E. Roanes-Lozano (Eds.): AISC 2000, LNAI 1930, pp. 65–77, 2001.
© Springer-Verlag Berlin Heidelberg 2001

Annotated logic systems are typically found towards the general side of the spectrum, in roles such as query distribution tools or, where paraconsistent reasoning is required, as mediators. Annotations are often derived from a specialized application domain, and there either from its object domain, e.g. as sets of objects to which a result applies, or from additional information such as (un-) certainties associated with results. Measures of certainty from different specialist systems are often incompatible, making a comparison on qualitative terms necessary as opposed to quantitative comparison between e.g. a fuzzy and a probabilistic value.

From this it follows that AL systems need to be flexible and extensible with respect to their annotations. It is well known that the size (for instance in terms of the number of grounded or of join-irreducible elements) of the annotation lattice is a major complexity parameter in AL proof procedures, therefore as more and more annotation domains are introduced into an AL system it would appear to become impractical. However, just as a logic program with a large number of predicate symbols typically uses only few of them in any single clause or in a set of closely related clauses, only a small number of annotation domains are connected by the literals in individual clauses and sets of closely related clauses. In other words, locally the annotation lattices do not grow boundlessly, while globally the annotation domain may grow almost proportionally to the number of specialized domains to be integrated. A traditional Generalized Annotated Logic Program (GAP) demands that all literals be annotated with elements of the same lattice, but from the preceding argument it would seem reasonable to demand that different lattices should be permitted in a single AL program. The only restriction is that because of the way satisfaction is defined in Def. 1(b), a single predicate symbol must always be annotated with elements of the same lattice.

The annotation values in a clause instance are not unrelated, even if they come from different lattices. For example, one would like to be able to pass a certainty value from one or more body literals to the head literal even if those literals have different annotation lattices. One method to accomodate this requirement is that of multiple annotations: a predicate symbol has not a single annotation lattice but a fixed tuple of annotation lattices. The sharing of annotation values through annotation variables would be permitted between same-lattice annotations of literals in a clause. However, the semantics of multiply annotated literals would remain to be defined. We present a more general solution, allowing various ways of composing and decomposing lattices and lattice values, along with one version of multiple annotations which is encompassed by our solution.

Another, unrelated reason to introduce generic structuring into annotation lattices is that some examples of complex AL applications naturally construct their annotation domains in a structured manner. As an example we will present how stable models of a logic program with negation would be translated to embed them into a mediatory AL system.

Following some definitions in Section 2, we present the free distributive product and a few other composition operations on the domains of complete bounded distributive lattices and bilattices in Section 3, and their implementation in KOMET in Section 4. The aforementioned example is presented in Section 5.

2 Definitions

Annotated logic has been studied since ca. 1989 as an outgrowth of multi-valued logics where the set of truth values is a lattice (e.g. [3,10,13]). They are set apart from general multi-valued logics by the use of bilattices, treatment of inconsistencies, and two types of negation, called *epistemic* and *ontologic* e.g. in [10], *explicit* and *default* in more recent literature. Usually, satisfaction of annotated atoms and therefore complex formulas are two-valued. A framework for clausal logic programs using this type of annotated logic was introduced under the term Generalized Annotated Logic (GAP) by Kifer and Subrahmanian [11]. At the same time, logic programs with multiply annotated atoms were proposed whose semantics were not based on lattice properties of the sets of annotations, e.g. [14]. The following definition already takes into account that we want to allow each predicate to carry a different annotation lattice.

Definition 1 (GAP).
(a) A GAP signature consists of a first-order predicate logic (PL1) signature (disjoint sets of symbols for predicates, object variables, functions and constants) and an annotation signature, which consists of a complete bounded distributive lattice L_p for each predicate symbol p, disjoint sets of symbols for annotation variables for each lattice, and annotation function symbols. All predicate and function symbols are considered to have unique arity, furthermore all annotation function symbols have fixed argument and result types.

The set $Ann(L)$ of annotations of an annotation signature that is associated with a lattice L is defined recursively:

- *Every lattice element $a \in L$ is a* constant *annotation.*
- *Every annotation variable is a* variable *annotation.*
- *For any k-ary annotation function symbol f with result type L and annotations $t_1 \ldots t_k$ from the appropriate $Ann(L_1) \ldots Ann(L_k)$, $f(t_1 \ldots t_k)$ is a* complex *annotation.*

An annotated literal $p : a$ consists of a PL1 atom p and an annotation $a \in Ann(L_p)$.

A negative annotated literal is **not** $p : a$ *where $p : a$ is a (positive) annotated literal. This negation is called* default *negation.*

If L is a bilattice, the explicit *negation of a literal is $\neg p : a := p : \neg a$ where the $\neg$ on the right side is the bilattice negation.*

Generalized annotated logic programs are built as conjunctions of clauses of annotated literals in the usual manner:

$$p : a \quad \leftarrow \quad q_1 : b_1 \wedge \cdots \wedge q_n : b_n \quad \wedge \quad \mathbf{not}\, r_1 : c_1 \wedge \cdots \wedge \mathbf{not}\, r_m : c_m \quad (1)$$

where $p : a$ is called head, $q_1 : b_1 \wedge \cdots \wedge \mathbf{not}\, r_m : c_m$ *the* body *of the clause. The head or body of a clause may be empty. All variables that appear in the head must also occur in the body. Unbound object and annotation variables are implicitly universally quantified.*

(b) An interpretation *of a GAP consists of a mapping $M : H \rightarrow \bigcup_p L_{p\,int}$ of the Herbrand universe H of the PL1 signature of the GAP into the interpretation lattice $L_{p\,int}$, such that $M(A) \in L_{p\,int}$ if p is the predicate symbol of the ground literal A. $L_{p\,int}$ may be L_p itself, the ideal lattice $I(L_p)$ or some other sublattice of the powerset lattice $P(L_p)$; the choice must be the same for all p. Furthermore, an interpretation maps every annotation function symbol to an evaluable function over the appropriate lattices. Satisfaction is defined as follows:*

$M \models p : a \qquad$ *iff* $a \leq M(p) \qquad$ *if* $L_{p\,int} = L_p$
$M \models p : a \qquad$ *iff* $a \in M(p) \qquad$ *if* $L_{p\,int} = I(L_p)$ *or* $L_{p\,int} = P \leq 2^{L_p}$
$M \models \neg p : a \qquad$ *iff* $M \models p : (\neg a)$
$M \models \mathbf{not}\, p : a$ *iff* $M \not\models p : a$

Satisfaction of a composite formula is defined recursively in the usual (two-valued) manner. Where function symbols appear in a formula, satisfaction is determined by evaluating the function which substantiates the function symbol.

The semantic resulting from $L_{int} = L$ was called restricted *by [11], the one with $L_{int} = I(L)$ is called* general.

Completeness and boundedness of annotation lattices are required in order for the restricted and general semantics of GAPs to remain closely related to those of non-annotated clausal logic programs [11].

Under the restricted or general semantics, some logic programs do not have a model. In the two-valued case, the well-founded semantics under which all safe logic programs have a three-valued model has gained wide acceptance. The extension of the well-founded semantics to GAPs has been described in [15].

Definition 2 (Well-Founded Semantics of GAPs).

1. *Given a set of annotated clauses P and an interpretation I, a set of transformed clauses (the Gelfond-Lifschitz transform) is defined as*

$G(P, I) = \{ p : a \leftarrow q_1 : b_1 \wedge \cdots \wedge q_n : b_n \,|$
$\qquad p : a \leftarrow q_1 : b_1 \wedge \cdots \wedge q_n : b_n \wedge \mathbf{not}\, r_1 : c_1 \wedge \cdots \wedge$
$\qquad \mathbf{not}\, r_m : c_m \in P$ *is ground instance of a clause in* P,
$\qquad \forall\, i = 1 \ldots m : I(r_i) \not\geq_{L_{r_i}} c_i \}$

$G(P, I)$ is an annotated clause set without negation, and the mapping $I \mapsto G(P, I)$ is antimonotonic. Therefore, $\mathcal{G}_P(I) := lfp(R_{G(P,I)})$ exists and $\mathcal{G}_P$ is antimonotonic.

2. *$\mathcal{G}_P^2$ is monotonic and has a least and a greatest fixpoint. The well-founded semantics of P is defined as*
$P \models_r^{wfs} p : a \qquad$ *iff* $a \leq_{L_p} lfp(\mathcal{G}_P^2)(p)$
$P \models_r^{wfs} \mathbf{not}\, p : a$ *iff* $a \not\leq_{L_p} gfp(\mathcal{G}_P^2)(p)$

A proof procedure which is complete and sound with respect to the well-founded semantics is SLG [7]. An annotated extension of SLG, called SLG-AL, has been described by P. Kullmann [12], who also provided the implementation KOMET which is the basis for an implementation of structured lattices presented here. The long description of the proof procedure as well as soundness and completeness proofs for the two-valued and annotated cases are omitted here and can be found in [7] and [12] respectively.

3   Lattice Products and Other Composite Lattices

The following sections introduce some lattice composition operations. The best known such operation is the cross product; however there are several other, at least equally important operations, among them the free distributive product.

3.1   Composition Operations

An annotated clause as in (1) can be read as

If $I(q_1) \geq b_1$ and ... and $I(q_n) \geq b_n$ and $I(r_1) \not\geq c_1$ and ... and $I(r_m) \not\geq c_m$, then $I(p) \geq a$.

Read as a condition on interpretations, $I(p), \ldots I(r_m)$ are merely placeholders, which we subsequently write as variables x_i. The set of annotation conditions $L_i' := \{(x \geq a)|a \in L_i\} \cup \{false\}$ on variables $x \in L_i - \{\top\}$ with reverse implication ordering is a lattice which is isomorphic to L.

Assume from now on that all lattices L_i are distributive and bounded. If a literal $p : a_1, \ldots, a_k$ is multiply annotated with $a_i \in L_{pi}$, the value assigned to p by an interpretation is a tuple $I(p) = (I_1(p), \ldots, I_k(p))$, $I_i(p) \in L_{pi}$ and the obvious extension of the above reading would be

If $I_1(p) \geq a_1$ and ... and $I_k(p) \geq a_k$ ...

The set of multiple annotation conditions $L^* = \{(x_1 \geq a_1) \wedge \cdots \wedge (x_k \geq a_k)|a_i \in L_{pi}, i = 1 \ldots k\}$ is no longer a lattice, but only a meet-semilattice. Still, this approach is frequently taken for the semantics of multiple annotations.

To extend L^* to a lattice, we add disjunctive expressions, i.e. L^{**} is the set of expressions recursively defined as

1. Every $(x_i \geq a_i)$ is a member of L^{**},
2. If $A, B \in L^{**}$, then $A \vee B \in L^{**}$ and $A \wedge B \in L^{**}$,
3. *true* and *false* are members of L^{**}.

The expressions thus defined may be simplified using absorption and distributive laws. Comparison is defined through subsumtion. The definition implies that each L_i is embedded homomorphically in L^{**} and $L_i \cap L_j = \{\bot, \top\}$ for $i \neq j$. Thus L^{**} is the free distributive lattice generated by the L_i', or the **free distributive product** of the L_i'.

The preceding transformation from lattices to sets of inequalities is not required for the construction of free distributive products, however it may be useful for understanding their application to annotated logic. We write

$$L = L_1 \otimes \cdots \otimes L_k \tag{2}$$

L is bounded and distributive and if all L_i are complete, L is also a complete lattice [9].

In free distributive lattice products (FDLP) there are strong, well known normal form properties (e.g. [9]). If an element $a \in L$ is given as a polynomial expression

$$a = \bigvee_{j=1}^{n} \bigwedge_{i \in I} a_{ji}, \quad a_{ji} \in L_i - \{\bot\}, \tag{3}$$

its unique conjunctive normal form consists of the join-terms of the fully expanded dual transform of the polynomial which are not subsumed by other terms. Its computation is closely related to that of reduced dual transforms in other domains, thus the same algorithmic techniques can be used and some algorithms can easily adapted for normal form computation in FDLPs. The exceptions are as follows: Algorithms that attempt to generate a minimal number of terms do not produce the unique normal form. Also, lattices admit simplification of terms by replacing sets of mutually incomparable elements of the same lattice with their join or meet resp. as appropriate, which is not possible in propositional logic and is therefore not exploited by most algorithms. On the other hand, elimination of tautological or paradoxical terms containing complementary literals, a standard optimisation in propositional or predicate logic, is generally not possible in non-boolean lattices.

The **cross product**, written as $\times$, is defined as $L_1 \times \cdots \times L_k = \{(a_1 \ldots a_k)|a_i \in L_i\}$ with $(a_i)_i \leq (b_i)_i \Leftrightarrow a_i \leq b_i \forall i = 1 \ldots k$, and meet and join are defined componentwise. Again, boundedness, completeness and distributivity of factors are preserved in the product. In the terminology of categories, the cross product is a *sum* or co-product, sometimes written as $\oplus$, and distributive laws hold for the free distributive product and the cross product on the class of bounded distributive lattices.

In the previous two definitions, all lattices have equal weight in their composite ordering. By contrast, the **lexicographic product** assigns strict priorities to its components.

Definition 3 (Lexicographic Product). *Given distributive bounded lattices $L_i = \langle M_i, \leq_i \rangle$, $i = 1 \ldots n$ and $M := M_1 \times \cdots \times M_n$ their cross product as sets. M is partially ordered by*

$$(a_1, \ldots, a_n) <_{\text{lex}} (b_1, \ldots, b_n) \text{ iff}$$
$$\exists i, 1 \leq i \leq n, \text{ such that } a_i < b_i \text{ and } \forall k, 1 \leq k < i : a_k = b_k \tag{4}$$

If all L_i are totally ordered, so is $\langle M, \leq_{\text{lex}} \rangle$.

Let $a = (a_1, \ldots, a_n), b = (b_1, \ldots, b_n) \in M$ *and* i *maximal*, $1 \leq i \leq n+1$, *such that* $\forall\, k, 1 \leq k < i : a_k = b_k$. *Then*

$$a \vee_{\mathrm{lex}} b = \begin{cases} a & \text{if } a \geq_{\mathrm{lex}} b \\ b & \text{if } a <_{\mathrm{lex}} b \\ (c_1, \ldots, c_n) & \text{if } a, b \text{ incomparable} \end{cases}$$

$$\text{with } c_k = \begin{cases} a_k & \text{if } 1 \leq k < i \\ a_k \vee b_k & \text{if } k = i \\ \bot & \text{if } i < k \leq n \end{cases}$$

and $\wedge$ *defined dually. Then* $L_{\mathrm{lex}} := \langle M, \leq_{\mathrm{lex}}, \vee_{\mathrm{lex}}, \wedge_{\mathrm{lex}} \rangle$ *is a lattice.*

A typical use of a lexicographic product occurs in mediators, where a best solution in some sense has to be selected from a set of possible solutions and it is therefore desirable to have an unique maximal element in any set.

Other operations on lattices which preserve completeness, distributivity and boundedness are order reversal and the addition of a new bottom element $\bot'$ below $\bot$. The latter is helpful if lattice values are to be imported from an external source but appearance of its $\bot$ in a term should not cause the whole term to be considered $\bot$ and therefore eliminated, as is the case with the standard embedding of components in FDLPs.

The operations on the domain of complete bounded distributive lattices (of which bilattices are an important subdomain) that have been introduced so far are connected in various ways. One of these is the aforementioned distributivity of FDLP and cross product. We briefly state a few other results; the proofs and further discussion can be found in [20].

Bilattices can be constructed uniquely as the cross product of two distributive lattices, and if the bilattice is symmetrical, the two lattices are isomorphic. In this case one writes $B = B(L)$.

Lemma 1. *A symmetrical bilattice* $B = B(L)$ *is isomorphic to the FDLP* $B \cong L \otimes \mathcal{FOUR}$, *where* $\mathcal{FOUR} = \{\bot, t, f, \top\}$ *with the ordering* $\bot < t, f < \top$.

Lemma 2. *The smallest distributive bounded lattice* $E = \{\bot, \top\}$ *is the neutral element of the domain of distributive bounded lattices with respect to the FDLP, i.e.* $E \otimes L \cong L$.

Lemma 3. *If* $M_i \leq L_i$ *are sublattices with* $\top_{M_i} = \top_{L_i}$, $\bot_{M_i} = \bot_{L_i}$, *then* $M = \otimes_{i=1}^n M_i \leq L = \otimes_{i=1}^n L_i$. *In particular, setting* $M_{k+1} \ldots M_n = E$ *gives* $\otimes_{i=1}^k L_i \leq L$.

The last statement means that the difference between the traditional definition of GAPs which demands the same lattice for all annotations and the relaxed form which allows each predicate symbol to be annotated with a different lattice may be reconciled, since the different lattices could be embedded into a single FDLP.

The latter remark raises a notational difficulty, namely identical and isomorphic components which are shared between composite annotation lattices need to be distinguished. In KOMET this is currently only possible on a clause by clause basis through variables which appear in several annotations in the clause.

Also, purely from the viewpoint of the lattices involved, it would appear possible to embed non-bilattices in bilattices, e.g. $L = E \otimes L \leq \mathcal{FOUR} \otimes L = B(L)$. However identifying E with $\{false, true\}$ leads to the embedding $true \mapsto \top$, which is semantically unreasonable. This coincides with the fact that an AL program with bilattice annotations needs to be written very differently from an AL program with non-bilattice annotations and it would therefore be questionable to mix those two types of annotations in an AL program. A more reasonable, if not highly complex embedding would take the three-valued well-founded model of a general two-valued logic program as an element of an interpretation bilattice.

3.2   Join-Irreducibility

Recent work on AL which focused on the computational complexity of inference has brought the concept of join-irreducibility to the front (e.g. [16]). Representing lattice elements as joins of join-irreducible elements has been recognized as helpful in reducing the computational complexity, therefore determining the number of such elements in a lattice and identifying the subset of all such elements becomes an implementation concern.

A lattice element $a \in L$ is *join-irreducible* iff for any $x, y \in L$, $x \vee y = a$ implies $x = a$ or $y = a$. The set of join-irreducible elements of a lattice L is written $JIR(L)$.

Lemma 4 (Birkhoff). *Every element of a distributive lattice with descending chain property has a unique non-redundant representation as the join of join-irreducible elements.*

The join-irreducible elements of elementary lattices are easily determined. In our situation, where lattices are built from the ground up from elementary lattices, the join-irreducible elements of some composites are also known. This knowledge is not exploited directly, since it does not appear to offer much advantage over the computation of normal forms based on the implementations of the component lattices. Elementary lattice implementations can be expected to take advantage of the join-irreducible representation of their elements. The following table lists some elementary and composite lattices and their join-irreducible subsets.

In particular, if the complexity of operations in lattices is assumed to be $O(|JIR(L)|)$, the complexity of FDLP operations is exponential in the number of components. The same can be achieved by an implementation of FDLP using normal form computations and $O(|JIR(L)|)$ implementations of the components.

Lattice L	$JIR(L)$	$|JIR(L)|$
Powerset $L = P(M)$	$\{\{m\} \mid m \in M\}$	$|M|$
Totally ordered set	$L - \{\bot\}$	$|L| - 1$
Cross product $L_1 \times \ldots \times L_n$	$\{(x, \bot \ldots \bot) \mid x \in JIR(L_1)\}$ $\cup \cdots \cup \{(\bot \ldots \bot, x) \mid x \in JIR(L_n)\}$	$\sum_{i=1}^{n} |JIR(L_i)|$
FDLP $L_1 \otimes \cdots \otimes L_n$	$\{x_1 \otimes \cdots \otimes x_n \mid x_i \in JIR(L_i)\}$	$\prod_{i=1}^{n} |JIR(L_i)|$
Raised lattice $L \cup \{\bot'\}, \bot' < \bot$	$JIR(L) \cup \{\bot\}$	$|JIR(L)| + 1$

Table 1. Join-Irreducible Elements of Elementary and Composite Lattices

The composition approach offered here does not suffer in situations where $JIR(L)$ is not easily obtainable. For example, the join-irreducible elements of an order-reversed lattice L_{rev} would correspond to the meet-irreducible elements of L, which may in general be a completely different set. In our elementary lattices, however, the sets of meet- and join-irreducible elements are closely related.

3.3   Constraint Representation of Composite Annotations

We still need to show how composition and decomposition of lattice *values* is carried out at the clause (instance) level. For this it is necessary to discuss the additional SLG-AL concepts of constraints and modes.

The original definition of lattice function instantiations as evaluable functions in GAPs was extended and reinterpreted in the KOMET version of SLG-AL through constraints. Besides lattices, built-in sorts (types for object variables) and connections with external information sources are represented as constraints within KOMET.

Constraints, like predicates, have a number of arguments, each of a fixed sort, and a lattice annotation. n-ary functions are regarded as two-valued (i.e. annotated with $TWO = \{\bot, \top\}$) $n+1$-ary constraints.

Polymorphism is not used in KOMET, however constraints can be poly-modal. A *mode* of an n-ary constraint is an n-tuple of elements of the set $\{bound, unbound, any\}$. A constraint in a clause instance is evaluable according to a mode when every argument of type *unbound* is a variable and every *bound* argument is ground. Evaluation of a constraint results in one or more answer substitutions which bind the variable arguments. Modes offer substantial, flexible control over the order of subgoal evaluation.

Returning to composite annotations, we again consider multiple annotations as a guiding example. If at the end of the evaluation of a subgoal $p : X_1 \wedge \ldots \wedge X_n$, an answer $p : a$ has been obtained and $a = \bigvee_{j=1}^{r} \bigwedge_{k=1}^{n} a_{jk}$ (where $a_{ji}, X_i \in L_i$), it is reasonable to return r answer substitutions, each of which extracts one of the conjunctive terms, i.e. $\langle X_i \rightarrow a_{ji}, i = 1 \ldots n \rangle$. More generally, a literal (partially instantiated) with a mixed variable and ground annotation, translated into FDLP notation, would look like $p : u$ where $u = a_{i_1} \wedge \cdots \wedge a_{i_k} \wedge X_{j_1} \wedge \cdots \wedge X_{j_l}$ and $a_i, X_i \in L_i$. If the term u includes more than one variable, a query may yield multiple, mutually incomparable answers, corresponding to conjunctive terms of a disjunctive normal form.

In KOMET, constant composite annotations can be specified using lattice class specific constructor syntaxes, variable composite annotations are obviously given as variable symbols, and mixed annotations are written using constructor constraints which are specific to each composite lattice type. The constructor constraints generally have two modes, one where all arguments are bound and the value is unbound (the constructor mode), and one where the value and some but not all arguments are bound and the remaining arguments are variable (deconstructor modes). As shown in the last paragraph, the FDLP deconstructor may return multiple answer substitutions.

Finding a workable but fully flexible syntax is a challenge. For instance, if $M = M_1 \otimes \cdots \otimes M_r$, $N = N_1 \otimes \cdots \otimes N_s$ and $L = M_1 \otimes \cdots \otimes M_r \otimes N_1 \otimes \cdots \otimes N_s$, then also $L \cong M \otimes N$. One would like to be able to write a single variable for each part, to avoid the mentioned multiple answers which may just be reduced to one in another literal, among other reasons. But a syntax which allows this and also handles cases where the components are arranged differently in L is probably cumbersome. This problem has not yet been solved completely.

4   Implementation

As discussed in the introduction, annotated logic with strong negation and composite annotation lattices is a formalism well suited for mediators and similar applications. For this reason it was implemented in KOMET, a modular program with an extensible type system at the base, an implementation of SLG-AL resolution with constraints at its core and multiple object and lattice classes on top of these components. Concerns with speed and software engineering, particularly the easier integration with common relational and object databases and more generally with arbitrary library and network interfaces led to the decision not to implement it in Prolog like many more logic-centered systems, but in C++, which was at the time (1994) chosen over Java for its maturity. KOMET's purpose is the investigation of mediator programming techniques and of the suitability of annotated logic for mediators. It is called a mediator shell because of its open architecture [4,5,6,12].

The first version of KOMET contained several simple lattices and the cross product, and the infrastructure to easily add more base lattices and lattice composition operations.

Of the lattice operations discussed in the previous section, only the FDLP presents notable challenges for an implementation. Elements are commonly represented as polynomials, either disjunctions of meet-terms or conjunctions of join-terms. For the comparison of FDLP elements, criteria are available which are based on either normal form [9,20]. The normal form computation is therefore the key element of an FDLP implementation.

A review of several dual transform algorithms of the matrix path search category (e.g. [17,18]) shows that after taking into account the aforementioned differences between propositional or predicate logic on one hand and non-boolean distributive lattices on the other, many refinements are lost. Therefore our example implementation contains only a relatively simple algorithm of this class.

A promising, very general dual transformation algorithm of a different type is presented in [2]. Its definition encompasses the FDLP case as well as propositional and predicate logics. Its strength comes from the fact that it organizes the propagation of partial dual terms in such a manner that it will not generate any terms that can be subsumed, thus avoiding the usual step of finding and eliminating subsumed terms. It does generate large numbers of identical terms, which are guaranteed to be collected in one node without prior comparison but still need to be removed, e.g. by hashing and common subexpression detection. Prolog provides this as part of most implementations, which facilitates the efficient implementation of this algorithm. As part of the work presented in this paper, the algorithm was implemented in a straightforward, non-optimized manner in C++ within KOMET. That implementation could be used to verify the applicability of the algorithm to FDLPs. Its performance was inferior to the simple matrix path search algorithm, however this is most likely due to the nonoptimal implementation and not the algorithm itself, so no conclusion could be reached about its claimed efficiency.

5   Example: Weighted Stable Models

A two-valued logic program with negation may in general have more than one stable model. These models represent interpretations which are consistent with the program, but a two-valued proof procedure is not able to infer any preference among those models. A hypothetical mediator would pass a program P to an external prover with stable semantics, returning a three-valued answer substitution of the result variables A, B, C for each of the models. Also, two weighting predicates assign preference to answers.

Constraint *StableProver* :: *Solve*(*String, THREE, THREE, THREE*) : [*TWO*]
Lattice *RFOUR* = *RAISE*(*FOUR*)
Lattice *Interpretation* = *FDLP*(*RFOUR, RFOUR, RFOUR*)
Lattice *WeightedInterpretation* = *LEXPR*(*REAL01, REAL01, Interpretation*)
Predicate *BestModel*(*void*) : [*WeightedInterpretation*]
Predicate *Pref1*(*FOUR, FOUR*) : [*REAL01*]
Predicate *Pref2*(*FOUR, FOUR*) : [*REAL01*]
Predicate *Raise*(*THREE*) : [*RFOUR*]
BestModel() : *LEXPR*($V1, V2$, *CTERM*(RA, RB, RC)) ←
    StableProver :: *Solve*(P, A, B, C) : [*true*],
        Pref1(A, B) : [$V1$],   *Pref2*(B, C) : [$V2$],
        Raise(A) : [RA],   *Raise*(B) : [RB],   *Raise*(C) : [RC]

The clause is first instantiated by answers from *StableProver::Solve* because the other constraints are initially not evaluable. For each answer, two weights and three raised-lattice versions of the values of the P-predicates are obtained as follows: *Pref1*, *Pref2* and *Raise* pass information between the object and annotation domains, in this case taking object values and binding annotation variables. *Raise* also embeds *THREE*, which is not a bounded distributive lattice, in *FOUR*. These simple predicates can be defined through lists of facts, e.g. $Pref1(t,f) : [0.5] \leftarrow$. Larger or changing tables might be implemented as constraints accessing external database tables.

The resulting *RFOUR* values are combined into a FDLP term which represents one model. *CTERM* (Conjunctive Term) is one of the constructor constraints for the FDLP *Interpretation* (the other being *DTERM*, Disjunctive Term). Its result is $RA \wedge RB \wedge RC$. These operations return one answer each so the search tree does not branch any further. The lexicographic product of the weights and the model is accumulated. Here, the constraint *LEXPR* is named identically to the lattice whose values it constructs.

The reduction rule which is part of SLG-AL computes the least upper bound of all weighted models. If exactly one model has the highest weight, this model and its weight are retained. If more than one model has equal highest weight, all of these models which are incomparable by the definition of the stable semantics are kept as terms of an *Interpretation* value.

6   Conclusion

Based on an analysis of the requirements of annotated logic (AL) in mediator applications, we have presented an approach to complex distributive lattice annotations in which large annotation lattices are built bottom-up from elementary lattices. The two new lattice composition operations introduced are the free distributive lattice product (FDLP) and the lexicographic product. The FDLP is a general solution to the multiple annotation problem, whereas the lexicographic product allows the selection of optimal answers. The complexity of an AL system is not affected prohibitively by the introduction of large numbers of component lattices into an AL program, as long as locally, i.e. within each literal, only a small number of the components are nontrivial.

Compared with results regarding join-irreducible representations (JIR) of lattice elements [16], the computational complexity of our approach is shown to be similar to the JIR approach. Furthermore the methods presented here do not require the join-irreducible elements of a lattice to be known.

All of the results presented here affect only the lattice component in an AL system. Therefore, they should transfer easily to lattice-based AL systems other than traditional GAPs, such as coherent well-founded AL programs [8] which have a different notion of default negation.

References

1. Bertoli, P.G.; Calmet, J.; Giunchiglia, F.; Homann, K.: Specification and Integration of Theorem Provers and Computer Algebra Systems. Fundamenta Informaticae 39(1,2):39–57, IOS Press 1999.

2. Bittencourt, Guilherme: Concurrent Inference through Dual Transformation. L.J. IGPL Vol. 6 No. 6 pp. 795–833. Oxford University Press 1998.
3. Blair, H.A.; Subrahmanian, V.S.: Paraconsistent Logic Programming. Theoretical Computer Science 68(2):135–154, 1989.
4. Calmet, J.; Jekutsch, S.; Kullmann, P.; Schü, J.; Svec, J.; Taneda, M.; Trcek, S.; Weißkopf, J.: KOMET. Interner Bericht, IAKS. Karlsruhe, 1997.
5. Calmet, J.; Kullmann, Peter.: A Data Structure for Subsumption-Based Tabling in Top-Down Resolution Engines for Data-Intensive Logic Applications. In Z.W.Ras, A.Skowron (Eds.), Foundations of Intelligent Systems, 11th International Symposium, ISMIS'99, Warsaw, Poland, June 8–1, 1999. Proceedings, Lecture Notes in Computer Science, Vol. 1609: 475–483, Springer 1999.
6. Calmet, J.; Kullmann, Peter.: Meta Web Search with KOMET. Workshop for Intelligent Information Integration (IJCAI-99), Stockholm, Sweden, July 1999.
7. Chen, Weidong; Warren, David S.: Tabled Evaluation with Delaying for General Logic Programs. Journal of the ACM, 43(1):20–74, 1996.
8. Damásio, C.V.; Pereira, L.M.; Swift, T.: Coherent Well-founded Annotated Logic Programs. In M.Gelfond, N.Leone, G.Pfeifer (Eds.), LPNMR 99, LNAI 1730:262–276, Springer 1999.
9. Graetzer, George: Lattice Theory. Freeman, San Francisco, 1971.
10. Kifer, M.; Lozinskii, E.L.: A Logic for Reasoning with Inconsistency. Journal of Automated Reasoning 9:179–215, 1992.
11. Kifer, M.; Subrahmanian, V.S.: Theory of Generalized Annotated Logic Programming. Journal of Logic Programming, 12(1):335–367, 1992.
12. Kullmann, Peter: Entwurf und Implementierung einer Expertensystem-Shell für annotierte Logik. Institut für Algorithmen und Kognitive Systeme, Dipl.-Arb., Karlsruhe, 1995.
13. Lu, J.; Murray, Neil V.; Rosenthal, Erik: Signed Formulas and Annotated Logics. Proc. 23rd International Symposium on Multiple-Valued Logics, 48–53. IEEE Press 1993.
14. Lu, J.; Nerode, Anil; Remmel, Jeffrey; Subrahmanian, V. S.: Towards a Theory of Hybrid Knowledge Bases. Technical Report, University of Maryland, 1993.
15. Lu, J.; Nerode, A.; Subrahmanian, V.S.: Hybrid knowledge bases. IEEE Transactions on Data and Knowledge Engineering 8(5):773–785, 1996.
16. Mobasher, Bamshad; Pigozzi, Don; Slutzki, Giora: Multi-valued logic programming semantics. An algebraic approach, Theoretical Computer Science 171:77–109, 1997.
17. Slagle, J.R.; Chang, C.; Lee, R.C.T.: A new algorithm for generating prime implicants. IEEE Transactions on Comp. 19(4):304–310, 1970.
18. Socher, Rolf: Optimizing the Clausal Normal Form Transformation. Journal of Automated Reasoning 7:325–336, 1991.
19. Subrahmanian, V. S.: Amalgamating Knowledge Bases. Technical Report, University of Maryland, CS-TR-2949, August 1992.
20. Taneda, Morio: Semantik mehrfacher Annotationen und Implementation in der Mediatorshell KOMET. Diplomarbeit, Fakultät für Informatik, University of Karlsruhe, June 1999.

A Proof Strategy Based on a Dual Representation

Guilherme Bittencourt and Isabel Tonin*

Departamento de Automação e Sistemas
Universidade Federal de Santa Catarina
88040-900 - Florianópolis - SC - Brazil
{ gb | isabel }@lcmi.ufsc.br

Abstract. In this paper we describe a first-order logic inference strategy based on information extracted from both conjunctive and disjunctive normal forms. We claim that the search problem for a proof can benefit from this further information, extending the heuristic possibilities of resolution and connection proof methods.
Keywords: First-order logic, theorem proving, inference strategy, dual transformation.
Topic: Logic and Symbolic Computing.

1 Introduction

There are two main families of automatic theorem proving systems: the systems based on the *Resolution* rule [17] and the systems based on the *Connection Method* [2]. Resolution is the best known and most widely used method for theorem proving. In recent years, several proof strategies have been proposed to improve its efficiency [11] and much of the development in expert systems [22] and logic programming [18] has been strongly influenced by it. Before a resolution method is applied, the negation of the theorem to be proved, along with the appropriate hypothesis, must be converted to conjunctive normal form. Resolution methods are characterized by a local inference rule – the *resolution rule* – able to generate new clauses – the *resolvents* – which are logical consequences of the clauses already admitted. The termination criterion is the generation of the *empty clause*. Its main disadvantage is that it retains the (non subsumed) newly inferred clauses, augmenting the search space at each successful application of the resolution rule.

The connection method has its roots in the *Semantic Tableaux* [20] and *Natural Deduction* methods [3] and inspired several theorem proving methods (e.g. the *Consolution Method* [10]). Although some methods of the connection family work on formulas represented in normal form (e.g., [1]), in general they can be applied on formulas expressed in full first-order logic language. The termination

---

* The authors express their thanks to the Brazilian research support agency "Fundação Coordenação de Aperfeiçoamento de Pessoal de Nível Superior (Capes)" for the partial support of this work.

criterion is the generation of a *spanning complementary mating* of the set of formulas. A mating is a set of connections, where a connection is an unordered pair of literals with the same predicate symbol but different signs. A connection is complementary if the two atomic formulas occurring in its literals unify. Finally, a complementary mating spans a set of formulas if each path through the formula literals contains a connection from the mating. Connection methods are high level proof methods, in the sense that they search for a global property of the set of formulas, the spanning matings. Differently from the resolution methods, no inferred formula is retained during the deduction process, although, for first-order logic, the set of formulas must be *expanded* (or *amplified*) by the duplication of some of its formulas. But this operation can be performed using only indices and not actually duplicating the formulas.

The proposed inference method for first-order logic has some characteristics that combine resolution and connection methods features. On the one hand, the proposed method presents the following properties in common with resolution: (i) it demands the problem to be transformed to a normal form, in fact to both dual normal forms, (ii) it retains (non subsumed) inferred theorems, (iii) it supports a refutation-based theorem proving method, and (iv) it is a local process. On the other hand, the proposed method presents the following properties in common with the connection method: (i) its proof strategy is based on the combination of substitutions associated with "connections", i.e., substitutions that unify two complementary literals in different clauses, (ii) it treats *linear chains* quite efficiently, (iii) it supports an affirmative theorem proving method, and (iv) it is a global process.

The particularity of the proposed method comes from the apparent paradox between the last properties of the above two lists. This paradox is explained by the "holographic" character of the proposed method. On the one hand, the atomic goals are identified through a local process applied to the conjunctive normal form. This process is analogous to the choice of candidate clauses in the resolution method and therefore can benefit from strategies such as the *set of support* [7]. On the other hand, once the goals are defined in the conjunctive normal form, the substitutions to be applied are calculated taking into account global properties about the contradictory character of the dual clauses that belong to the disjunctive normal form, analogously to the connection method. This second aspect of the inference process can benefit from the linear chains and hinged loops treatment available in the connection method [3].

The main idea of the proposed inference method is to use the information about the occurrence of literals within clauses and dual clauses, along with the subsumption relation among them, to guide the inference process, when a specific goal is given. Intuitively, if we have a clause that contains a literal that unifies with the given goal, then it is enough to eliminate all other literals of the clause and the goal will be proved, provided that all the involved substitutions combine. To eliminated some literal from a clause, we have to find a substitution that turns into contradictions all the dual clauses that are represented, in that clause, by the given literal.

The paper is organized as follows. In Section 2, we present the adopted representation for conjunctive and disjunctive normal forms, where the relations between literals that occur in both forms are explicitly stored. In Section 3, the proposed proof strategy is described. In Section 4, some results of the strategy implementation are presented. Finally, in Section 5, we draw some conclusions and comment upon future perspectives.

2   Dual Representation

Consider the first-order language $L(P, F, C)$, where P, F and C are finite or countable sets of predicate, function and constant symbols, respectively. Following the usual definition of *terms, atomic formulas,* and *formulas* (e.g., [12]), and given the formulas $X_1, ..., X_n$, we define a *generalized disjunction* and a *generalized conjunction* as $[X_1, ..., X_n] \equiv X_1 \vee ... \vee X_n$ and $\langle X_1, ..., X_n \rangle \equiv X_1 \wedge ... \wedge X_n$ respectively. A *literal* is an atomic formula or the negation of an atomic formula, or one of the constants *True* or *False*. A *clause* is a generalized disjunction in which each member is a literal. A *dual clause* is a generalized conjunction in which each member is a literal.

A first-order formula W_c is in *conjunctive normal form* or is in *clause form* if it is a generalized conjunction $\langle C_1, ..., C_n \rangle$ in which each member is a clause. A first-order formula W_d is in *disjunctive normal form* or is in *dual clause form* if it is a generalized disjunction $[D_1, ..., D_n]$ in which each member is a dual clause. Given an *ordinary formula* W, i.e., one not restricted to generalized conjunctions and generalized disjunctions, there are algorithms for converting it into a formula W_c, in clause form, and into a formula W_d, in dual clause form, such that $W \Leftrightarrow W_c \Leftrightarrow W_d$ (e.g., [16], [19], [21]). To transform a formula from one clause form to the other, what we here call the *dual transformation*, only the distributivity of the logical operators $\vee$ and $\wedge$ is needed.

The proposed proof procedure needs, beside the two canonical forms W_c and W_d, some information about the relation between the literals in one form and the literals in the other form. The clauses in W_c and the dual clauses in W_d are a kind of "holographic" representation of each other. Each clause in W_c consists of a combination of all dual clauses in W_d and, conversely, each dual clause in W_d consists of a combination of all clauses in W_c. They are combinations in the sense that each literal in a clause belongs to a different dual clause. If all literals in the clause set are ground than each dual clause will contain exactly one literal of each clause, but if we have variables and some literals subsume some others, than a single literal in one dual clause may represent more than one clause. This *representation* relation that must be captured for our purposes. This is done through the introduction of the notion of *quantum*[1].

A quantum is a mathematical object that consists of three elements: a literal ϕ and two sets of integers F, S. If the quantum belongs to W_c, its set F of *fixed*

---

[1]   The metaphoric notations adopted to name the defined mathematical objects (quantum, coordinates, etc.) are intended to facilitate the understanding of the algorithm and do not have any further significance.

coordinates indicates which dual clauses contain the literal ϕ in their definition. The set S of *subsumed coordinates* indicates which dual clauses contain (in their definitions) literals ϕ' such that ϕ subsumes ϕ'. A quantum is noted by $\phi^{F,S}$. The representation is symmetric, i.e., the F and S sets associated with the quanta that belong to dual clauses represented the information about the presence of the respective literals (and their subsumed literals) in the clause form.

The dual transformation is a very expensive procedure and can be traced back to Quine [15, 16], who first proposed an algorithm to solve the problem of reducing an arbitrary truth-functional formula to a shortest equivalent in normal form. This problem, also referred to in the literature as the prime implicants/implicates determination problem, finds applications in the minimization of switching circuits [23] and has since been the subject of several publications [19, 13, 14, 21]. The proposed representation is not only useful in the framework of the proposed proof strategy, but can also be used as a base for an efficient procedure to calculate the dual transformation [5]. This procedure is part of the theorem proving system whose strategy is presented below.

3   Strategy

The proposed strategy can be divided into three steps. Initially, it is necessary to determine which dual clauses are eliminated by each substitution that unify atomic formulas of literals occurring in different clauses. Next, for each literal, it is necessary to determine all the different ways that it can be eliminated, i.e., all the different ways the dual clauses it represents can be turned into contradictions. This part of the calculation depends only on the theory and not on the specific goal to be proven. Finally, given a goal to be proven and the clauses containing literals that unify with its negation, it is necessary to combine the different ways the other literals of the clauses can be eliminated. These three steps are presented in the following subsections.

3.1   Elimination Set

Given both conjunctive and disjunctive normal forms of a theory, it is necessary to calculate the set: $\Theta = \{(\theta, P_\theta, E_\theta)\}$. It contains all substitutions that are able to turn into explicit contradictions one or more dual clauses in W_d. Each element of the set Θ contains the following elements: (i) θ - a substitution, (ii) $P_\theta = \{(\phi_i^{F_i,S_i}, \phi_i'^{F_i',S_i'})\}$ - the set of pairs of quanta in the conjunctive normal form of the theory such that: $\phi_i\theta = \neg\phi_i'\theta$, and (iii) E_θ - a set of integers containing the numbers of the dual clauses eliminated[2] by θ.

The problem is how to calculate E_θ given P_θ, the set of conjunctive normal form quanta pairs. Clearly, we must have that: $\bigcup_i (F_i \cap F_i') \subseteq E_\theta$ because, by definition of the set F, both literals – ϕ_i and ϕ_i' – are present in all dual clauses

---

[2] We use the terms kill and eliminate as abbreviations of "turn into explicit contradictions".

k, such that $k \in \cup_i (F_i \cap F_i')$. But there are cases where the S sets should also play a role; this happens when ϕ_i or ϕ_i' are ground. The final expression for E_θ is: $E_\theta = \bigcup_{i \notin G, i \notin G'} (F_i \cap F_i') \cup \bigcup_{i \in G, i \notin G'} ((F_i \cup S_i) \cap F_i') \cup \bigcup_{i \notin G, i \in G'} (F_i \cap (F_i' \cup S_i')) \cup \bigcup_{i \in G, i \in G'} ((F_i \cup S_i) \cap (F_i' \cup S_i'))$ where G contains the indices of the quanta containing ground or linear[3] literals in the left hand side of the pairs in P_θ and G' contains the indices of the quanta containing ground or linear literals in the right hand side of these pairs.

The reason why we can include these extra dual clauses into the set of dual clauses that θ is able to eliminate is the following. Suppose a dual clause k and a pair $(\phi_j^{F_j, S_j}, \phi_j'^{F_j', S_j'}) \in P_\theta$ where ϕ_j is ground or linear (i.e., $j \in G$), ϕ_j' is not ground (i.e., $j \notin G'$), $k \in S_j$ and $k \in F_j'$. In this case, dual clause k contains literal ϕ_j', because $k \in F_j'$, but does not contain literal ϕ_j, because it is only present in dual clauses whose numbers are in F_j. But, by the definition of the set S, there is surely some non ground literal $\varphi(\boldsymbol{x})$ in clause k and substitution σ, such that $\varphi(\boldsymbol{x})\sigma = \phi_j$, where $\boldsymbol{x}$ stands for the set of variables that occur in the literal φ. So we can write dual clause k as: $\langle \dots, \phi_j', \varphi(\boldsymbol{x}), \dots \rangle$ and, if we apply θ to it, we obtain: $\langle \dots, \phi_j'\theta, \varphi(\boldsymbol{x})\theta, \dots \rangle$ which might not be explicitly contradictory because we only have that: $\phi_j = \neg\phi_j'\theta$ and not that $\phi_j = \neg\varphi(\boldsymbol{x})\theta$.

But we can freely introduce a ϕ_j in dual clause k, before we apply θ to it. The reason why we can do that is the idempotency of the $\wedge$ operator, even inside a disjunctive formula: $[F_1[\boldsymbol{x}], \dots, F_n[\boldsymbol{x}], \langle \varphi(\boldsymbol{x}), \dots \rangle] \Leftrightarrow [F_1[\boldsymbol{x}], \dots, F_n[\boldsymbol{x}], \langle \varphi(\boldsymbol{x}), \varphi(\boldsymbol{y}), \dots \rangle]$ given that the set $\boldsymbol{y}$ of variables contains only new variables that do not appear anywhere else. It is easy to see, by the definition of the set S, that there is a substitution ω such that: $\varphi(\boldsymbol{y})\omega = \phi_j$. If we apply this substitution ω to the modified dual clause, we obtain: $\langle \dots, \phi_j', \varphi(\boldsymbol{x}), \phi_j, \dots \rangle$ which is equivalent to the original dual clause k but, nevertheless, is contradictory under θ.

The most interesting point is that we do not need to transform the substitution θ into $\theta\omega$ in the set Θ, because, as the variables in ω do not occur anywhere else in the problem, they have no effect in the global solution (in this case, what we are looking for).

Example 1. Consider the theory:

$$W_c = \langle\ 0 : [P_1(x_0)^{\{1,3\},\emptyset}, Q(f(x_0))^{\{0,2\},\emptyset}], \qquad 1 : [\neg Q(f(a))^{\{2,3\},\{0,1\}}],$$
$$\qquad 2 : [\neg Q(x_1)^{\{0,1\},\emptyset}, P_2(x_1)^{\{2,3\},\emptyset}]\ \rangle$$
$$W_d = [\ 0 : \langle Q(f(x_0))^{\{0\},\emptyset}, \neg Q(x_1)^{\{2\},\{1\}}\rangle, \qquad 1 : \langle P_1(x_0)^{\{0\},\emptyset}, \neg Q(x_1)^{\{2\},\{1\}}\rangle$$
$$\qquad 2 : \langle P_2(x_1)^{\{2\},\emptyset}, \neg Q(f(a))^{\{1\},\emptyset}, Q(f(x_0))^{\{0\},\emptyset}\rangle,$$
$$\qquad 3 : \langle P_1(x_0)^{\{0\},\emptyset}, P_2(x_1)^{\{2\},\emptyset}, \neg Q(f(a))^{\{1\},\emptyset}\rangle\ ]$$

Here, in this case, the set Θ contains the elements: $(\{x_1/f(x_0)\}, \{(\neg Q(x_1)^{\{0,1\},\emptyset}, Q(f(x_0))^{\{0,2\},\emptyset})\}, \{0\})$ and $(\{x_0/a\}, \{\ (\neg Q(f(a))^{\{2,3\},\{0,1\}}, Q(f(x_0))^{\{0,2\},\emptyset})\}, \{0,2\})$. The first element corresponds to the simple case where both literals in the pair are non ground, and the eliminated dual clause is just that where

---

[3] We call a non ground literal *linear* in a clause if its variables don't occur anywhere else in the clause (and because of renaming of variables, also in the theory).

both are present (dual clause 0). The pair of the second element presents one ground literal $\neg Q(f(a))$ and the associated quantum has a non empty S set – $\{0,1\}$. In this case, additionally to the dual clause where both literals of the pair are present (dual clause 2), the substitution also eliminates dual clause 0: $\langle Q(f(x_0)), \neg Q(x_1)\rangle$, because this dual clause is equivalent to the following dual clause: $\langle Q(f(x_0)),\ \neg Q(x_1),\ \neg Q(y)\rangle$. And, if we apply the substitution $\omega = \{y/f(a)\}$ to this clause, we obtain: $\langle Q(f(x_0)), \neg Q(x_1), \neg Q(f(a))\rangle$ which becomes contradictory after the application of the substitution $\theta = \{x_0/a\}$: $\langle Q(f(a)), \neg Q(x_1), \neg Q(f(a))\rangle$. It should be noted that these results would be exactly the same if the literal in the second clause – $\neg Q(f(a))$ – was linear, e.g., $\neg Q(f(z))$, instead of being ground.

In this same case, where ϕ_j is a ground literal and ϕ_j' is a non ground and non linear one, it should be noted that the dual clause numbers in the S set associated with the non ground literal – ϕ_j' – are not taken into account in the calculation of the elimination set. This is because, to apply the same duplication trick to this literal, we have to identify a literal $\varphi'(\boldsymbol{x})$ that subsumes ϕ_j' and that is present in dual clause k. But if we duplicate this literal, obtaining $\varphi'(\boldsymbol{z})$, and apply to it a suitable substitution ω', we will obtain, as expected, literal ϕ_j', which has variables that are not linear in the dual clause set (because, by hypothesis, ϕ_j' is non ground and non linear) and this would change the semantics of the theory. Nevertheless, the S set in this case has a role to play in the proposed strategy (see example 3 in section 3.2). If both ϕ_j and ϕ_j' were non ground literals, then $\varphi(\boldsymbol{y})\omega\theta$ and $\phi_j'\theta$ would be identical and non ground, therefore they would contain the same variables. But a dual clause with two literals containing the same variables has no meaning, because each literal must represent different clauses and therefore, because of renaming of variables, must have distinct variables. If we transform the modified theory to conjunctive normal form and rename the variables in the clauses, we will obtain a theory with a different semantic than the original one.

Example 2. Consider now the theory:

$$W_c = \langle\ 0 : [Q(f(x_0))^{\{0,1,2\},\emptyset}], \qquad 1 : [\neg Q(f(x_1))^{\{1\},\{0\}}, P_1(x_1)^{\{2\},\emptyset}],$$
$$2 : [\neg Q(x_2)^{\{0\},\emptyset}, P_2(x_2)^{\{1,2\},\emptyset}]$$
$$W_d = [\ 0 : \langle Q(f(x_0))^{\{0\},\emptyset}, \neg Q(x_2)^{\{2\},\{1\}}\rangle,$$
$$1 : \langle P_2(x_2)^{\{2\},\emptyset}, \neg Q(f(x_1))^{\{1\},\emptyset}, Q(f(x_0))^{\{0\},\emptyset}\rangle,$$
$$2 : \langle P_1(x_1)^{\{1\},\emptyset}, P_2(x_2)^{\{2\},\emptyset}, Q(f(x_0))^{\{0\},\emptyset}\rangle\ ]$$

In this case, the set Θ contains the elements: ($\{x_2/f(x_0)\}$, $\{$ $(\neg Q(x_2)^{\{0\},\emptyset}, Q(f(x_0))^{\{0,1,2\},\emptyset})\}, \{0\}$), and $(\{x_1/x_0\}, \{(\neg Q(f(x_1))^{\{1\},\{0\}}, Q(f(x_0))^{\{0,1,2\},\emptyset})\}, \{1\})$. The first element corresponds again to the simple case where both literals in the pair are non ground and have empty S sets. Then the eliminated dual clause is just that where both are present (dual clause 0). The pair in the second element presents one literal $\neg Q(f(x_1))$ with a non empty S set – $\{0\}$ –, but now it is non ground and non linear. If we try to apply the literal duplication trick to dual clause 0, we obtain: $\langle Q(f(x_0)), \neg Q(x_2), \neg Q(y) \rangle$. And, if we apply the substitution $\omega = \{y/f(x_0)\}$ to this clause, we obtain: $\langle Q(f(x_0)), \neg Q(x_2), \neg Q(f(x_0)) \rangle$ which is indeed contradictory, but is non linear with respect to variable x_0, something that is not allowed in a dual clause.

3.2   Elimination Graph

Suppose a unitary goal $\neg \varphi$ is to be proven, given a theory represented by both normal forms W_c and W_d. Initially, we look for the set of clauses in W_c that contain one quantum $\phi^{F,S}$ such that there exists a substitution θ that verifies: $\phi\theta = \varphi\theta$. Call this set C_φ, and consider one of the clauses belonging to it: $[\phi_1^{F_1, S_1}, \dots, \phi_k^{F_k, S_k}, \phi^{F,S}]$.

Each of the quanta in this clause represents, through its F and S sets, some set of dual clauses in W_d. If we find a substitution, call it σ, that turns into contradictions all the dual clauses whose numbers are in the set $\cup_{i=1}^{k}(F_i \cup S_i)$, than all the quanta $\phi_i^{F_i, S_i}, i = 1, \dots, k$ can be eliminated from the clause. If we apply substitution σ to W_c and W_d, we obtain a new theory where the original clause has been reduced to: $[\phi^{F,S}\sigma]$.

If it is possible to combine substitutions θ and σ, then we have: $\phi\theta\sigma = \varphi\theta\sigma$ and the substitution $\theta\sigma$ is an answer to the given goal. Otherwise, $\phi\sigma$ can be used as a new hypothesis that should be incorporated into the theory. Therefore, the proposed strategy is designed to find an adequate combination of substitutions, belonging to the elements of the set Θ, that eliminates all dual clauses $k \in \cup_{i=1}^{k}(F_i \cup S_i)$, for each clause in the set C_φ.

Analogously to all proof procedures, what we have is a standard state space search problem. The originality of the proposed strategy can be stated in two points: (i) the information contained in the set Θ comes from both normal forms, allowing a combination of techniques from the resolution and connection families of theorem proving methods, and (ii) the search for the set of combined substitutions that eliminate one literal in a clause is independent of the eventual goal to be proven and, given a theory, can be performed only once and stored for later use.

Consider first the problem of eliminating one literal in a clause. In this case we have two types of information to explore. On the one hand, the substitutions to be combined should span an acyclic graph in the conjunctive normal form, with the root node in the clause to which the literal to be eliminated belongs. The elements in the Θ set contains the necessary information to construct such graph in their lists of pairs. In fact, each pair contains two quanta, and although these are conjunctive normal form quanta, whose F and S sets contain information about the disjunctive normal form, their correspondent quanta in the disjunctive normal form (which we call their *mirror* quanta) contain in their F and S sets the information about the clauses where the respective literals occur. This information reduces drastically the number of elements of the Θ set that should be tested, because we just have to consider those that connect the present clause to another one.

On the other hand, the F and S sets of the quanta that contains the literal to be eliminated determine the set of numbers that designate the dual clauses that should be eliminated. Each element in the set Θ also contains a set of numbers associated with the dual clauses it eliminates. The match between these two sets of numbers can also be used to reduce the search space of Θ elements whose substitutions should be tested for combination. Elements of Θ that don't eliminate any of the dual clauses we want to kill should not be considered in the search.

The search problem can be defined as the process of combining graph fragments where the nodes of each of these graph fragments are labelled with clause numbers and the edges with elements of the set Θ. Formally, each of these fragments is represented by a pair (E, G), where the set E contains the number of the dual clauses eliminated by the fragment and $G = (N, A)$ is a graph with N, the set of nodes, a subset of the set of clause numbers and A, the set of edges, given by tuples (n_1, n_2, c_θ) such that n_1 and n_2 are clause numbers and $c_\theta \in \Theta$.

Initially, it is necessary to construct a set of basic fragments. The basic fragments are constructed from the elements of Θ. The dual clauses eliminated by each basic fragment are the same eliminated by the associated Θ element, and the graph associated with it is constructed according to the pair set of the Θ element. Each pair in this set gives rise to one or more edges, depending on the F sets of the mirror quanta associated with the quanta in the pair. More formally, given $c_\theta = (\theta, P_\theta, E_\theta) \in \Theta$ with $P_\theta = \{(\phi^{F,S}, \phi'^{F',S'})\}$, let $\{(\phi^{F_m,S_m}, \phi'^{F'_m,S'_m})\}$ be the set of mirror quanta associated with the quanta in P_θ. In this case, c_θ gives rise to the following basic fragment: $(E_\theta, (F_m \cup F'_m, \{(n, m, c_\theta), (m, n, c_\theta) \mid n \in F_m \text{ and } m \in F'_m\}))$.

If there are more than one element in Θ that eliminate the same set of dual clauses, i.e., that have the same E_θ set, we combine their graphs together, by making the union of the node and edge sets, respectively, and generate just one basic fragment. This further reduces the search space. In the basic fragment graphs, all edges are bidirectional, except those that have one node associated with a ground clause, in this case the edge is directed to the ground clause.

A special case occurs when, in one element of P_θ, ϕ is ground, ϕ' is not ground and the intersection of $F \cup S$ and S' is not empty. In this case, it is necessary to find all literals φ such that φ subsumes ϕ'. Let $\{\varphi^{F_\varphi,S_\varphi}\}$ be their associated quanta and $\{\varphi^{F_{\varphi,m},S_{\varphi,m}}\}$ the respective mirror quanta. For each φ, we should verify if the intersection: $(F \cup S) \cap S' \cap F_\varphi$ is not empty and, if it is the case, then we should include the following edges in the graph associated with the basic fragment: $\{(n, m, c_\theta), (m, n, c_\theta) \mid n \in F_m \text{ and } m \in F_{\varphi,m}\}$.

Example 3. Consider the theory given by:

$$W_c = \langle\, 0 : [P(f(a))^{\{0,2\},\{1\}}],$$
$$1 : [\neg P(f(x_0))^{\{2\},\{0\}}, P(x_0)^{\{1\},\emptyset}],$$
$$2 : [\neg P(x_1)^{\{0\},\emptyset}, R(x_1)^{\{1,2\},\emptyset}] \,\rangle$$
$$W_d = [\, 0 : \langle P(f(a))^{\{0\},\emptyset}, \neg P(x_1)^{\{2\},\{1\}}\rangle,$$
$$1 : \langle R(x_1)^{\{2\},\emptyset}, P(x_0)^{\{1\},\{0\}}\rangle,$$
$$2 : \langle R(x_1)^{\{2\},\emptyset}, \neg P(f(x_0))^{\{1\},\emptyset}, P(f(a))^{\{0\},\emptyset}\rangle \,]$$

In this case, the set Θ contains the elements: ($\{x_1/f(a)\}, \{$ ($\neg P(x_1)^{\{0\},\emptyset}$, $P(f(a))^{\{0,2\},\{1\}})\}, \{0\}$) and ($\{x_0/a\}, \{(\neg P(f(x_0))^{\{2\},\{0\}}, P(f(a))^{\{0,2\},\{1\}}))\}$, $\{0,2\})$. The second element of the set Θ is an instance of this special case. There we have $\phi^{F,S} = P(f(a))^{\{0,2\},\{1\}}$, $\phi'^{F',S'} = \neg P(f(x_0))^{\{2\},\{0\}}$, $\varphi^{F_\varphi,S_\varphi} = \neg P(x_1)^{\{0\},\emptyset}$ and $\varphi^{F_\varphi,m,S_\varphi,m} = \neg P(x_1)^{\{2\},\{1\}}$. In this case, we have: $F \cap S' \cap F_\varphi = \{0,2\} \cap \{0\} \cap \{0\} = \{0\}$ and we should include the following edges: $\{(1,2,c_{\{x_0/a\}}),(2,1,c_{\{x_0/a\}})\}$.

The search begins with the basic fragments that contain, in the pair set of the Θ elements occurring in the edges of their graph, quanta associated with the literal we want to eliminate. These basic fragments are joined together into a general fragment. The only difference between basic and general fragments is that general fragment graphs only have unidirectional edges. Initially, all the edges begin in the clause associated with the literal we want to eliminate, the ends of all these edges determine the *fringe* of the graph.

Given a general fragment, the search proceeds by looking into the basic fragment set for one that has a graph with a node labelled with the same clause as those labelling the fringe nodes of the graph in the given general fragment. Augmenting the graph only from its fringe nodes guarantees that we will get an acyclic graph as a result. Once an adequate basic fragment is found its graph is joined with the graph in the given general fragment, if the substitutions associated with the Θ elements in neighbor edges combine. It is important to note that, when an adequate basic fragment is found, only the edges of its graph that begin in clauses of the fringe of the graph in the current general fragment are included, keeping the general fragment graph acyclic and only containing unidirectional edges.

Each graph may store several different substitutions, because only the substitutions in the same path can be combined together. Another advantage is that the complete substitutions associated with each path don't need to be stored during the search; it is only necessary that the substitution associated with a new edge combine with the substitution in the previous edge. This is so because, the graph being acyclic, the paths in it never return to the same clause and therefore variables are never repeated along one path.

Besides the fact that we only choose basic fragments that lead to the expansion of the fringe nodes of the graph, the search for basic fragments is further restricted by the number of dual clauses that these basic fragments eliminate.

Once an adequate basic fragment is found and its graph is joined with the graph of the given general fragment, a new general fragment is generated. These new general fragments will eliminate dual clauses corresponding to the union of those eliminated by the given general fragment and those eliminated by the basic fragment found. From this new general fragment the search can proceed, until a general fragment is generated that kills all the dual clauses we want to eliminate, finishing the search.

Example 4. Consider the theory:

$$W_c = \langle\ 0 : [P(a)^{\{0,1,2,3,4,5\},\emptyset}], \qquad 1 : [P(b)^{\{0,1,2,3,4,5\},\emptyset}],$$
$$2 : [\neg P(x_0)^{\{0,2\},\emptyset}, Q(f(x_0))^{\{1,3,4,5\},\emptyset}],$$
$$3 : [\neg P(f(x_1))^{\{1,5\},\{0,2\}}, R(x_1)^{\{3,4\},\emptyset}],$$
$$4 : [\neg Q(f(x_2))^{\{0,1,3\},\emptyset}, R(x_2)^{\{2,4,5\},\emptyset}]\ \rangle$$
$$W_d = [\ 0 : \langle P(a)^{\{0\},\emptyset}, P(b)^{\{1\},\emptyset}, \neg Q(f(x_2))^{\{4\},\emptyset}, \neg P(x_0)^{\{2\},\{3\}}\rangle$$
$$1 : \langle Q(f(x_0))^{\{2\},\emptyset}, \neg Q(f(x_2))^{\{4\},\emptyset}, \neg P(f(x_1))^{\{3\},\emptyset}, P(b)^{\{1\},\emptyset}, P(a)^{\{0\},\emptyset}\rangle$$
$$2 : \langle P(a)^{\{0\},\emptyset}, P(b)^{\{1\},\emptyset}, R(x_2)^{\{4\},\emptyset}, \neg P(x_0)^{\{2\},\{3\}}\rangle$$
$$3 : \langle P(a)^{\{0\},\emptyset}, P(b)^{\{1\},\emptyset}, R(x_1)^{\{3\},\emptyset}, \neg Q(f(x_2))^{\{4\},\emptyset}, Q(f(x_0))^{\{2\},\emptyset}\rangle$$
$$4 : \langle P(a)^{\{0\},\emptyset}, P(b)^{\{1\},\emptyset}, R(x_1)^{\{3\},\emptyset}, R(x_2)^{\{4\},\emptyset}, Q(f(x_0))^{\{2\},\emptyset}\rangle$$
$$5 : \langle P(a)^{\{0\},\emptyset}, P(b)^{\{1\},\emptyset}, \neg P(f(x_1))^{\{3\},\emptyset}, R(x_2)^{\{4\},\emptyset}, Q(f(x_0))^{\{2\},\emptyset}\rangle\ ]$$

In this case, the set Θ contains the elements:
$c_1 = (\{x_0/a\}, \{(\neg P(x_0)^{\{0,2\},\emptyset}, P(a)^{\{0,1,2,3,4,5\},\emptyset})\}, \{0,2\})$,
$c_2 = (\{x_0/b\}, \{(\neg P(x_0)^{\{0,2\},\emptyset}, P(b)^{\{0,1,2,3,4,5\},\emptyset})\}, \{0,2\})$ and
$c_3 = (\{x_2/x_0\}, \{(\neg Q(f(x_2))^{\{0,1,3\},\emptyset}, Q(f(x_0))^{\{1,3,4,5\},\emptyset})\}, \{1,3\})$.

These three elements give rise to two basic fragments, because we join in a single basic fragment all the substitutions that eliminate the same set of dual clauses. The two basic fragments (see figure 1 (a) and (b)) are given by:
$f_1 = (\{0,2\}, (\{0,1,2\}, \{(2,0,c_1), (2,1,c_2)\}))$ and
$f_2 = (\{1,3\}, (\{2,4\}, \{(2,4,c_3), (4,2,c_3)\}))$.

Given a goal $\varphi = \neg R(y)$, the set C_φ would contain clauses 3 and 4. To eliminate literal $\neg P(f(x_1))$ in clause 3, it would be necessary to eliminate dual clauses $\{0,1,2,5\}$, according to the F and S sets of the quantum associated with it. It is easy to see that no combination of elements of Θ can kill these dual clauses. On the other hand, the literal $\neg Q(f(x_2))$ in clause 4 can be eliminated. The search begins with fragment f_2, because Θ element c_3, which occurs in the edge of its graph, contains, in its pair set, the quantum $\neg Q(f(x_2))^{\{0,1,3\},\emptyset}$ associated with the literal $\neg Q(f(x_2))$ we want to kill. In this case the search is trivial and the solution is given by the following general fragment (see figure 1 (c)): $(\{0,1,2,3\}, (\{0,1,2,4\}, \{(4,2,c_3), (2,0,c_1), (2,1,c_2)\}))$ which is the combination of basic fragments f_1 and f_2 and eliminates dual clauses $\{0,1,2,3\}$. The resulting substitutions are obtained by combining the goal substitution $\theta = \{y/x_2\}$ with the substitutions in the Θ elements in each of the paths in the graph of the solution fragment: $\{y/x_2, x_2/x_0, x_0/a\}$ and $\{y/x_2, x_2/x_0, x_0/b\}$.

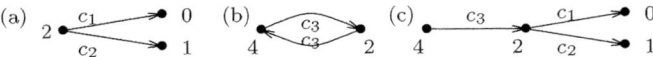

Fig. 1. Graphs: (a) fragment f_1, (b) fragment f_2, (c) solution fragment

3.3 Combining Elimination Graphs

Consider the following clause containing the literal ϕ we want to eliminate: $[\phi_1^{F_1,S_1}, \ldots, \phi_k^{F_k,S_k}, \phi^{F,S}]$. If, differently from the examples presented in the previous section, $k > 1$, then the elimination graphs associated with literals $\phi_1, \ldots, \phi_k$ must be combined in a single elimination graph. This combined elimination graph differs from the elimination graphs associated with single fragments in that, in the latter, each path corresponds to one substitution and, in the former, a single substitution may be represented by more than one path, because of this, combined elimination graphs have more than one fringe, each fringe associated with one substitution.

The simplest case is when the graphs to be combined involve different variables. In this case, the graphs have only to be merged together. The only difficulty is the definition of the set of fringes: each fringe of the combined graph is composed by exactly one path of each of the elimination graphs to be combined.

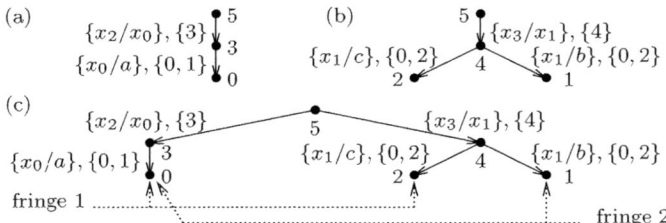

Fig. 2. Elimination Graphs: (a) $\neg P(f(x_2))$, (b) $\neg Q(f(x_3))$, (c) combined graph

Example 5. Consider the following theory:

$$W_c = \langle\ 0 : [P(a)^{\{0,1,2,3,4,5\},\emptyset}]$$
$$1 : [Q(b)^{\{0,1,2,3,4,5\},\emptyset}]$$
$$2 : [Q(c)^{\{0,1,2,3,4,5\},\emptyset}]$$
$$3 : [\neg P(x_0)^{\{0,1\},\emptyset}, P(f(x_0))^{\{2,3,4,5\},\emptyset}]$$
$$4 : [\neg Q(x_1)^{\{0,2\},\emptyset}, Q(f(x_1))^{\{1,3,4,5\},\emptyset}]$$
$$5 : [\neg P(f(x_2))^{\{3\},\{0,1\}}, \neg Q(f(x_3))^{\{4\},\{0,2\}}, R(x_2,x_3)^{\{5\},\emptyset}]\ \rangle$$

$W_d = [$
$0 : \langle P(a)^{\{0\},\emptyset}, Q(b)^{\{1\},\emptyset}, Q(c)^{\{2\},\emptyset}, \neg P(x_0)^{\{3\},\{5\}}, \neg Q(x_1)^{\{4\},\{5\}} \rangle$
$1 : \langle P(a)^{\{0\},\emptyset}, Q(b)^{\{1\},\emptyset}, Q(c)^{\{2\},\emptyset}, Q(f(x_1))^{\{4\},\emptyset}, \neg P(x_0)^{\{3\},\{5\}} \rangle$
$2 : \langle Q(b)^{\{1\},\emptyset}, Q(c)^{\{2\},\emptyset}, P(a)^{\{0\},\emptyset}, P(f(x_0))^{\{3\},\emptyset}, \neg Q(x_1)^{\{4\},\{5\}} \rangle$
$3 : \langle Q(b)^{\{1\},\emptyset}, Q(c)^{\{2\},\emptyset}, Q(f(x_1))^{\{4\},\emptyset}, \neg P(f(x_2))^{\{5\},\emptyset}, P(f(x_0))^{\{3\},\emptyset}, P(a)^{\{0\},\emptyset} \rangle$
$4 : \langle Q(b)^{\{1\},\emptyset}, Q(c)^{\{2\},\emptyset}, Q(f(x_1))^{\{4\},\emptyset}, \neg Q(f(x_3))^{\{5\},\emptyset}, P(f(x_0))^{\{3\},\emptyset}, P(a)^{\{0\},\emptyset} \rangle$
$5 : \langle R(x_2, x_3)^{\{5\},\emptyset}, Q(f(x_1))^{\{4\},\emptyset}, Q(c)^{\{2\},\emptyset}, Q(b)^{\{1\},\emptyset}, P(a)^{\{0\},\emptyset}, P(f(x_0))^{\{3\},\emptyset} \rangle]$
and a goal to be proven of the form $\neg R(a, y)$. In this case, we should eliminate
the literals $\neg P(f(x_2))$ and $\neg Q(f(x_3))$. The search for elimination fragments for
these two literals results in two fragments. The first one, see figure 2 (a), has
a graph with only one path and the other, see figure 2 (b), has a graph with
two paths. The combination is shown in figure 2 (c); it has two fringes (1 and
2) which correspond to two different solutions: $\{x_3/x_1, x_2/x_0, x_0/a, x_1/b\}$ and
$\{x_3/x_1, x_2/x_0, x_0/a, x_1/c\}$.

If the graphs to be combined share variables, there are two possibilities: if the
variables in each of the edges contain all the variables of a given clause, then
these edges can be maintained in parallel, because they represent two different
instances of the covered clause. Otherwise the substitutions in each edge should
be combined and the parallel edges should be considered as part of the same
path.

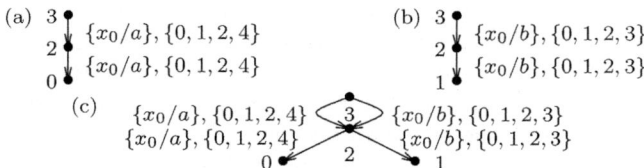

Fig. 3. Elimination Graphs: (a) $\neg Q(a)$, (b) $\neg Q(b)$, (c) combined graph

Example 6. Consider the theory given by:
$W_c = \langle\ 0 : [P(a)^{\{0,1,2,3,4,5\},\emptyset}],$
$\qquad 1 : [P(b)^{\{0,1,2,3,4,5\},\emptyset}],$
$\qquad 2 : [\neg P(x_0)^{\{0,1,2\},\emptyset}, Q(x_0)^{\{3,4,5\},\emptyset}],$
$\qquad 3 : [\neg Q(a)^{\{2,4\},\emptyset}, \neg Q(b)^{\{1,3\},\emptyset}, R(c)^{\{0,5\},\emptyset}]\ \rangle$
$W_d = [\ 0 : \langle R(c)^{\{3\},\emptyset}, \neg P(x_0)^{\{2\},\emptyset}, P(b)^{\{1\},\emptyset}, P(a)^{\{0\},\emptyset} \rangle,$
$\qquad 1 : \langle \neg P(x_0)^{\{2\},\emptyset}, \neg Q(b)^{\{3\},\emptyset}, P(b)^{\{1\},\emptyset}, P(a)^{\{0\},\emptyset} \rangle,$
$\qquad 2 : \langle \neg P(x_0)^{\{2\},\emptyset}, \neg Q(a)^{\{3\},\emptyset}, P(b)^{\{1\},\emptyset}, P(a)^{\{0\},\emptyset} \rangle,$
$\qquad 3 : \langle Q(x_0)^{\{2\},\emptyset}, \neg Q(b)^{\{3\},\emptyset}, P(b)^{\{1\},\emptyset}, P(a)^{\{0\},\emptyset} \rangle,$
$\qquad 4 : \langle Q(x_0)^{\{2\},\emptyset}, \neg Q(a)^{\{3\},\emptyset}, P(b)^{\{1\},\emptyset}, P(a)^{\{0\},\emptyset} \rangle,$
$\qquad 5 : \langle P(b)^{\{1\},\emptyset}, Q(x_0)^{\{2\},\emptyset}, R(c)^{\{3\},\emptyset}, P(a)^{\{0\},\emptyset} \rangle]$

and a goal to be proven of the form $\neg R(c)$. In this case, we should eliminate the literals $\neg Q(a)$ and $\neg Q(b)$. The search for elimination fragments for these two literals results in two fragments and each of them has only one path (see figure 3 (a) and (b). The combination is shown in figure 3 (c), it has one fringe that correspond to two different solutions: $\{x_0/a\}$ and $\{x_0/b\}$.

4   Results

To test the proposed strategy, the LOGIK system has been implemented. The system is an object-oriented laboratory for first-order logic, written in Common Lisp/CLOS, which includes the proposed proof strategy and a dual transformation algorithm. All entities of the logical syntax – variables, functions, predicates, terms, literals, clauses, dual clauses and substitutions – have been implemented as classes with their associated manipulation methods – substitution, unification and subsumption. To test the strategy, we used examples from the TPTP[4] problem library. The results obtained, even with this experimental implementation where no special concern was taken over performance, are promising.

5   Conclusion

We have presented a proof strategy for first-order logic based on the complementary information that can be extracted from both conjunctive and disjunctive normal forms and mainly from the relations between literals that appear in each one of them. This strategy integrates a theorem proving system [6] that is part of a more ambitious cognitive modeling project [4], [9], [8]. The originality of the proposed strategy lies in the fact that, because it uses the information contained in both normal forms, it allows a combination of techniques from the resolution and connection families of theorem proving methods. Besides that, the results obtained during the search for combined substitutions that eliminate each literal in a clause are independent of the eventual theorem to be proven and, given a theory, need be performed only once and stored for later use. The proposed strategy has been implemented and tested, showing promising results.

References

[1] P.B. Andrews. Theorem proving via general matings. *Journal of the ACM*, 28(2):193–214, 1981.
[2] W. Bibel. *Automated Theorem Proving*. Vieweg Verlag, 1987.
[3] W. Bibel and E. Eder. Methods and calculi for deduction. In Dov M. Gabbay, C.J. Hogger, and J.A. Robinson, editors, *Handbook of Logic in Artificil Intelligence and Logic Programming - Logical Foundations*, volume 1, pages 67–182. Oxford University Press, 1993.

---

[4] The TPTP (Thousands of Problems for Theorem Provers) Problem Library is a collection of test problems for Automated Theorem Proving systems. It can be obtained at: `coral.cs.jcu.edu.au:/pub/research/tptp-library/` [137.219.17.4].

[4] G. Bittencourt. In the quest of the missing link. In *Proceedings of IJCAI 15, Nagoya, Japan, August 23-29*, pages 310–315. Morgan Kaufmann (ISBN 1-55860-480-4), 1997.

[5] G. Bittencourt. Concurrent inference through dual transformation. *Logic Journal of the Interest Group in Pure and Applied Logics (IGPL)*, 6(6):795–834, 1998.

[6] G. Bittencourt and I. Tonin. A multi-agent approach to first-order logic. In *Proceedings of 8th Portuguese Conference on Artificial Intelligence (EPIA'97), Coimbra, Portugal, October 6-9*. Springer-Verlag, 1997.

[7] C.-L. Chang and R.C.-T. Lee. *Symbolic Logic and Mechanical Theorem Proving*. Academic Press, Computer Science Classics, 1973.

[8] A. C. P. L. da Costa and G. Bittencourt. From a concurrente agent architecture to a autonomous concurrente agent architecture. *IJCAI'99 Third RoboCup Workshop*, Stockolm, Sweden, July 31-August 1st 1999.

[9] A.C.P.L. da Costa and G. Bittencourt. Dynamic social knowledge: A cooperation strategie for cognitive multi-agent systems. *Third International Conference on Multi-Agent Systems (ICMAS'98)*, pages 415–416, Paris, France, July 2-7 1998. IEEE Computer Society.

[10] E. Eder. Consolation and its relation to resolution. In *Proceedings of IJCAI 12*, 1991.

[11] N. Eisinger and H.J. Ohlbach. Deduction systems based on resolution. In Dov M. Gabbay, C.J. Hogger, and J.A. Robinson, editors, *Handbook of Logic in Artificil Intelligence and Logic Programing - Logical Foundations*, volume 1, pages 183–271. Oxford University Press, 1993.

[12] M. Fitting. *First-Order Logic and Automated Theorem Proving*. Springer-Verlag, New York, 1990.

[13] P. Jackson. Computing prime implicants. In *Proceedings of the 10th International Conference on Automatic Deduction, Kaiserslautern, Germany, Springer-Verlag LNAI No. 449*, pages 543–557, 1990.

[14] A. Kean and G. Tsiknis. An incremental method for generating prime implicants/implicates. *Journal of Symbolic Computation*, 9:185–206, 1990.

[15] W.V.O. Quine. The problem of simplifying truth functions. *American Mathematics Monthly*, 59:521–531, 1952.

[16] W.V.O. Quine. On cores and prime implicants of truth functions. *American Mathematics Monthly*, 66:755–760, 1959.

[17] J.A. Robinson. A machine-oriented logic based on the resolution principle. *Journal of the ACM*, 12(1):23–41, January 1965.

[18] J.A. Robinson. Logic and logic programming. *Communications of the ACM*, 35(3):41–65, March, 1992.

[19] J.R. Slagle, C.L. Chang, and R.C.T. Lee. A new algorithm for generating prime implicants. *IEEE Transactions on Computing*, 19(4):304–310, 1970.

[20] R.M. Smullyan. *First Order Logic*. Springer-Verlag, 1968.

[21] R. Socher. Optimizing the clausal normal form transformation. *Journal of Automated Reasoning*, 7(3):325–336, 1991.

[22] A.C. Stylianou, G.R. Madey, and R.D. Smith. Selection criteria for expert system shells: A socio-technical framework. *Communications of the ACM*, 35(10):32–48, October 1992.

[23] P. Tison. Generalized consensus theory and application to the minimization of boolean functions. *IEEE Transactions on Computing*, 16(4):446–456, 1967.

Formalizing Rewriting in the ACL2 Theorem Prover*

José-Luis Ruiz-Reina, José-Antonio Alonso, María-José Hidalgo, and
Francisco-Jesús Martín-Mateos

Departamento de Ciencias de la Computación e Inteligencia Artificial.
Facultad de Informática y Estadística, Universidad de Sevilla
Avda. Reina Mercedes, s/n. 41012 Sevilla, Spain
{jruiz,jalonso,mjoseh,fjesus}@cica.es

Abstract. We present an application of the ACL2 theorem prover to
formalize and reason about rewrite systems theory. This can be seen
as a first approach to apply formal methods, using ACL2, to the design of symbolic computation systems, since the notion of rewriting or
simplification is ubiquitous in such systems. We concentrate here on formalization and representation aspects of abstract reduction and term
rewriting systems, using the first-order, quantifier-free ACL2 logic based
on Common Lisp.

1 Introduction

We report in this paper the status of our work on the application of the ACL2
theorem prover to reason about rewrite systems theory: confluence, local confluence, noetherianity, normal forms and other related concepts have been formalized in the ACL2 logic and some results about abstract reduction relations
and term rewriting systems have been mechanically proved, including Newman's
lemma and the Knuth-Bendix critical pair theorem.

ACL2 is both a logic and a mechanical theorem proving system supporting
it. The ACL2 logic is an existentially quantifier-free, first-order logic with equality. ACL2 is also a programming language, an applicative subset of Common
Lisp. The system evolved from the Boyer-Moore theorem prover, also known as
Nqthm.

A formal proof using a theorem proving environment provides not only formal
verification of mathematical theories, but allows us to understand and examine
their theorems with much greater detail, rigor and clarity. On the other hand, the
notion of rewriting or simplification is a crucial component in symbolic computation: simplification procedures are needed to transform complex objects in order
to obtain equivalent but simpler objects and to compute unique representations
for equivalence classes (see, for example, [4] or [9]).

---

* This work has been supported by DGES/MEC: Projects PB96-0098-C04-04 and
PB96-1345

J.A. Campbell and E. Roanes-Lozano (Eds.): AISC 2000, LNAI 1930, pp. 92–106, 2001.
© Springer-Verlag Berlin Heidelberg 2001

Since ACL2 is also a programming language, this work can be seen as a first step to obtain verified executable Common Lisp code for components of symbolic computation systems. Although a fully verified implementation of such a system is currently beyond our possibilities, several basic algorithms can be mechanically "certified" and integrated as part of the whole system.

We also show here how a weak logic like the ACL2 logic (no quantification, no infinite objects, no higher order variables, etc.) can be used to represent, formalize, and mechanically prove non-trivial theorems. In this paper, we place emphasis on describing the formalization and representation aspects of our work. Due to the lack of space, we will skip details of the mechanical proofs. The complete information is available on the web at `http://www-cs.us.es/~jruiz/acl2-rewr`.

1.1   The ACL2 System

We briefly describe here the ACL2 theorem prover and its logic. The best introduction to ACL2 is [6]. To obtain more background on ACL2, see the ACL2 user's manual in [7]. A description of the main proof techniques used in Nqthm, also used in ACL2, can be found in [3].

ACL2 stands for A Computational Logic for Applicative Common Lisp. The ACL2 logic is a quantifier-free, first-order logic with equality, describing an applicative subset of Common Lisp. The syntax of terms is that of Common Lisp [14] (we will assume that the reader is familiar with this language). The logic includes axioms for propositional logic and for a number of Lisp functions and data types. Rules of inference include those for propositional calculus, equality, and instantiation. By the *principle of definition*, new function definitions (using `defun`) are admitted as axioms only if there exists an ordinal measure in which the arguments of each recursive call decrease. This ensures that no inconsistencies are introduced by new definitions. The theory has a constructive definition of the ordinals up to ε_0, in terms of lists and natural numbers, given by the predicate `e0-ordinalp` and the order `e0-ord-<`. One important rule of inference is the *principle of induction*, which permits proofs by induction on ε_0.

In addition to the definition principle, the encapsulation mechanism (using `encapsulate`) allows the user to introduce new function symbols by axioms constraining them to have certain properties (to ensure consistency, a witness local function having the same properties has to be exhibited). Inside an `encapsulate`, properties stated with `defthm` need to be proved for the local witnesses, and outside, those theorems work as assumed axioms. The functions partially defined with `encapsulate` can be seen as second-order variables, representing functions with those properties. A derived rule of inference, functional instantiation, allows some kind of second-order reasoning: theorems about constrained functions can be instantiated with function symbols known to have the same properties.

The ACL2 theorem prover is inspired by Nqthm, but has been considerably improved. The main proof techniques used by the prover are simplification and induction. Simplification is a combination of decision procedures, mainly term rewriting, using the rules previously proved by the user. The command `defthm` starts a proof attempt, and, if it succeeds, the theorem is stored as a rule. The theorem prover is automatic in the sense that once **defthm** is invoked, the user can no longer interact with the system. However, in a deeper sense, the system is interactive. Very often, non-trivial proofs are not found by the system in the first attempt. The user has to guide the prover by adding lemmas and definitions, used in subsequent proofs as rules. The role of the user is important: a typical proof effort consists of formalizing the problem in the logic and helping the prover to find a preconceived proof by means of a suitable set of rewrite rules.

1.2   Abstract Reductions and Term Rewriting Systems

This section provides a short introduction to basic concepts and definitions from rewriting theory used in this paper. A complete description can be found in [1].

An *abstract reduction* is simply a binary relation $\rightarrow$ defined on a set A. We will denote as $\leftarrow$, $\leftrightarrow$, $\overset{*}{\rightarrow}$ and $\overset{*}{\leftrightarrow}$ respectively the inverse relation, the symmetric closure, the reflexive-transitive closure and the equivalence closure. The following concepts are defined with respect to a reduction relation $\rightarrow$. An element x is in *normal form* (or *irreducible*) if there is no z such that $x \rightarrow z$. We say that x and y are joinable (denoted as $x \downarrow y$) if it exists u such that $x \overset{*}{\rightarrow} u \overset{*}{\leftarrow} y$. We say that x and y are *equivalent* if $x \overset{*}{\leftrightarrow} y$.

An important property to study about reduction relations is the existence of unique normal forms for equivalent objects. A reduction relation has the *Church-Rosser property* if every two equivalent objects are joinable. An equivalent property is *confluence*: for all x, u, v such that $u \overset{*}{\leftarrow} x \overset{*}{\rightarrow} v$, then $u \downarrow v$. If a reduction has the Church-Rosser property, then two distinct normal forms cannot be equivalent. If in addition the relation is *normalizing* (i.e. every element has a normal form, noted as $x \downarrow$) then $x \overset{*}{\leftrightarrow} y$ iff $x \downarrow = y \downarrow$. Provided normal forms are computable and identity in A is decidable, then the equivalence relation $\overset{*}{\leftrightarrow}$ is decidable in this case, using a test for equality of normal forms.

Another important property is termination: a reduction relation is *terminating* (or *noetherian*) if there is no infinite reduction sequence $x_0 \rightarrow x_1 \rightarrow x_2 \rightarrow \ldots$. Obviously, every noetherian reduction is normalizing. The Church-Rosser property can be localized when the reduction is terminating. In that case an equivalent property is *local confluence*: for all x, u, v such that $u \leftarrow x \rightarrow v$, then $u \downarrow v$. This result is known as **Newman's lemma**.

One important type of reduction relations is defined in the set $T(\Sigma, X)$ of first order terms in a given language, where Σ is a set of function symbols, and X is a set of variables. In this context, an *equation* is a pair of terms $l = r$. The reduction relation induced by a set of equations E is defined as follows: $s \rightarrow_E t$ if there exist $l = r \in E$ and a substitution σ of the variables in l (the *matching* substitution) such that $\sigma(l)$ is a subterm of s and t is obtained from s by replacing the subterm $\sigma(l)$ by $\sigma(r)$. This reduction relation is of great interest in universal algebra because it can be proved that $E \models s = t$ iff $s \overset{*}{\leftrightarrow}_E t$. This implies decidability of every equational theory defined by a set of axioms E such that $\rightarrow_E$ is terminating and locally confluent. To emphasize the use of the equation $l = r$ from left to right as described above, we write $l \rightarrow r$ and talk about *rewrite rules*. A *term rewriting system* (TRS) is a set of rewrite rules. Unless denoted otherwise, E is always a set of equations (equational axioms) and R is a term rewriting system.

Local confluence is decidable for finite and terminating TRSs: joinability has only to be checked for a finite number of pair of terms, called *critical pairs*, accounting for the most general forms of local divergence. The **critical pair theorem** states that a TRS is locally confluent iff all its critical pairs are joinable. Thus, Church-Rosser property of terminating TRSs is a decidable property: it is enough to check if every critical pair has a common normal form. If a TRS R has a critical pair with different normal forms, there is still a chance to obtain a decision procedure for the equational theory of R, adjoining that equation as a new terminating rewrite rule. This is the basis for the well-known *completion algorithms* (see [1] for details).

In the sequel, we describe the formalization of these properties in the ACL2 logic and a proof of them using the theorem prover. For the rest of the paper, when we talk about "prove" we mean "mechanically prove using ACL2".

2   Formalizing Abstract Reductions in ACL2

Our first attempt to represent abstract reduction relations in the ACL2 logic was simply to define them as binary boolean functions, using `encapsulate` to state their properties. Nevertheless, if $x \rightarrow y$, more important than the relation between x and y is the fact that y is obtained from x by applying some kind of transformation or *operator*. In its most abstract formulation, we can view a reduction as a binary function that, given an element and an operator, returns another object, performing a *one-step reduction*. Consider, for example, equational reductions: elements in that case are first-order terms and operators are the objects constituted by a position (indicating the subterm replaced), an equation (the rule applied) and a substitution (the matching substitution).

Of course not any operator can be applied to any element. Thus, a second component in this formalization is needed: a boolean binary function to test if it is *legal* to apply an operator to an element. Finally, a third component is introduced: since computation of normal forms requires searching for legal operators to apply, we will need a unary function such that when applied to an element, it returns a legal operator, whenever it exists, or `nil` otherwise (a *reducibility* test).

The above considerations lead us to formalize the concept of abstract reductions in ACL2, using three partially defined functions: `reduce-one-step`, `legal` and `reducible`. This can be done with the following `encapsulate` (dots are used to omit technical details, as in the rest of the paper):

```
(encapsulate
  ((legal (x u) t) (reduce-one-step (x u) t) (reducible (x) t))
  ...
  (defthm legal-reducible-1
    (implies (reducible x) (legal x (reducible x)))))
```

```
(defthm legal-reducible-2
  (implies (not (reducible x)) (not (legal x op))))
 ...)
```

The first line of every `encapsulate` is a signature description of the non-local functions partially defined. The two theorems assumed above as axioms are minimal requirements for every reduction we defined: if further properties (for example, local confluence, confluence or noetherianity) were assumed, they had to be stated inside the `encapsulate`. This is a very abstract framework to formalize reductions in ACL2. We think that these three functions capture the basic abstract features every reduction has. On the one hand, a procedural aspect: the computation of normal forms, applying operators until irreducible objects are obtained. On the other hand, a declarative aspect: every reduction relation describes its equivalence closure. Representing reductions in this way, we can define concepts like Church-Rosser property, local confluence or noetherianity and even prove non-trivial theorems like Newman's lemma, as we will see.

To instantiate this general framework, concrete instances of `reduce-one--step`, `legal` and `reducible` have to be defined and the properties assumed here as axioms must be proved for those concrete definitions. By functional instantiation, results about abstract reductions can then be easily exported to concrete cases (as we will see for the equational case).

2.1   Equivalence and Proofs

Due to the constructive nature of the ACL2 logic, in order to define $x \overset{*}{\leftrightarrow} y$ we have to include an argument with a sequence of steps $x = x_0 \leftrightarrow x_1 \leftrightarrow x_2 \dots \leftrightarrow x_n = y$. This is done by the function `equiv-p` defined in figure 1. (`equiv-p x y p`) is `t` if `p` is a proof justifying that $x \overset{*}{\leftrightarrow} y$. A *proof*[1] is a sequence of legal steps and each proof step is a structure `r-step` with four fields: `elt1`, `elt2` (the elements connected), `direct` (the direction of the step) and `operator`. Two proofs justifying the same equivalence will be said to be *equivalent*. A proof step is *legal* (as defined by `proof-step-p`) if one of its elements is obtained applying the (legal) operator to the other (in the sense indicated by `direct`).

Church-Rosser property and local confluence can be redefined with respect to the form of a proof (subsections 2.2 and 2.3). For that purpose, we define (omitted here) functions to recognize proofs with particular shapes (*valleys* and *local peaks*): `local-peak-p` recognizes proofs of the form $v \leftarrow x \rightarrow u$ and `steps-valley` recognizes proofs of the form $v \overset{*}{\rightarrow} x \overset{*}{\leftarrow} u$.

2.2   Church-Rosser Property and Decidability

We describe how we formalized and proved the fact that every Church-Rosser and normalizing reduction relation is decidable. Valley proofs can be used to

---

[1] Do not confuse with proofs done using the ACL2 system.

```
(defstructure r-step direct operator elt1 elt2)

(defun proof-step-p (s)
  (let ((e1 (elt1 s)) (e2 (elt2 s)) (op (operator s)) (dt (direct s)))
    (and
      (r-step-p s)
      (implies dt (and (legal e1 op)
                       (equal (reduce-one-step e1 op) e2)))
      (implies (not dt) (and (legal e2 op)
                             (equal (reduce-one-step e2 op) e1))))))

(defun equiv-p (x y p)
  (if (endp p) (equal x y)
    (and (proof-step-p (car p)) (equal x (elt1 (car p)))
         (equiv-p (elt2 (car p)) y (cdr p)))))
```

Fig. 1. Definition of proofs and equivalence

reformulate the definition of the Church-Rosser property: a reduction is Church-Rosser iff for every proof there exists an equivalent valley proof. Since the ACL2 logic is quantifier-free, the existential quantifier in this statement has to be replaced by a Skolem function, which we call `transform-to-valley`. The concept of being normalizing can also be reformulated in terms of proofs: a reduction is normalizing if for every element there exists a proof to an equivalent irreducible element. This proof is given by the (Skolem) function `proof-irreducible` (note that we are not assuming noetherianity yet). Properties defining a Church-Rosser and normalizing reduction are `encapsulated` as shown in figure 2, item (a).

The function `r-equiv` tests if normal forms are equal. Note that the normal form of an element x is the last element of (`proof-irreducible` x):

```
(defun normal-form (x)
  (last-of-proof x (proof-irreducible x)))

(defun r-equiv (x y)
  (equal (normal-form x) (normal-form y)))
```

To prove decidability of a Church-Rosser and normalizing relation, it is enough to prove that `r-equiv` is a complete and sound algorithm deciding the equivalence relation associated with the reduction relation. See figure 2, item (b). We also include the main lemma used, stating that there are no distinct equivalent irreducible elements. Note also that soundness is expressed in terms of a Skolem function `make-proof-common-normal-form` (definition omitted), which constructs a proof justifying the equivalence. These theorems are proved easily, without much guidance from the user. See the web page for details.

```
;;; (a) Definition of Church-Rosser and normalizing reduction:
(encapsulate
  ((legal (x u) t) (reduce-one-step (x u) t) (reducible (x) t)
   (transform-to-valley (x) t) (proof-irreducible (x) t))
  .....
  (defthm Church-Rosser-property
    (let ((valley (transform-to-valley p)))
      (implies (equiv-p x y p)
               (and (steps-valley valley) (equiv-p x y valley)))))
  .....
  (defthm normalizing
    (let* ((p-x-y (proof-irreducible x))
           (y (last-of-proof x p-x-y)))
      (and (equiv-p x y p-x-y) (not (reducible y))))))

;;; (b) Main theorems proved:
(defthm if-C-R--two-ireducible-connected-are-equal
  (implies
    (and (equiv-p x y p) (not (reducible x)) (not (reducible y)))
    (equal x y)))

(defthm r-equiv-sound
  (implies (r-equiv x y) (equiv-p x y (make-proof-common-n-f x y))))

(defthm r-equiv-complete
  (implies (equiv-p x y p) (r-equiv x y)))
```

Fig. 2. Church-Rosser and normalizing implies decidability

2.3   Noetherianity, Local Confluence, and Newman's Lemma

A relation is *well founded* on a set A if every non-empty subset has a minimal element. A restricted notion of well-foundedness is built into ACL2, based on the following meta-theorem: a relation on a set A is well-founded iff there exists a function $F : A \to Ord$ such that $x < y \Rightarrow F(x) < F(y)$, where Ord is the class of all ordinals (axiom of choice needed). In ACL2, once a relation is proved to satisfy these requirements, it can be used in the admissibility test for recursive functions. A general well-founded partial order rel can be defined in ACL2 as shown in item (a) of figure 3. Since only ordinals up to ε_0 are formalized in the ACL2 logic, a limitation is imposed in the maximal order type of well-founded relations that can be represented. Consequently, our formalization suffers from the same restriction. Nevertheless, no particular properties of ε_0 are used in our proofs, except well-foundedness, so we think the same formal proofs could be carried out if higher ordinals were involved.

```
;;; (a) A well-founded partial order:
(encapsulate
 ((rel (x y) t) (fn (x) t))
 ...
 (defthm rel-well-founded-relation
   (and (e0-ordinalp (fn x))
        (implies (rel x y) (e0-ord-< (fn x) (fn y))))
   :rule-classes :well-founded-relation)

 (defthm rel-transitive
   (implies (and (rel x y) (rel y z)) (rel x z))))

;;; (b) A noetherian and locally confluent reduction relation:
(encapsulate
 ((legal (x u) t) (reduce-one-step (x u) t)
  (reducible (x) t) (transform-local-peak (x) t))
 ....
 (defthm locally-confluent
   (let ((valley (transform-local-peak p)))
     (implies (and (equiv-p x y p) (local-peak-p p))
              (and (steps-valley valley)
                   (equiv-p x y valley)))))

 (defthm noetherian
   (implies (legal x u) (rel (reduce-one-step x u) x))))

;;; (c) Definition of transform to valley:
(defun transform-to-valley (p)
  (declare (xargs :measure (proof-measure p)
                  :well-founded-relation mul-rel))
  (if (not (exists-local-peak p))
      p
      (transform-to-valley (replace-local-peak p))))

;;; (d) Main theorems proved:
(defthm transform-to-valley-admission
  (implies (exists-local-peak p)
           (mul-rel (proof-measure (replace-local-peak p))
                    (proof-measure p))))

(defthm Newman-lemma
  (let ((valley (transform-to-valley p)))
    (implies (equiv-p x y p)
             (and (steps-valley valley)
                  (equiv-p x y valley)))))
```

Fig. 3. Newman's lemma

In item (b) of figure 3, a general definition of a noetherian and locally confluent reduction relation is presented. Local confluence is easily expressed in terms
of the shape of proofs involved: a relation is locally confluent iff for every local
peak proof there is an equivalent valley proof. This valley proof is given by the
partially defined function `transform-local-peak`. As for noetherianity, our formalization relies on the following meta-theorem: a reduction is noetherian if and
only if it is contained in a well-founded partial ordering (AC). Thus, the general
well-founded relation `rel` previously defined is used to justify noetherianity of
the general reduction relation defined: for every element x such that a `legal`
operator u can be applied, then `reduce-one-step` obtains an element less than
x with respect to `rel`.

The standard proof of Newman's lemma found in the literature (see [1]) shows
confluence by noetherian induction based on the reduction relation. The proof
we obtained in ACL2 differs from the standard one and it is based on the proof
given in [8]. In our formalization, we have to show that the reduction relation
has the Church-Rosser property by defining a function `transform-to-valley`
and proving that for every proof p, (`transform-to-valley` p) is an equivalent
valley proof.

This function is defined to iteratively apply `replace-local-peak` (which
replaces a local peak subproof by the equivalent proof given by `transform-lo-
cal-peak`) until there are no local peaks. See definition in item (c) of figure 3.

Induction used in the standard proof is hidden here by the termination proof
of `transform-to-valley`, needed for admission. The main proof effort was to
show that in each iteration, some measure on the proof, `proof-measure`, de-
creases with respect to a well-founded relation, `mul-rel`. This can be seen as a
normalization process acting on proofs. The measure `proof-measure` is the list
of elements involved in the proof and the relation `mul-rel` is defined to be the
multiset extension of `rel`. We needed to prove in ACL2 that the multiset extension of a well-founded relation is also well-founded, a result interesting in its own
right (see the web page for details). Once `transform-to-valley` is admitted, it
is relatively easy to show that it always returns an equivalent valley proof. See
item (d) of figure 3.

Note that we gave a particular "implementation" of `transform-to-valley`
and proved as theorems the properties assumed as axioms in the previous subsection. The same was done with `proof-irreducible`. Decidability of noetherian
and locally confluent reduction relations can now be easily deduced by functional
instantiation from the general results proved in the previous subsection, allowing some kind of second-order reasoning. Name conflicts are avoided by using
Common Lisp packages that are capable of removing them.

3   Formalizing Rewriting in ACL2

We defined in the previous section a very general formalization of reduction
relations. The results proved can be reused for every instance of the general

framework. As a major example, we describe in this section how we formalized and reasoned about term rewriting in ACL2.

Since rewriting is a reduction relation defined on the set of first order terms, we needed to use a library of definitions and theorems formalizing the lattice theoretic properties of first-order terms: in particular, matching and unification algorithms are defined and proved correct. See [12] for details of this work. Some functions of this library will be used in the following. Although definitions are not given here, their names suggest what they do.

The very abstract concept of operator can be instantiated for term rewriting reductions. Equational operators are structures with three fields, containing the rewriting rule to apply, the position of the subterm to be replaced and the matching substitution: (defstructure eq-operator rule pos matching).

As we said in section 2, every reduction relation is given by concrete versions of legal, reduce-one-step and reducible. In the equational case:

– (eq-legal term op R) tests if the rule of the operator op is in R, and can be applied to term at the position indicated by the operator (using the matching in op).
– (eq-reduce-one-step term op) replaces the subterm indicated by the position of the operator op, by the corresponding instance (using matching) of the right-hand side of the rule of the operator.
– (eq-reducible term R) returns a legal equational operator to apply, whenever it exists, or nil otherwise.

Note that for every fixed term rewriting system R a particular reduction relation is defined. The rewriting counterpart of the abstract equivalence equiv-p can be defined in an analogous way: (eq-equiv-p t1 t2 p R) tests if p is a proof of the equivalence of t1 and t2 in the equational theory of R. Due to the lack of space, we do not give the definitions here. Recall also from section 2 that two theorems (assumed as axioms in the general framework) have to be proved to state the relationship between eq-legal and eq-reducible. We proved them:

```
(defthm eq-reducible-legal-1
   (implies (eq-reducible term R)
            (eq-legal term (eq-reducible term R) R)))

(defthm eq-reducible-legal-2
   (implies (not (eq-reducible term R))
            (not (eq-legal term op R))))
```

Formalizing term rewriting in this way, we proved a number of results about term rewriting systems. In the following subsections, two relevant examples are sketched.

```
(defthm eq-equiv-p-reflexive (eq-equiv-p term term nil E))

(defthm eq-equiv-p-symmetric
  (implies (eq-equiv-p t1 t2 p E)
           (eq-equiv-p t2 t1 (inverse-proof p) E)))

(defthm eq-equiv-p-transitive
  (implies (and (eq-equiv-p t1 t2 p E) (eq-equiv-p t2 t3 q E))
           (eq-equiv-p t1 t3 (proof-concat p q) E)))

(defthm eq-equiv-p-stable
  (implies (eq-equiv-p t1 t2 p E)
           (eq-equiv-p (instance t1 sigma) (instance t2 sigma)
                       (eq-proof-instance p sigma) E)))

(defthm eq-equiv-p-compatible
  (implies (and (eq-equiv-p t1 t2 p E) (positionp pos term))
       (eq-equiv-p (replace-term term pos t1)
                   (replace-term term pos t2)
                   (eq-proof-context p term pos) E)))
```

Fig. 4. Congruence: an algebra of proofs

3.1   Equational Theories and an Algebra of Proofs

An equivalence relation on first-order terms is a congruence if it is stable (closed under instantiation) and compatible (closed under inclusion in contexts). Equational consequence, $E \models s = t$, can be alternatively defined as the least congruence relation containing E. In order to justify that the above described representation is appropriate, it would be suitable to prove that, for a given E, the relation established by (eq-equiv-p t1 t2 p E), is the least congruence containing E (formally speaking, p has to be understood as existentially quantified).

We proved it in ACL2. In figure 4 we sketch part of our formalization showing that eq-equiv-p is a congruence. The ACL2 proof obtained is a good example of the benefits gained by considering proofs as objects that can be transformed to obtain new proofs. Following Bachmair [2], we can define an "algebra" of proofs, a set of operations acting on proofs: proof-concat to concatenate proofs, inverse-proof to obtain the reverse proof, eq-proof-instance, to instantiate the elements involved in the proof and eq-proof-context to include the elements of the proof as subterms of a common term. The empty proof nil can be seen as a proof constant. Each of these operations corresponds with one of the properties needed to show that eq-equiv-p is a congruence. The theorems are proved easily by ACL2, with minor help from the user.

```
;;; (a) A TRS with joinable critical pairs
(encapsulate
 ((RLC () t) (transform-cp (l1 r1 pos l2 r2) t))
  ...
 (defthm RLC-joinable-critical-pairs
   (implies
     (and (member (cons l1 r1) (RLC)) (member (cons l2 r2) (RLC))
          (positionp pos l1) (not (variable-p (occurence l1 pos))))
     (let* ((cp-r (cp-r l1 r1 pos l2 r2)))
       (implies cp-r
         (and (eq-equiv-p (lhs cp-r) (rhs cp-r)
                          (transform-cp l1 r1 pos l2 r2) (RLC))
              (steps-valley (transform-cp l1 r1 pos l2 r2)))))))))

;;; (b) Theorem proved:
(defun transform-eq-local-peak (p) ...)

(defthm critical-pair-theorem
   (let ((valley (transform-eq-local-peak p)))
     (implies (and (eq-equiv-p t1 t2 p (RLC)) (local-peak-p p))
              (and (steps-valley valley)
                   (eq-equiv-p t1 t2 valley (RLC))))))
```

Fig. 5. The critical pair theorem

3.2   The Critical Pair Theorem

The main result we have proved is the critical pair theorem: a rewrite system R is locally confluent iff every critical pair obtained with rules in R is joinable. This result is formalized in our framework and proved guiding the system to the classical proof given in the literature (see [1] for example).

In item (a) of figure 5, a term rewriting system (RLC) is partially defined assuming the property of joinability of its critical pairs. The partially defined function (transform-cp l1 r1 pos l2 r2) is assumed to obtain a valley proof for the critical pair determined by the rules (l1 . r1) and (l2 . r2) at the non-variable position pos of l1. The function (cp-r l1 r1 pos l2 r2) computes such a critical pair, whenever it exists (after prior renaming of the variables of the rules, in order to get them standardized apart).

To prove the critical pair theorem in our formalization, we have to define a function transform-eq-local-peak and prove that it transforms every equational local peak proof to an equivalent valley proof. The final theorem is shown in item (b) of figure 5. The ACL2 proof of this theorem is the largest proof we developed. Due to the lack of space, we cannot describe here the proof effort. We urge the interested reader to see the web page.

This theorem and the theorems described in Section 2 for abstract reduction relations were used to prove that equational theories described by a terminating TRS such that every critical pair has a common normal form are decidable. This result (which some authors call the Knuth-Bendix theorem) is easily obtained by functional instantiation from the abstract case, taking advantage of the fact that the whole formalization is done in the same framework. Note how this last result can be used to "certify" decision procedures for equational theories defined by confluent and terminating TRSs.

4   Conclusions and Further Work

We have presented a case study of using the ACL2 system as a metalanguage to formalize properties of object proof systems (abstract reductions and equational logic) in it. It should be stressed that the task of proving in ACL2 is not trivial. As claimed in [6], difficulties come from "the complexity of the whole enterprise of formal proofs", rather than from the complexity of ACL2. A typical proof effort consists of formalizing the problem, and guiding the prover to a preconceived "hand proof", by decomposing the proof into intermediate lemmas. Most of our lemmas are proved mainly by simplification and induction, without hints from the user. If one lemma is not proved in a first attempt, then additional lemmas are often needed, as suggested by inspecting the failed proof (for example, the proof of the critical pair theorem needed more than one hundred lemmas and fifty auxiliary definitions). Nevertheless, proofs can be simpler if a good library of previous results (*books* in the ACL2 terminology) is used. We think our work provides a good collection of books to be reused in further verification efforts.

Our formalization has the following main features:

- Reduction relations and their properties are stated in a very general framework, as explained in section 2.
- The concept of proof is a key notion in our formalization. Proofs are treated as objects that can be transformed to obtain new proofs.
- Functional instantiation is extensively used as a way of exporting results from the abstract case to the concrete case of term rewriting systems.

Some related work has been done in the formalization of abstract reduction relations in other theorem proving systems, mostly as part of formalizations on the λ-calculus. For example, Huet [5] in the Coq system or Nipkow [11] in Isabelle/HOL. A comparison is difficult because our goal was different and, more important, the logics involved are significantly different: ACL2 logic is a much weaker logic than those of Coq or HOL. A more related work is Shankar [13], using Nqthm. Although his work is on the concrete reduction relation from the λ-calculus and he does not deal with the abstract case, some of his ideas are reflected in our work.

To our knowledge, no formalization of term rewriting systems has been done yet and consequently the formal proofs of their properties presented here are the first ones we know that have been performed using a theorem prover.

We think the results presented here are important for two reasons. From a theoretical point of view, it is shown that a very weak logic can be used to formalize properties of TRSs. From a practical point of view, this is an example of how formal methods can help in the design of symbolic computation systems. Usually, algebraic techniques are applied to the design of proof procedures in automated deduction. We show how benefits can be obtained in the reverse direction: automated deduction used as a tool to "certify" components of symbolic computation systems. Since ACL2 is also a programming language, this paper shows how computing and proving tasks can be mixed. Although a fully verified computer algebra system is currently beyond our possibilities, the guard verification mechanism [6] can be used to obtain verified Lisp code (executable in *any* compliant Common Lisp) for some basic procedures of term rewriting systems. There are also several ways in which this work can be extended. For example:

- In order to obtain certified decision procedures for some equational theories (or for the word problem of some finitely presented algebras) work has to be done to formalize in ACL2 well-known terminating term orderings (recursive path orderings, Knuth-Bendix orderings, etc.). Maybe some problems will arise due to the restricted notion of noetherianity supported by ACL2.
- The work presented in [10] suggests another application of this work: other theorem provers can be combined with ACL2 in order to obtain mechanically verified decision algorithms for some equational theories.
- Our goal in the long term is to obtain a certified completion procedure written in Common Lisp. Although for the moment this may be far from the current status of our development, we think the work presented here is a good starting point.

References

1. F. Baader and T. Nipkow. *Term rewriting and all that*. Cambridge University Press, 1998.
2. L. Bachmair. *Canonical equational proofs*. Birkhäuser, 1991.
3. R. Boyer and J S. Moore. *A Computational Logic Handbook*. Academic Press, 2nd edition, 1998.
4. B. Buchberger and R. Loos. Algebraic simplification. In *Computer Algebra, Symbolic and Algebraic Computation. Computing Supplementum 4*. SV, 1982.
5. G. Huet. Residual theory in λ-calculus: a formal development. *Journal of Functional Programming*, (4):475–522, 1994.
6. M. Kaufmann, P. Manolios and J S. Moore. *Computer-Aided Reasoning: An Approach*. Kluwer Academic Publishers, 2000.
7. M. Kaufmann and J S. Moore. http://www.cs.utexas.edu/users/moore/acl2/. ACL2 Version 2.5, 2000.
8. J.W. Klop. Term rewriting systems. *Handbook of Logic in Computer Science*, Clarendon Press, 1992.
9. P. Le Chenadec. *Canonical forms in finitely presented algebras*. Pitman-Wiley, London, 1985.

10. W. McCune and O. Shumsky. Ivy: A preprocessor and proof checker for first-order logic. In *Computer-Aided Reasoning: ACL2 Case Studies*, Kluwer Academic Publishers, 2000, ch. 16.
11. T. Nipkow. More Church-Rosser proofs (in Isabelle/HOL). In *13th International Conference on Automated Deduction*, LNAI 1104, pages 733–747. Springer-Verlag, 1996.
12. J.L. Ruiz-Reina, J.A. Alonso, M.J. Hidalgo, and F.J. Martín. Mechanical verification of a rule based unification algorithm in the Boyer-Moore theorem prover. In *AGP'99 Joint Conference on Declarative Programming*, pages 289–304, 1999.
13. N. Shankar. A mechanical proof of the Church-Rosser theorem. *Journal of the ACM*, 35(3):475–522, 1988.
14. G.L. Steele. *Common Lisp the Language, 2nd edition*. Digital Press, 1990.

Additional Comments on Conjectures, Hypotheses, and Consequences in Orthocomplemented Lattices[*]

Angel Fernandez Pineda[1], Enric Trillas[1], and Claudio Vaucheret[2]

[1] Universidad Politécnica de Madrid
Departmento de Inteligencia Artificial
28660 Madrid, Spain
etrillas@fi.ump.es
[2] Universidad Nacional del Comahue
Departamento de Informática,
Q8300azn Neuquén, Argentina
cvaucher@uncoma.edu.ar

Abstract. This paper is a brief continuation of earlier work by the same authors [4] and [5] that deals with the concepts of *conjecture, hypothesis* and *consequence* in orthocomplemented complete lattices. It considers only the following three points:

1. Classical logic theorems of both *deduction* and *contradiction* are reinterpreted and proved by means of one specific operator $C_\wedge$ defined in [4].
2. Having shown that there is reason to consider the set $C_\wedge(P)$ of consequences of a set of premises P as too large, it is proven that $C_\wedge(P)$ is the largest set of consequences that can be assigned to P by means of a Tarski's consequences operator, provided that L is a Boolean algebra.
3. On the other hand, it is proven that, also in a Boolean algebra, the set $\Phi_\wedge(P)$ of strict conjectures is the smallest of any $\Phi(P)$ such that $P \subseteq \Phi(P)$ and that if $P \subseteq Q$ then $\Phi(Q) \subseteq \Phi(P)$.

Keywords Conjectures, Hypotheses, Consequences, Boolean Algebras, Deduction, Contradiction.

1 Introduction

Let L be an orthocomplemented complete lattice with operations $\cdot$ (meet), $+$ (join), $'$ (complement), minimum 0 and maximum 1. We will consider subsets (of premises) $P \subseteq L$ such that $\wedge P \neq 0$[1] and designate the respective family $\{P \in \mathcal{P}(L); p_\wedge \neq 0\}$ as $\mathcal{P}_0(L)$. It is obvious that there are no contradictory pairs

---

[*] Paper partially supported by Spanish Ministry of Education and Culture under projects PB98-1379-C02-C02 and CICYT-TIC99-1151
[1] $\wedge P = Inf(P) = p_\wedge$ and $\vee P = Sup(P) = p_\vee$.

p_1, p_2 in P, that is pairs p_1, p_2 such that $p_1 \leq p_2'$. Pursuant to [4], we will deal with the operators $\Phi_\vee : \mathcal{P}_0(L) \to \mathcal{P}(L)$, $\Phi_\wedge : \mathcal{P}_0(L) \to \mathcal{P}(L)$, $H_\wedge : \mathcal{P}_0(L) \to \mathcal{P}(L)$ and $C_\wedge : \mathcal{P}_0(L) \to \mathcal{P}_0(L)$ defined respectively, as:

$$\Phi_\vee(P) = \{q \in L - \{0\}; p_\vee \leq q'\}^c \quad \text{(loose conjectures of } P\text{)}$$
$$\Phi_\wedge(P) = \{q \in L - \{0\}; p_\wedge \leq q'\}^c \quad \text{(strict conjectures of } P\text{)}$$
$$H_\wedge(P) = \{q \in L - \{0\}; q \leq p_\wedge\} \quad \text{(hypotheses of } P\text{)}$$
$$C_\wedge(P) = \{q \in L - \{0\}; p_\wedge \leq w\} \quad \text{(consequences of } P\text{)}$$

for any $P \in \mathcal{P}_0(L)$ and where c is the complement of subsets in $P(L)$.

As it was proven in [4], $P \subseteq C_\wedge(P) \subseteq \Phi_\wedge(P) \subseteq \Phi_\vee(P)$, $H_\wedge(P) \subseteq \Phi_\wedge(P) \subseteq \Phi_\vee(P)$ and $C_\wedge(P) \cap H_\wedge(P) = \{p_\wedge\}$. If $P \subseteq Q$, $\Phi_\vee(P) \subseteq \Phi_\vee(Q)$, $\Phi_\wedge(Q) \subseteq \Phi_\wedge(P)$, $H_\wedge(Q) \subseteq H_\wedge(P)$ and $C_\wedge(P) \subseteq C_\wedge(Q)$; that is, loose conjectures and consequences are monotonic, but strict conjectures and hypotheses are anti-monotonic.

For each $P \in \mathcal{P}_0(L)$, the set $P' = \{p'; p \in P\}$ verifies $Inf(P') = (Sup(P))' \neq 0$ if and only if $Sup(P) \neq 1$. Let $\mathcal{P}_{01}(L) = \{P \in \mathcal{P}_0(L); Sup(P) \neq 1\}$ and suppose we designate the restrictions on $\mathcal{P}_{01}$ of $\Phi_\vee$, $\Phi_\wedge$, $H_\wedge$ and $C_\wedge$ using the same symbols. Obviously, $H_\wedge(P) = \{q \in L - \{0\}; Sup(P') = (Inf(P))' \leq q'\}$.

Remark. It is quite clear that $C_\wedge(P)$ is always a *filter* that, generally, is not prime, and that $H_\wedge(P) \cup \{0\} = \{q \in L : q \leq p_\wedge\}$ is always an *ideal* that, generally, is not prime either. Neither $\Phi_\vee(P)$ nor $\Phi_\wedge(P)$ are either filters or ideals.

2   The Classical Theorems of Deduction and Contradiction Reconsidered

In this section, the classical theorems of deduction and contradiction are reviewed and it is found that only one part of each one depends on both distributivity and a specific operation of implication.

2.1   Theorem of Deduction

Let L be an orthocomplemented complete lattice, $P \in \mathcal{P}_0(L)$, and a operation $\to$ of *implication* in L. So, $\to$ is such that $a \cdot (a \to b) \leq b$ for all $a, b \in L$. If L is a Boolean algebra, this property is equivalent to $a \to b \leq a' + b$, and the material implication $a \to b = a' + b$ is the greatest.

Under such conditions, we will split this theorem into two parts that require different additional conditions:

Lemma 1 $\forall a, b \in L : a \to b \in C_\wedge(P) \implies b \in C_\wedge(P \cup \{a\})$,

which stands for "b is a consequence of P extended with a if $a \to b$ is a consequence of P".

Proof: Provided that $a \rightarrow b \in C_\wedge(P)$ then:

$p_\wedge \leq a \rightarrow b$ (by definition),

$a \cdot p_\wedge \leq a \cdot (a \rightarrow b)$ (monotonicity of $\cdot$),

$a \cdot (a \rightarrow b) \leq b$ (definition of $\rightarrow$),

$a \cdot p_\wedge \leq b$ (transitivity of $\leq$).

As $\wedge(P \cup \{a\}) = a \cdot p_\wedge$, it follows that $\wedge(P \cup \{a\}) \leq b$, which is the definition for $b \in C_\wedge(P \cup \{a\})$

Lemma 2 *If L is a distributive lattice and the operator $\rightarrow$ is defined as $a \rightarrow b = a' + b$ (material implication) , then:*

$$\forall a, b \in L : b \in C_\wedge(P \cup \{a\}) \Longrightarrow a \rightarrow b \in C_\wedge(P),$$

which stands for "$a \rightarrow b$ is a consequence of P if b is a consequence of P extended with a".

Proof: Provided that $b \in C_\wedge(P \cup \{a\})$, then:

$a \cdot p_\wedge \leq b$ (by definition),

$a' + (a \cdot p_\wedge) \leq a' + b$ (monotonicity of $+$),

$(a' + a) \cdot (a' + p_\wedge) \leq a' + b$ (distributivity),

$a' + p_\wedge \leq a' + b$ (since $a' + a = 1$),

$p_\wedge \leq a' + b$ (transitivity and $p_\wedge \leq a' + p_\wedge$),

$p_\wedge \leq a \rightarrow b$ (by definition of $\rightarrow$),

$a \rightarrow b \in C_\wedge(P)$ (by definition).

Remark. Lemma 2 requires the *implication $a \rightarrow b = a' + b$*. Section 2.1 shows that it cannot be generalized to all implications.

Theorem 1 *Let L be a complete Boolean algebra and $a \rightarrow b = a' + b$, then:*

$$\forall a, b \in L : a \rightarrow b \in C_\wedge(P) \Longleftrightarrow b \in C_\wedge(P \cup \{a\}),$$

which is a generalization of the the first-order logic *theorem of deduction*.

Proof:
Immediate, as a complete Boolean algebra is nothing other than a distributive orthocomplemented and complete lattice.

Lemma 2 Is Not Valid for All Implications. Let L be the Boolean algebra shown in Figure 1 and consider the following *implications*:

$$a \rightarrow_1 b = \begin{cases} 1 \ if a \leq b \\ 0 \ otherwise \end{cases}, a \rightarrow_2 b = a \cdot b$$

Let $P = \{a\}$, so $C_\wedge(P) = \{a, b', c', 1\}$. Now, let's consider $C_\wedge(P \cup \{b'\}) = \{a, b', c', 1\}$. If lemma 2 were true, "$a \in C_\wedge(P \cup \{b'\}) \Longrightarrow b' \rightarrow_1 a \in C_\wedge(P)$", but $b' \rightarrow_1 a = 0 \notin C_\wedge(P)$.

Similarly, if $P = \{a'\}$, $C_\wedge(P) = \{a', 1\}$ and $C_\wedge(P \cup \{c'\}) = \{b, a', c', 1\}$ but $b \in C_\wedge(P \cup \{c'\})$ and $c' \rightarrow_2 b = c' \cdot b = b \notin C_\wedge(P)$.

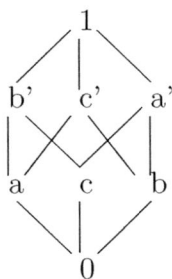

Fig. 1. The 2^3 Boolean Algebra

2.2   Theorem of Contradiction

Let L be an orthocomplemented complete lattice and $P \in \mathcal{P}_0(L)$.

Lemma 3 $\forall a \in L : a \in C_\wedge(P) \Longrightarrow a' \cdot p_\wedge = 0,$

which stands for "a' *is incompatible with* P *if* a *is a consequence of* P".

   Proof: if $a \in C_\wedge(P)$, as it means $p_\wedge \leq a$, it follows that $a' \cdot p_\wedge \leq a' \cdot a = 0$, and $a' \cdot p_\wedge = 0$.

Lemma 4 *If* L *is a distributive lattice, then:*

$$\forall a \in L : a' \cdot p_\wedge = 0 \Longrightarrow a \in C_\wedge(P),$$

which stands for "a *is a consequence of* P *if* a' *is incompatible with* P".

   Proof: If $a' \cdot p_\wedge = 0$, it follows:
$a + (a' \cdot p_\wedge) = a + 0$ (monotonicity of $+$),
$(a + a') \cdot (a + p_\wedge) = a$ (distributivity),
$a + p_\wedge = a$ (since $a + a' = 1$),
$p_\wedge \leq a$ (transitivity, and $p_\wedge \leq a + p_\wedge$). Hence $a \in C_\wedge(P)$.

Theorem 2 *If* L *is a complete Boolean algebra:*

$$\forall a \in L : a \in C_\wedge(P) \Longleftrightarrow a' \cdot p_\wedge = 0,$$

which is a generalization of the first-order logic *theorem of contradiction*.

   Proof: Immediate, since a complete Boolean algebra is a distributive orthocomplemented complete lattice.

3   Another Look at $H_\wedge$ and $C_\wedge$

3.1   The Dualized Operator

For each operator $F : \mathcal{P}_{01}(L) \to \mathcal{P}(L)$ let the dualized operator be δF, defined by $\delta F = c \circ F \circ '$, where $' : \mathcal{P}_{01}(L) \to \mathcal{P}_{01}$ is given by $'(P) = P'$, and $c : \mathcal{P}_{01} \to \mathcal{P}(L)$ is $c(P) = P^c = \{q \in L; q \notin P\}$. If $P \in \mathcal{P}_{01}(L)$, $\delta F(P) = F(P')^c$ and, obviously:

- $\delta(\delta F) = \delta(c \circ F \circ ') = c \circ (c \circ F \circ ')\circ ' = (c \circ c) \circ F \circ ('\circ ') = F$
- Defining $F \le G$ by $F(P) \subseteq G(P)$ for all $P \in \mathcal{P}_{01}(L)$, $F \le G$ implies $F(P') \subseteq G(P')$ and $G(P')^c \subseteq F(P')^c$, or $\delta G \le \delta F$.
- The selfdualized operators or operators F such that $\delta F = F$ verify $c \circ F = F \circ '$, and these operators are neither strictly monotonic nor strictly anti-monotonic. For example, if F is strictly anti-monotonic and selfdualized, it follows from $P \subset Q$ that $F(Q) \subset F(P)$ and then $F(P)^c \subset F(Q)^c$. This is equivalent to $F(P') \subset F(Q')$, against $P' \subset Q'$.

Hence, neither $\Phi_\vee, \Phi_\wedge, H_\wedge$ nor $C_\wedge$ are, generally, selfdualized operators. Let's look at what they change into.

3.2   Are $H_\wedge$ and $C_\wedge$ Too Restrictive?

$\delta\Phi_\vee(P) = \Phi_\vee(P')^c = (\{q \in L - \{0\}; Sup(P') \le q'\}^c)^c = \{q \in L - \{0\}; SupP' \le q'\} = \{q \in L - \{0\}; q \le (SupP')'\} = \{q \in L - \{0\}; q \le InfP\} = H_\wedge(P)$. So $\delta\Phi_\vee = H_\wedge$ and $\delta H_\wedge = \Phi_\vee$.

Similarly, $\delta\Phi_\wedge(P) = \Phi_\wedge(P')^c = \{q \in L - \{0\}; InfP' \le q'\} = \{q \in L - \{0\}; q \le SupP\}$. This set, $H_\vee(P) = \{q \in L - \{0\}; a \le SupP\}$, which contains P and $H_\wedge(P)$, verifies $H_\vee(P) \cap C_\wedge(P) = \{q \in L - \{0\}; Inf(P) \le q \le Sup(P)\} = C(P)$. The set of restricted consequences of P (see [4] and [5]) was not considered in [4] as a set of hypotheses of P, because it is uncommon to accept the premises and consequences of P other than $Inf(P)$ as hypotheses of P.

However, the operator $H_\vee$ now appears as the dualized operator of $\Phi_\wedge$, like $H_\wedge$ of $\Phi_\vee$. The set $C_\wedge(P)$ could be too large or the set $H_\wedge(P)$ could be too small. Are there operators $\hat{C}$ and $\hat{H}$ such that

1. $\hat{C}(P) \subseteq C_\wedge(P)$,
2. $H_\wedge(P) \subseteq \hat{H}(P)$, and
3. $\hat{C}(P) \cap \hat{H}(P) = \{Inf(P)\}$?

Neither Tarski's consequences operator C of restricted consequences answers this question, nor will this paper solve the problem. The only new thing related to (1) is that $C_\wedge$ is the largest of the Tarski's consequences operators in $\mathcal{P}_0(L)$, where L is a Boolean algebra.

4   Some New Properties of Operators $C_\wedge$ and $\varPhi_\wedge$

4.1 $C_\wedge$ Is the Largest Tarski's Operator

Lemma 5 *Let $P \in \mathcal{P}(L)$ and $x \in L$, where L is a Boolean algebra. Then:*

$$x \notin C_\wedge(P) \Rightarrow P \cup \{x'\} \in \mathcal{P}_0(L)$$

Proof:
If $P \cup \{x'\} \notin \mathcal{P}_0(L)$, then $p_\wedge \cdot x' = 0$ and, hence, $(p_\wedge \cdot x') + x = x$. As L is distributive, $(p_\wedge + x) \cdot (x' + x) = x$, thus $(p_\wedge + x) \cdot 1 = x$ and $p_\wedge + x = x$. Then $p_\wedge \leq x$ and, therefore, $x \in C_\wedge(P)$.

Theorem 3 *Let L be a Boolean algebra. For every function $C : \mathcal{P}_0(L) \to \mathcal{P}_0(L)$ for which the following properties hold:*

- *$P \subseteq C(P)$ (expansion), and*
- *$P \subseteq Q \Rightarrow C(P) \subseteq C(Q)$ (monotonicity),*

$$C(P) \subseteq C_\wedge(P), \quad \forall P \in \mathcal{P}_0(L).$$

Proof:
Suppose that there is $P \in \mathcal{P}_0(L)$ such that $C(P) \not\subseteq C_\wedge(P)$. Then there exists $x \in C(P)$ such that $x \notin C_\wedge(P)$. Because of lemma 5, we have $P \cup \{x'\} \in \mathcal{P}_0(L)$ and, therefore, it is in the domain of C. However, $x \in C(P \cup \{x'\})$ by monotonicity and $x' \in C(P \cup \{x'\})$ by expansion. Hence, $\wedge C(P \cup \{x'\}) = 0$ and $C(P \cup \{x'\}) \notin \mathcal{P}_0(L)$, which is contradictory. Therefore, $C(P) \subseteq C_\wedge(P)$.

Corollary 1 *Let L be a Boolean algebra. Any Tarski's consequences operator $C : \mathcal{P}_0(L) \to \mathcal{P}_0(L)$ verifies:*

$$C(P) \subseteq C_\wedge(P), \quad \forall P \in \mathcal{P}_0(L).$$

Remarks:

1. Distributivity is necessary as the following example shows. Given the typical non distributive hexagonal lattice L shown in Figure 2, define the following consequences operator: $C(\{1\}) = \{1\}$, $C(\{a\}) = C(\{b\}) = C(\{a,b\}) = C(\{a,1\}) = C(\{b,1\}) = C(\{a,b,1\}) = \{a,b,1\}$ and $C(\{a'\}) = C(\{b'\}) = C(\{a',b'\}) = C(\{a',1\}) = F(\{b',1\}) = F(\{a',b',1\}) = \{a',b',1\}$. It verifies the properties of expansion and monotonicity. However we have

$$C(\{b\}) \not\subseteq C_\wedge(\{b\}) \text{ and } C(\{a'\}) \not\subseteq C_\wedge(\{a'\})$$

2. Note the importance of C being defined as a mapping $\mathcal{P}_0(L) \to \mathcal{P}_0(L)$. $\varPhi_\vee$, which is defined between only $\mathcal{P}_0(L)$ and $\mathcal{P}(L)$, verifies $P \subseteq \varPhi_\vee(P)$ and $P \subseteq Q \Rightarrow \varPhi_\vee(P) \subseteq \varPhi_\vee(Q)$. However, $C_\wedge(P) \subseteq \varPhi_\vee(P)$.
3. If the lattice is distributive but not orthocomplemented, theorem 3 is no longer valid, as there can be consequences operators that are greater, lesser and even incomparable with $C_\wedge$ (see [2]).

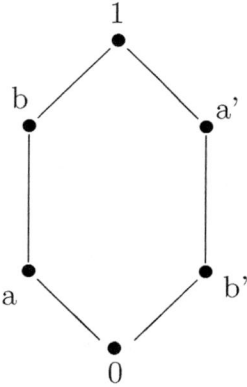

Fig. 2. Hexagonal Lattice

4.2 $\Phi_\wedge$ Is the Smallest Expansive and Anti-monotonic Operator

Lemma 6 *Let $P \in \mathcal{P}_0(L)$ and $x \in L$, where L is a Boolean algebra. Then:*

$$x \in \Phi_\wedge(P) \Rightarrow P \cup \{x\} \in \mathcal{P}_0(L)$$

Proof:
Suppose that $P \cup \{x\} \notin \mathcal{P}_0(L)$. Then $p_\wedge \cdot x = 0$ and, therefore, $(p_\wedge \cdot x) + x' = x'$. Because L is distributive, $(p_\wedge + x') \cdot (x + x') = x'$ and, thus, $(p_\wedge + x') \cdot 1 = x'$ and $p_\wedge + x' = x'$. Then, $p_\wedge \leq x'$ and, hence, $x \notin \Phi_\wedge(P)$.

Theorem 4 *Let L be a Boolean algebra. For any function $\Phi : \mathcal{P}_0(L) \to \mathcal{P}(L)$ such that:*

- *$P \subseteq \Phi(P)$ (expansion), and*
- *$P \subseteq Q \Rightarrow \Phi(Q) \subseteq \Phi(P)$ (anti-monotonicity),*

$$\Phi_\wedge(P) \subseteq \Phi(P), \quad \forall P \in \mathcal{P}_0(L).$$

Proof:
Let $x \in \Phi_\wedge(P)$. Because of lemma 6, $P \cup \{x\} \in \mathcal{P}_0(L)$ and, therefore, it is in the domain of Φ. Thus, by expansion and anti-monotonicity, we have

$$x \in P \cup \{x\} \subseteq \Phi(P \cup \{x\}) \subseteq \Phi(P).$$

Hence, $\Phi_\wedge(P) \subseteq \Phi(P)$

Remarks:

1. Distributivity is necessary as the following example shows. Given the non distributive lattice L shown in Figure 2, define the following operator:

$$\Phi(\{a\}) = \{a, b, 1\} \qquad\qquad \Phi(\{b'\}) = \{a', b', 1\}$$
$$\Phi(\{b\}) = \{a, b, 1\} \qquad\qquad \Phi(\{a'\}) = \{a', b', 1\}$$
$$\Phi(\{1\}) = \{a, b, 1, a', b'\}$$
$$\Phi(\{a, b\}) = \{a, b, 1\} \qquad\qquad \Phi(\{a', b'\}) = \{a', b', 1\}$$
$$\Phi(\{a, 1\}) = \{a, b, 1\} \qquad\qquad \Phi(\{b', 1\}) = \{a', b', 1\}$$
$$\Phi(\{b, 1\}) = \{a, b, 1\} \qquad\qquad \Phi(\{a', 1\}) = \{a', b', 1\}$$
$$\Phi(\{a, b, 1\}) = \{a, b, 1\} \qquad \Phi(\{a', b', 1\}) = \{a', b', 1\}$$

It verifies the properties of expansion and anti-monotonicity. However,

$$\Phi_\wedge(\{b\}) \not\subseteq \Phi(\{b\}).$$

2. Note the importance of the property $P \subseteq \Phi(P)$. $H_\wedge$ is anti-monotonic, but $H_\wedge \subseteq \Phi_\wedge$, because $P \not\subseteq H_\wedge(P)$.

Acknowledgement

The authors are indebted to both Professor Michio Sugeno (Brain Science Institute, Hirosawa) for suggesting the grounds of section 3.2 and to two anonymous referees, whose comments were of help in improving the final version of this paper.

References

[1] G Birkhoff. *Lattice Theory*. American Mathematical Society, 3rd edition, 1967.
[2] E. Castiñeira, S. Cubillo, A. Pradera, and E. Trillas. On conjectures and consequences in fuzzy logic. In *Proceedings of IEEE*. NAFIPS'2000, forthcoming.
[3] K. R. Popper. *Conjectures and Refutations*. Routledge & Kegan Paul, London, 1963.
[4] E. Trillas, S. Cubillo, and E. Castiñeira. On conjectures in orthocomplemented lattices. *Artificial Intelligence*, 117:255–275, 2000.
[5] E. Trillas, S. Cubillo, and E. Castiñeira. Averaging premises. *Mathware & Softcomputing*, forthcoming.
[6] S Watanabe. *Knowing & Guessing*. John Wiley & Sons, New York, 1969.

Reasoning about the Elementary Functions of Complex Analysis*

Robert M. Corless[1], James H. Davenport[2]**, David J. Jeffrey[1], Gurjeet Litt[1],
and Stephen M. Watt[1]

[1] Ontario Research Centre for Computer Algebra, University of Western Ontario
www.orcca.on.ca
[2] Dept. Mathematical Sciences, University of Bath, Bath BA2 7AY, England
jhd@maths.bath.ac.uk

Abstract. There are many problems with the simplification of elementary functions, particularly over the complex plane. Systems tend to make major errors, or not to simplify enough. In this paper we outline the "unwinding number" approach to such problems, and show how it can be used to prevent errors and to systematise such simplification, even though we have not yet reduced the simplification process to a complete algorithm. The unsolved problems are probably more amenable to the techniques of artificial intelligence and theorem proving than the original problem of complex-variable analysis.
Keywords: Elementary functions; Branch cuts; Complex identities.
Topics: AI and Symbolic Mathematical Computing; Integration of Logical Reasoning and Computer Algebra.

1 Introduction

The elementary functions are traditionally thought of as log, exp and the trigonometric and hyperbolic functions (and their inverses). This list should include powering (to non-integral powers) and also the n-th root. These functions are built in, to a greater or lesser extent, to many computer algebra systems (not to mention other programming languages [8,12]), and are heavily used. However, reasoning with them is more difficult than is usually acknowledged, and all algebra systems have one, sometimes both, of the following defects:

– they make mistakes, be it the traditional schoolchild one

$$1 = \sqrt{1} = \sqrt{(-1)^2} = -1 \qquad (1)$$

or more subtle ones (see footnote 6);

---

* The authors are grateful to Mrs. A. Davenport for her help with the original of [3], and to Dr. D. E. G. Hare of Waterloo Maple for many discussions.
** This work was performed while this author held the Ontario Research Chair in Computer Algebra at the University of Western Ontario. Background work was supported by the European Commission under Esprit project OpenMath (24.969).

J.A. Campbell and E. Roanes-Lozano (Eds.): AISC 2000, LNAI 1930, pp. 115–126, 2001.
© Springer-Verlag Berlin Heidelberg 2001

– they fail to perform obvious simplifications, leaving the user with an impossible mess when there "ought" to be a simpler answer. In fact, there are two possibilities here: maybe there is a simpler equivalent that the system has failed to find, but maybe there isn't, and the simplification that the user wants is not actually valid, or is only valid outside an exceptional set. In general, the user is not informed what the simplification might have been, nor what the exceptional set is.

Faced with these problems, the user of the algebra system is not convinced that the result is correct, or that the algebra system in use understands the functions with which it is reasoning. An ideal algebra system would never generate incorrect results, and would simplify the results as much as practicable, even though perfect simplification is impossible, and not even totally well-defined: is $1 + x + \cdots + x^{1000}$ "simpler" than $(x^{1001} - 1)/(x - 1)$?

Throughout this paper, z and its decorations indicate a complex variable, while x, y and t indicate real variables. The symbol $\Im$ denotes the imaginary part, and $\Re$ the real part, of a complex number. For the purposes of this paper, the precise definitions of the inverse elementary functions in terms of log are those of [4]: these are reproduced in Appendix A for ease of reference.

2 The Problem

The fundamental problem is that log is multi-valued: since $\exp(2\pi i) = 1$, its inverse is only valid up to adding any multiple of $2\pi i$. This ambiguity is traditionally resolved by making a *branch cut*: usually [1, p. 67] the branch cut $(-\infty, 0]$, and the rule (4.1.2) that

$$- \pi < \Im \log z \leq \pi. \tag{2}$$

This then completely specifies the behaviour of log: on the branch cut it is continuous with the positive imaginary side of the cut, i.e. counter-clockwise continuous in the sense of [10].

What are the consequences of this definition[1]? From the existence of branch cuts, we get the problem of a lack of continuity:

$$\lim_{y \to 0^-} \log(x + iy) \neq \log x : \tag{3}$$

for $x < 0$ the limit is $\log x - 2\pi i$. Related to this is the fact that

$$\log \overline{z} \neq \overline{\log z} \tag{4}$$

---

[1] Which we do not contest: it seems that few people today would support the rule one of us (JHD) was taught, viz. that $0 \leq \Im \log z < 2\pi$. The placement of the branch cut is "merely" a notational convention, but an important one. If we wanted a function that behaves like log but with this cut, we could consider $\underset{[0,2\pi)}{\underbrace{\log}} (z) = \log(-1) - \log(-1/z)$ instead. We note that, until 1925, astronomers placed the branch cut between one day and the next at noon [7, vol. 15 p. 417].

on the branch cut: instead $\log \overline{z} = \overline{\log z} + 2\pi i$ on the cut. Similarly,

$$\log \left(\frac{1}{z} \right) \neq -\log z \tag{5}$$

on the branch cut: instead $\log(1/z) = -\log z + 2\pi i$ on the cut.

Although not normally explained this way, the problem with (1) is a consequence of the multi-valued nature of log: if we define (as for the purposes of this paper we do)

$$\sqrt{z} = \exp \left(\frac{1}{2} \log z \right), \tag{6}$$

then $-\pi/2 < \Im \sqrt{z} \leq \pi/2$. On the real line, this leads to the traditional resolution of (1), namely that $\sqrt{x^2} = |x|$.

Three families of solutions have been proposed to these problems.

- Prof. W. Kahan points out that the concept of a "signed zero"[2] [9] (for clarity, we write the positive zero as 0^+ and the negative one as 0^-) can be used to solve the above problems, if we say that, for $x < 0$, $\log(x + 0^+ i) = \log x + \pi i$ whereas $\log(x + 0^- i) = \log x - \pi i$. Equation (3) then becomes an equality for all x, interpreting the x on the right as $x + 0^- i$. Similarly, (4) and (5) become equalities throughout. Attractive though this proposal is, it does not answer the fundamental question as far as the designer of a computer algebra system is concerned: what to do if the user types $\log(-1)$.
- The authors of [5] point out that most "equalities" do not hold for the complex logarithm, e.g. $\log(z^2) \neq 2\log z$ (try $z = -1$), and its generalisation

$$\log(z_1 z_2) \neq \log z_1 + \log z_2. \tag{7}$$

The most fundamental of all non-equalities is $z \overset{?}{=} \log \exp z$, whose most obvious violation is at $z = 2\pi i$. (A similar point was made in [2], where the correction term is called the "adjustment".) They therefore propose to formalise the violation of this equality by introducing the *unwinding number* $\mathcal{K}$, defined[3] by

$$\mathcal{K}(z) = \frac{z - \log e^z}{2\pi i} = \left\lceil \frac{\Im z - \pi}{2\pi} \right\rceil \in \mathbf{Z} \tag{8}$$

(note that the apparently equivalent definition $\lfloor \frac{\Im z + \pi}{2\pi} \rfloor$ differs precisely on the branch cut for log as applied to $\exp z$).

---

[2] One could ask why zero should be special and have two values. The answer seems to be that all the branch cuts we need to consider are on either the real or imaginary axes, so the side to which the branch cut adheres depends on the sign of the imaginary or real part, including the sign of zero. To handle other points similarly would require the arithmetic of non-standard analysis.

[3] Note that the sign convention here is the opposite to that of [5], which defined $\mathcal{K}(z)$ as $\lfloor \frac{\pi - \Im z}{2\pi} \rfloor$: the authors of [5] recanted later to keep the number of -1s occurring in formulae to a minimum. We could also change "unwinding" to "winding" when we make that sign change; but "winding number" is in wide use for other contexts, and it seems best to keep the existing terminology.

This definition has several attractive features: $\mathcal{K}(z)$ is integer-valued, and familiar in the sense that "everyone knows" that the multivalued logarithm can be written as the principal branch "plus $2\pi i k$ for some integer k"; it is single-valued; and it can be computed by a formula not involving logarithms. It does have a numerical difficulty, namely that you must decide if the imaginary part is an odd integer multiple of π or not, and this can be hard (or impossible in some exact arithmetic contexts), but the difficulty is inherent in the problem and cannot be repaired e.g. by putting the branch cuts elsewhere.

Some correct identities for elementary functions using $\mathcal{K}$ are given in Table 1.

1. $z = \log e^z + 2\pi i\mathcal{K}(z)$.
2. $\mathcal{K}(a\log z) = 0 \,\forall z \in \mathbf{C}$ if and only if $-1 < a \le 1$.
3. $\log z_1 + \log z_2 = \log(z_1 z_2) + 2\pi i\mathcal{K}(\log z_1 + \log z_2)$.
4. $a\log z = \log z^a + 2\pi i\mathcal{K}(a\log z)$.
5. $z^{ab} = (z^a)^b e^{2\pi ib\mathcal{K}(a\log z)}$.

Table 1. Some correct identities for logarithms and powers using $\mathcal{K}$.

(7) can then be rescued as

$$\log(z_1 z_2) = \log z_1 + \log z_2 - 2\pi i\mathcal{K}(\log z_1 + \log z_2). \tag{9}$$

Similarly (4) can be rescued as

$$\log \overline{z} = \overline{\log z} - 2\pi i\mathcal{K}\left(\overline{\log z}\right). \tag{10}$$

Note that, as part of the algebra of $\mathcal{K}$, $\mathcal{K}(\overline{\log z}) = \mathcal{K}(-\log z) \ne \mathcal{K}(\log 1/z)$. $\mathcal{K}(z)$ depends only on the imaginary part of z.
- Although not formally proposed in the same way in the computational community, one possible solution, often found in texts in complex analysis, is to accept the multi-valued nature of these functions (we adopt the common convention of using capital letters, e.g. Ln, to denote the multi-valued function), defining, for example

$$\text{Arcsin}\, z = \{y|\sin y = z\}.$$

This leads to $\sqrt{z^2} = \{\pm z\}$, which has the advantage that it is valid throughout $\mathbf{C}$. Equation 7 is then rewritten as

$$\text{Ln}(z_1 z_2) = \text{Ln}\, z_1 + \text{Ln}\, z_2 , \tag{11}$$

where addition is addition of sets ($A + B = \{a + b : a \in A, b \in B\}$) and equality is set equality[4].

---

[4] "The equation merely states that the sum of one of the (infinitely many) logarithms of z_1 and one of the (infinitely many) logarithms of z_2 can be found among the

However, it seems to lead in practice to very large and confusing formulae. More fundamentally, this approach does not say what will happen when the multi-valued functions are replaced by the single-valued ones of numerical programming languages.

A further problem that has not been stressed in the past is that this approach suffers from the same aliasing problem that naïve interval arithmetic does [6]. For example,

$$\mathrm{Ln}(z^2) = \mathrm{Ln}\, z + \mathrm{Ln}\, z \neq 2\,\mathrm{Ln}\, z \,,$$

since $2\,\mathrm{Ln}(z) = \{2\log(z) + 4k\pi i : k \in \mathbf{Z}\}$, but $\mathrm{Ln}(z) + \mathrm{Ln}(z) = \{2\log(z) + 2k\pi i : k \in \mathbf{Z}\}$: indeed if $z = -1$, $\log(z^2) \notin 2\,\mathrm{Ln}(z)$. Hence this method is unduly pessimistic: it may fail to prove some identities that are true.

3   The Rôle of the Unwinding Number

We claim that the unwinding number provides a convenient formalism for reasoning about these problems. Inserting the unwinding number systematically allows one to make "simplifying" transformations that *are* mathematically valid. The unwinding number can be evaluated at any point, either symbolically or via guaranteed arithmetic: since we know it is an integer, in practice little accuracy is necessary. Conversely, removing unwinding numbers lets us genuinely "simplify" a result. We describe insertion and removal as separate steps, but in practice every unwinding number, once inserted by a "simplification" rule, should be eliminated as soon as possible. We have thus defined a concrete goal for mathematically valid simplification.[5]

The following section gives examples of reasoning with unwinding numbers. Having motivated the use of unwinding numbers, the subsequent sections deal with their insertion (to preserve correctness) and their elimination (to simplify results).

4   Examples of Unwinding Numbers

This section gives certain examples of the use of unwinding numbers. We should emphasise our view that an ideal computer algebra system should do this manipulation for the user: certainly inserting the unwinding numbers where necessary, and preferably also removing/simplifying them where it can.

4.1   Forms of arccos

The following example is taken from [4], showing that two alternative definitions of arccos are in fact equal:

(infinitely many) logarithms of $z_1 z_2$, and conversely every logarithm of $z_1 z_2$ can be represented as a sum of this kind (with a suitable choice of [elements of] $\mathrm{Ln}\, z_1$ and $\mathrm{Ln}\, z_2$)." [3, pp. 259–260] (our notation).

[5] Just to remove the terms with unwinding numbers, as is done in some software systems, could be called "over-simplification."

Theorem 1.

$$\frac{2}{i}\log\left(\sqrt{\frac{1+z}{2}}+i\sqrt{\frac{1-z}{2}}\right) = -i\log\left(z+i\sqrt{1-z^2}\right). \tag{12}$$

First we prove the correct (and therefore containing unwinding numbers) version of $\sqrt{z_1 z_2}\overset{?}{=}\sqrt{z_1}\sqrt{z_2}$.

Lemma 1.

$$\sqrt{z_1 z_2} = \sqrt{z_1}\sqrt{z_2}(-1)^{\mathcal{K}(\log z_1 + \log z_2)}. \tag{13}$$

Proof.

$$\sqrt{z_1 z_2} = \exp\left(\frac{1}{2}\left(\log(z_1 z_2)\right)\right)$$
$$= \exp\left(\frac{1}{2}\left(\log z_1 + \log z_2 - 2\pi i\mathcal{K}(\log z_1 + \log z_2)\right)\right)$$
$$= \sqrt{z_1}\sqrt{z_2}\exp\left(-\pi i\mathcal{K}(\log z_1 + \log z_2)\right)$$
$$= \sqrt{z_1}\sqrt{z_2}(-1)^{\mathcal{K}(\log z_1 + \log z_2)}$$

Lemma 2. *Whatever the value of z,*

$$\sqrt{1-z}\sqrt{1+z} = \sqrt{1-z^2}.$$

This is a classic example of a result that is "obvious": the schoolchild just squares both sides, but in fact that loses information, and the identity requires proof. To show this, consider the apparently similar "result"[6]:

$$\sqrt{-i-z}\sqrt{-i+z}\overset{?}{=}\sqrt{-1-z^2}.$$

If we take $z = i/2$, the left-hand side becomes $\sqrt{-3i/2}\sqrt{-i/2}$: the inputs to the square roots[7] have arg $= -\pi/2$, so the square roots themselves have arg $= -\pi/4$, and the product has arg $= -\pi/2$, and therefore is $-i\sqrt{3}/2$. The right-hand side is $\sqrt{-3/4} = i\sqrt{3}/2$.

Proof. It is sufficient to show that the unwinding number term in lemma 1 is zero. Whatever the value of z, $1+z$ and $1-z$ have imaginary parts of opposite signs. Without loss of generality, assume $\Im z \geq 0$. Then $0 \leq \arg(1+z) \leq \pi$ and $-\pi < \arg(1-z) \leq 0$. Therefore their sum, which is the imaginary part of $\log(1+z)+\log(1-z)$, is in $(-\pi, \pi]$. Hence the unwinding number is indeed zero.

---

[6] Maple V.5, in the absence of an explicit declaration that z is complex, will say that the two are almost never equal, with the difference being $-2i\sqrt{1-z^2}$, but in fact at $z = 2i$, the two are equal.

[7] One is tempted to say "arguments of the square root", but this is easily confused with the function arg; we use 'inputs' instead.

Proof of Theorem 1. Now

$$\left(\sqrt{\frac{1+z}{2}}+i\sqrt{\frac{1-z}{2}}\right)^2 = z+i\sqrt{1-z}\sqrt{1+z}=z+i\sqrt{1-z^2}$$

by the previous lemma. Also $2\log a = \log(a^2)$ if $\mathcal{K}(2\log a)=0$, so we need only show this last stipulation, i.e. that

$$-\frac{\pi}{2} < \arg\left(\sqrt{\frac{1+z}{2}}+i\sqrt{\frac{1-z}{2}}\right) \le \frac{\pi}{2}.$$

This is trivially true at $z=0$. If it is false at any point, say z_0, then a path from z_0 to 0 must pass through a z where $\left|\arg\left(\sqrt{(1+z)/2}+i\sqrt{(1-z)/2}\right)\right| = \pi/2$, i.e. $\sqrt{(1+z)/2}+i\sqrt{(1-z)/2} = it$ for $t \in \mathbf{R}$, because, first, arg is continuous for $|z| \le \pi/2$, and indeed for $|z| < \pi$, and, second, that the inputs to arg are themselves discontinuous only on $z > 1$ and $z < -1$, and on these half-lines, the arguments in question are 0 and $\pi/2$, which are acceptable. Coming back to the continuity along the path, we find that by squaring both sides, $z+i\sqrt{1-z^2} = -t^2$, i.e. $(z+t^2)^2 = -(1-z^2)$. Hence $2zt^2+t^4 = -1$, so $z = -(1+t^4)/(2t^2) \le -1$, and in particular is real. On this half-line, as stated before, the argument in question is $+\pi/2$, which is acceptable. Hence the argument never leaves the desired range, and the theorem is proved.

4.2   arccos and arccosh

$\cos(z) = \cosh(iz)$, so we can ask whether the corresponding relation for the inverse functions, $\operatorname{arccosh}(z) = i\operatorname{arccos}(z)$, holds. This is known in [4] as the "couthness" of the arccos/arccosh definitions. The problem reduces, using equations (20) and (26), to

$$2\log\left(\sqrt{\frac{z+1}{2}}+\sqrt{\frac{z-1}{2}}\right) \stackrel{?}{=} i\left(\frac{2}{i}\log\left(\sqrt{\frac{1+z}{2}}+i\sqrt{\frac{1-z}{2}}\right)\right),$$

i.e.

$$\log\left(\sqrt{\frac{z+1}{2}}+\sqrt{\frac{z-1}{2}}\right) \stackrel{?}{=} \log\left(\sqrt{\frac{1+z}{2}}+i\sqrt{\frac{1-z}{2}}\right).$$

Since $\log a = \log b$ implies $a = b$ (n.b. this is not true for exp, which is part of the point of this paper), this reduces to

$$\sqrt{\frac{z-1}{2}} \stackrel{?}{=} i\sqrt{\frac{1-z}{2}} = \sqrt{-1}\sqrt{\frac{1-z}{2}}.$$

By lemma 1, the right-hand side reduces to $\sqrt{\frac{z-1}{2}}(-1)^{-\mathcal{K}(\log(-1)+\log(\frac{z-1}{2}))}$. Hence the two are equal if, and only if, the unwinding number is even (and therefore zero). This will happen if, and only if, $\arg\left(\frac{z-1}{2}\right) \le 0$, i.e. $\Im z < 0$ or $\Im z = 0$ and $z > 1$.

4.3   arcsin and arctan

The aim of this section is to prove the correct expression for arcsin in terms of arctan. We note that we need to add unwinding number terms to deal with the two cuts $\Re z < -1$, $\Im z = 0$ and $\Re z > 1$, $\Im z = 0$.

Theorem 2.

$$\arcsin z = \arctan \frac{z}{\sqrt{1-z^2}} + \pi\mathcal{K}(-\log(1+z)) - \pi\mathcal{K}(-\log(1-z)). \qquad (14)$$

We start from equations (19) and (21). Then

$$\begin{aligned}
2i \; \arctan \frac{z}{\sqrt{1-z^2}} &= \log\left(1 + i\frac{z}{\sqrt{1-z^2}}\right) - \log\left(1 - i\frac{z}{\sqrt{1-z^2}}\right) \\
&= \log\left([1 + i\frac{z}{\sqrt{1-z^2}}]/[1 - i\frac{z}{\sqrt{1-z^2}}]\right) \\
&\quad + 2\pi i\mathcal{K}\left(\log(1 + i\frac{z}{\sqrt{1-z^2}}) - \log(1 - i\frac{z}{\sqrt{1-z^2}})\right) \\
&= \log[iz + \sqrt{1-z^2}]^2 \\
&\quad + 2\pi i\mathcal{K}\left(\log(1 + i\frac{z}{\sqrt{1-z^2}}) - \log(1 - i\frac{z}{\sqrt{1-z^2}})\right) \\
&= 2i \arcsin(z) \\
&\quad - 2\pi i\mathcal{K}\left(2\log(iz + \sqrt{1-z^2})\right) \\
&\quad + 2\pi i\mathcal{K}\left(\log(1 + i\frac{z}{\sqrt{1-z^2}}) - \log(1 - i\frac{z}{\sqrt{1-z^2}})\right)
\end{aligned}$$

The tendency for $\mathcal{K}$ factors to proliferate is clear. To simplify we proceed as follows. Consider first the term

$$\mathcal{K}\left(2\log(iz + \sqrt{1-z^2})\right) \; .$$

For $|z| < 1$, the real part of the input to the logarithm is positive and hence has argument in $(-\pi/2, \pi/2)$; therefore $\mathcal{K} = 0$. For $|z| > 1$, we solve for the critical case in which the input to $\mathcal{K}$ is $-i\pi$ and find only $z = r\exp(i\pi)$, with $r > 1$. Therefore

$$\mathcal{K}(2\log(iz + \sqrt{1-z^2})) = \mathcal{K}(-\log(1+z)) \; .$$

Repeating the procedure with

$$\mathcal{K}\left(\log(1 + iz/\sqrt{1-z^2}) - \log(1 - iz/\sqrt{1-z^2})\right)$$

shows that $\mathcal{K} \neq 0$ only for $z > 1$. Therefore

$$\mathcal{K}\left(\log(1 + iz/\sqrt{1-z^2}) - \log(1 - iz/\sqrt{1-z^2})\right) = \mathcal{K}(-\log(1-z))$$

and so finally we get

$$\arctan \frac{z}{\sqrt{1-z^2}} = \arcsin(z) - \pi\mathcal{K}(-\log(1+z)) + \pi\mathcal{K}(-\log(1-z)) \; , \qquad (15)$$

and this cannot be simplified further.

5   The Unwinding Number: Insertion

We have seen that the systematic insertion of unwinding numbers while applying many "simplification" rules is necessary for mathematical correctness.

Unwinding numbers are normally inserted by use of equation (9) and its converse:

$$\log\left(\frac{z_1}{z_2}\right) = \log z_1 - \log z_2 - 2\pi\mathcal{K}\left(\log z_1 - \log z_2\right). \tag{16}$$

Equation (10) may also be used, as may its close relative (also a special case of (16))

$$\log\left(\frac{1}{z}\right) = -\log z - 2\pi\mathcal{K}\left(-\log z\right). \tag{17}$$

In practice, results such as lemma 1 would also be built in to a simplifier.

The definition of $\mathcal{K}$ gives us

$$\log(e^z) = z - 2\pi i\mathcal{K}(z), \tag{18}$$

which is another mechanism for inserting unwinding numbers while "simplifying". The formulae for other inverse functions are given in appendix B.

Many other "identities" among inverse functions require unwinding numbers. For example,

$$\arctan x + \arctan y = \arctan\left(\frac{x+y}{1-xy}\right) + \pi\mathcal{K}\left(2i(\arctan x + \arctan y)\right).$$

6   The Unwinding Number: Removal

It is clearly easier to insert unwinding numbers than to remove them. There are various possibilities for the values of unwinding numbers.

- An unwinding number may be identically zero. This is the case in lemma 2 and theorem 1. The aim is then to prove this.
- An unwinding number may be zero everywhere except on certain branch cuts in the complex plane. This is the case in equation (10), and its relative $\log(1/z) = -\log z - 2\pi i\mathcal{K}(-\log z)$. A less trivial case of this can be seen in equation (14). Derive has a different definition of arctan to eliminate this, so that, for Derive, $\arcsin(z) = \underbrace{\arctan}_{\text{Derive}} \frac{z}{\sqrt{1-z^2}}$. This definition can be related to ours either via unwinding numbers or via $\underbrace{\arctan}_{\text{Derive}}(z) = \overline{\arctan \overline{z}}$. It is often possible to disguise this sort of unwinding number, which is often of the form $\mathcal{K}(-\log(\ldots))$ or $\mathcal{K}(\overline{\log z})$, by resorting to such a "double conjugate" expression, though as yet we have no algorithm for this. Equally, we have no algorithm as yet for the sort of simplification we see in section 4.3.

- An unwinding number may divide the complex plane into two regions, one where it is non-zero and one where it is zero. A typical case of this is given in section 4.2. Here the proof methodology consists in examining the critical case, i.e. when the input to $\mathcal{K}$ has imaginary part $\pm\pi$, and examining when the functions contained in the input to $\mathcal{K}$ themselves have discontinuities.
- An unwinding number may correspond to the usual $+n\pi$: $n \in \mathbf{Z}$ of many trigonometric identities: examples of this are given in appendix B.

7 Conclusion

Unwinding number insertion permits the manipulation of logarithms, square roots etc., as well as the cancellation of functions and their inverses, while retaining mathematical correctness. This can be done completely algorithmically, and we claim this is one way, the only way we have seen, of guaranteeing mathematical correctness while "simplifying".

Unwinding number removal, where it is possible, then simplifies these results to the expected form. This is not a process that can currently be done algorithmically, but it is much better suited to current artificial intelligence techniques than the general problems of complex analysis.

When the unwinding numbers cannot be eliminated, they can often be converted into a case analysis that, while not ideal, is at least comprehensible while being mathematically correct.

More generally, we have reduced the analytic difficulties of simplifying these functions to more algebraic ones, in areas where we hope that artificial intelligence and theorem proving stand a better chance of contributing to the problem.

References

1. Abramowitz,M. & Stegun,I., Handbook of Mathematical Functions with Formulas, Graphs, and Mathematical Tables. US Government Printing Office, 1964. 10th Printing December 1972.
2. Bradford,R.J., Algebraic Simplification of Multiple-Valued Functions. Proc. DISCO '92 (Springer Lecture Notes in Computer Science 721, ed. J.P. Fitch), Springer, 1993, pp. 13–21.
3. Carathéodory,C., *Theory of functions of a complex variable* (trans. F. Steinhardt), 2nd. ed., Chelsea Publ., New York, 1958.
4. Corless,R.M., Davenport,J.H., Jeffrey,D.J. & Watt,S.M., "According to Abramowitz and Stegun". To appear in SIGSAM Bulletin. OpenMath Project (Esprit 24969) deliverable 1.4.6.
5. Corless,R.M. & Jeffrey,D.J., The Unwinding Number. SIGSAM Bulletin **30** (1996) 2, pp. 28–35.
6. Davenport,J.H. & Fischer,H.-C., Manipulation of Expressions. *Improving Floating-Point Programming* (ed. P.J.L. Wallis), Wiley, 1990, pp. 149–167.
7. Encyclopedia Britannica, 15th. edition. Encyclopedia Britannica Inc., Chicago etc., 15th ed., 1995 printing.
8. IEEE Standard Pascal Computer Programming Language. IEEE Inc., 1983.

9. IEEE Standard 754 for Binary Floating-Point Arithmetic. IEEE Inc., 1985.
10. Kahan,W., Branch Cuts for Complex Elementary Functions. The State of Art in Numerical Analysis (ed. A. Iserles & M.J.D. Powell), Clarendon Press, Oxford, 1987, pp. 165–211.
11. Litt,G., Unwinding numbers for the Logarithmic, Inverse Trigonometric and Inverse Hyperbolic Functions. M.Sc. project, Department of Applied Mathematics, University of Western Ontario, December 1999.
12. Steele,G.L.,Jr., Common LISP: The Language, 2nd. edition. Digital Press, 1990.

A   Definition of the Elementary Inverse Functions

These definitions are taken from [4]. They agree with [1, ninth printing], but are more precise on the branch cuts, and agree with Maple with the exception of arccot, for the reasons explained in [4].

$$\arcsin z = -i \log \left(\sqrt{1 - z^2} + iz \right). \tag{19}$$

$$\arccos(z) = \frac{\pi}{2} - \arcsin(z) = \frac{2}{i} \log \left(\sqrt{\frac{1+z}{2}} + i\sqrt{\frac{1-z}{2}} \right). \tag{20}$$

$$\arctan(z) = \frac{1}{2i} \left(\log(1 + iz) - \log(1 - iz) \right). \tag{21}$$

$$\operatorname{arccot} z = \frac{1}{2i} \log \left(\frac{z+i}{z-i} \right) = \arctan \left(\frac{1}{z} \right). \tag{22}$$

$$\operatorname{arcsec}(z) = \arccos(1/z) = -i \log(1/z + i\sqrt{1 - 1/z^2}), \tag{23}$$

with $\operatorname{arcsec}(0) = \frac{\pi}{2}$.

$$\operatorname{arccsc}(z) = \arcsin(1/z) = -i \log(i/z + \sqrt{1 - 1/z^2}), \tag{24}$$

with $\operatorname{arccsc}(0) = 0$.

$$\operatorname{arcsinh}(z) = \log \left(z + \sqrt{1 + z^2} \right). \tag{25}$$

$$\operatorname{arccosh}(z) = 2 \log \left(\sqrt{\frac{z+1}{2}} + \sqrt{\frac{z-1}{2}} \right). \tag{26}$$

$$\operatorname{arctanh}(z) = \frac{1}{2} \left(\log(1 + z) - \log(1 - z) \right). \tag{27}$$

$$\operatorname{arccoth}(z) = \frac{1}{2} \left(\log(-1 - z) - \log(1 - z) \right). \tag{28}$$

$$\operatorname{arcsech}(z) = 2 \log \left(\sqrt{\frac{z+1}{2z}} + \sqrt{\frac{1-z}{2z}} \right). \tag{29}$$

$$\operatorname{arccsch}(z) = \log \left(\frac{1}{z} + \sqrt{1 + \left(\frac{1}{z} \right)^2} \right), \tag{30}$$

B   Formulae for Inverse Functions

These formulae are taken from [11]. They make use of the secondary function csgn, which we define below in terms of $\mathcal{K}$ and was first defined by Dr. D. E. G. Hare as the piecewise function on the right hand side[8]:

$$\operatorname{csgn}(z) = (-1)^{\mathcal{K}(2\log(z))} = \begin{cases} +1 & \Re(z) > 0 \text{ or } \Re(z) = 0; \Im(z) \geq 0 \\ -1 & \Re(z) < 0 \text{ or } \Re(z) = 0; \Im(z) < 0 \end{cases}.$$

$$\arcsin(\sin(z)) = \begin{cases} z - 2\pi\mathcal{K}(zi) & \operatorname{csgn}(\cos z) = 1 \\ \pi - z - 2\pi\mathcal{K}(i(\pi - z)) & \operatorname{csgn}(\cos z) = -1 \end{cases}. \tag{31}$$

$$\arccos(\cos z) = \begin{cases} z - 2\pi\mathcal{K}(zi) & \operatorname{csgn}(\sin z) = 1 \\ -z - 2\pi\mathcal{K}(-zi) & \operatorname{csgn}(\sin z) = -1 \end{cases}. \tag{32}$$

$$\arctan(\tan z) = z + \pi\left(\mathcal{K}(-zi - \log\cos z) - \mathcal{K}(zi - \log\cos z)\right) \tag{33}$$

provided $z \neq \frac{\pi}{2} + n\pi$: $n \in \mathbf{Z}$.

$$\operatorname{arcsinh}(\sinh(z)) = \begin{cases} z - 2\pi i\mathcal{K}(z) & \operatorname{csgn}(\cosh z) = 1 \\ i\pi - z - 2\pi i\mathcal{K}(i\pi - z)) & \operatorname{csgn}(\cosh z) = -1 \end{cases}. \tag{34}$$

$$\operatorname{arccosh}(\cosh z) = \begin{cases} z - 2\pi\mathcal{K}(z) & \operatorname{csgn}(\sinh z)\cos(n\pi) = 1 \\ -z - 2\pi i\mathcal{K}(-z) & \operatorname{csgn}(\sinh z)\cos(n\pi) = -1 \end{cases} \tag{35}$$

where $n = \mathcal{K}\left(\log(\cosh(z) - 1) + \log(\cosh(z) + 1)\right)$.

$$\operatorname{arctanh}(\tanh z) = z + i\pi\left(\mathcal{K}(z - \log\cosh z) - \mathcal{K}(z - \log\cosh z)\right) \tag{36}$$

provided $z \neq \frac{\pi}{2}i + in\pi$: $n \in \mathbf{Z}$.

---

[8] This function simplifies $\sqrt{z^2}$ to $z\operatorname{csgn}(z)$. Dr. J. Carette observed that if we put $\omega = \exp(2\pi i/n)$, then the function defined by $\omega^{\mathcal{K}(n\log z)}$ and sometimes abbreviated by $C_n(z)$, that generalizes csgn, is useful in simplifying $(z^n)^{1/n}$ (private communication).

Solving Nonlinear Systems by Constraint Inversion and Interval Arithmetic

Martine Ceberio and Laurent Granvilliers

IRIN, Université de Nantes
B.P. 92208, F-44322 Nantes Cedex 3, France
{Martine.Ceberio, Laurent.Granvilliers}@irin.univ-nantes.fr

Abstract A reliable symbolic-numeric algorithm for solving nonlinear systems over the reals is designed. The symbolic step generates a new system, where the formulas are different but the solutions are preserved, through partial factorizations of polynomial expressions and constraint inversion. The numeric step is a branch-and-prune algorithm based on interval constraint propagation to compute a set of outer approximations of the solutions. The processing of the inverted constraints by interval arithmetic provides a fast and efficient method to contract the variables' domains. A set of experiments for comparing several constraint solvers is reported.

Keywords: AI and symbolic mathematical computing, constraint solving, nonlinear system, symbolic-numeric algorithm, interval arithmetic.

1 Introduction

Symbolic-numeric algorithms (solvers) processing sets of formulas over the reals (constraints) have been widely studied in the last years. In this framework, the symbolic algorithms are mainly devoted to polynomial (or quasi-polynomial) systems, like Gaussian elimination, Simplex, Gröbner bases [5], CAD [7], resultants [10], and triangular set-based techniques [2]. The numeric methods, like Gauss-Seidel, Newton-Raphson or optimisation techniques [15], compute sequences of approximate solutions or tightened variables' domains until some convergence properties or required distances to solutions are verified. The combination of both kinds of constraint solving techniques is a promising approach to prevent the drawbacks of each individual solver, namely the complexity and the lack of expressiveness of symbolic algorithms, the approximate nature of numeric solutions and the restricting convergence properties of numeric algorithms.

The core constraint solver developed in this work is a numeric branch-and-prune algorithm [9, 17] iterating two steps: first, an interval-based pruning operator associated with each constraint of the system to be solved discards from the variables' domains some of the values that are inconsistent with the constraint (technique related to local consistency methods such as arc consistency [11]); the

J.A. Campbell and E. Roanes-Lozano (Eds.): AISC 2000, LNAI 1930, pp. 127–141, 2001.
© Springer-Verlag Berlin Heidelberg 2001

domain modifications are then propagated to the other constraints for reinvocation of their pruning operator. When quiescence is reached, a bisection stage occurs—splitting of the domains—to separate solutions and obtain tighter domains that otherwise could have been obtained by constraint propagation alone. The use of interval arithmetic (IA) [13] permits the computation of outer approximations (supersets) of the relations defined by the constraints (solution set). Let us note that reliability is mandatory for various applications, for instance in order to ensure the physical meaning of a solution. Further, when techniques from interval computations are applied, the result of a procedure can be guaranteed also in the presence of rounding errors inherent in digital computations [1].

The symbolic part of the solving process is a preprocessing step implementing constraint factorization and inversion. The factorization of syntactically equivalent sub-expressions of a constraint permits us to partially tackle the dependency problem of IA: the multiple occurrences of a variable are considered as different variables during interval evaluation. As a consequence, interval computations often over-estimate the real quantities to be approximated. For this purpose, we define the *cross nested form* of a quasi-polynomial expression (polynomial where the coefficients can be complex expressions), which is intuitively a kind of Horner form of a multivariate quasi-polynomial. It is computed by an algorithm that iterates the choice of a variable w.r.t. which the Horner form is computed, until reaching a fixed-point, *i.e.*, no sub-expression can be factorized. With respect to the nested form [14] that aims at minimizing the total degree of an expression, the cross nested form aims at reducing the multiple occurrences of the variables.

The inversion of a constraint w.r.t. a variable consists in generating a new (syntactically different) equivalent constraint (*i.e.* with the same solution set) where this variable is expressed according to the rest of the constraint. For instance, the constraint obtained from the inversion of $x^3 - y = 2$ w.r.t. x is $x = \sqrt[3]{2 + y}$. The aim is to have a form of constraint that can be processed easily by IA to contract the domain of the considered variable. The associate pruning operator first approximates the range of the right-hand expression (using IA) and then interprets the relation symbol so as to keep all the solutions of the constraint (in the case of equality, the domains are intersected). The inversion of arbitrary constraints extends the framework presented in [6, 16] to process primitive constraints (constraints with one operation at most). The main idea is to introduce a new operation symbol to invert each real operation in the constraint, associated with a method to evaluate it over the intervals. To illustrate this process, one may cite the Gauss-Seidel method that inverts each row (a constraint) of a linear system with respect to a column (a variable) to contract the variables' domains.

The contribution of this paper is twofold: the definition of the cross nested form and the design of a branch-and-prune algorithm based on constraint inversion and interval arithmetic. Some experimental results show the algorithm's efficiency as well as the possibility for solving difficult nonlinear systems modelling real applications.

The outline of this paper is as follows: Section 2 introduces some notions from IA and presents the cross nested form. Constraint inversion is described in Section 3. The constraint solving algorithm is devised in Section 4. The experimental results are discussed in Section 5, and some conclusions are stated in Section 6.

2   Interval Arithmetic

Interval Arithmetic (IA) has been designed by R. E. Moore [13] for automatically computing roundoff error bounds of numerical computations. In this paper, it is used to compute a superset of the range of a real function over a domain. In this section, some notions from IA are presented, and the *cross nested form* of a real function is defined.

2.1   Preliminaries

Let $\mathbb{R}$ denote the set of real numbers. Let $\Sigma = \langle \mathbb{R}, \mathcal{F}, \{=, \leqslant, \geqslant\} \rangle$ be a real-based structure, where $\mathcal{F}$ is a set of operation symbols, and consider a set of real-valued variables. A *term* is a syntactic expression built from constants, operations and variables. Let V_f denote the set of variables occurring in a term f. A *constraint* is a first order formula built from terms and relation symbols—a nonlinear equality/inequality over the reals. Given an n-ary constraint c, let ρ_c denote the relation defined by c in the standard interpretation of Σ, and V_c the set of variables occurring in c. Two constraints c and c' are said to be *equivalent* if $\rho_c = \rho_{c'}$. Let $c \equiv c'$ denote the equivalence of c and c'.

Given $a, b \in \mathbb{R}$, the set of reals $I = \{x \in \mathbb{R} \mid a \leqslant x \leqslant b\}$ is an *interval*, denoted $[a, b]$ or $[\underline{I}, \overline{I}]$. Practical experiments are often based on the set $\mathbb{I}$ of machine-representable intervals whose bounds are floating-point numbers. Let $\mathbb{U}$ denote the set of unions of intervals from $\mathbb{I}$. Given a subset ρ of $\mathbb{R}$, let $\mathsf{Hull}(\rho)$ denote the smallest element of $\mathbb{I}$ (with respect to set inclusion) enclosing ρ.

The main notion from IA is the notion of interval extension of a real function/relation [13, 17].

Definition 1 (Interval extension). *An* interval extension *(also called interval form or inclusion function) of a function $f : \mathbb{R}^n \to \mathbb{R}$ is a function $F : \mathbb{I}^n \to \mathbb{I}$ such that for any tuple of intervals $(I_1, \ldots, I_n)$ in the domain of f, we have the following property:*

$$\{f(a_1, \ldots, a_n) \mid \exists a_1 \in I_1, \ldots, \exists a_n \in I_n\} \subseteq F(I_1, \ldots, I_n)$$

An interval extension of a relation $\rho \subseteq \mathbb{R}^n$ is a relation $\Gamma \subseteq \mathbb{I}^n$ such that for any tuple of intervals $(I_1, \ldots, I_n)$, we have the following property:

$$\exists a_1 \in I_1, \ldots, \exists a_n \in I_n, (a_1, \ldots, a_n) \in \rho \Rightarrow (I_1, \ldots, I_n) \in \Gamma$$

In other words, an interval extension represents a superset of the associated real quantity. This property is commonly called the inclusion property or fundamental theorem of IA. IA operations are set theoretic extensions of the real ones; given $I, J \in \mathbb{I}$ and an operation $\diamond$, we have:

$$I \diamond J = \mathsf{Hull}(\{a \diamond b \mid \exists a \in I, \exists b \in J\}).$$

In practice, these operations are evaluated by floating point computations over the bounds of intervals; for instance, we have $[a, b] + [c, d] = [a + c, b + d]$, and $[a, b] - [c, d] = [a - d, b - c]$, provided the resulting bounds are rounded towards the infinities. In such an approach, there are many ways to extend a real function.

The *natural interval extension* is a componentwise extension of a term representing the real function: each constant a of the term is replaced with $\mathsf{Hull}(\{a\})$, each variable with an interval variable, and each operation by the corresponding interval one. Let us remark that the natural extensions obtained from different terms of one function are generally different. For instance, let us consider $f(x) = x^2 - x$ and $g(x) = x(x - 1)$ and their natural extensions F and G. The range of f (and g) is $[-0.25, 90]$ over the domain $[0, 10]$, though $F([0, 10]) = [-10, 100]$ and $G([0, 10]) = [-10, 90]$. The inclusion property is preserved, but the over-estimation of the range cannot be anticipated. This weakness of IA is known as the *dependency problem*, which comes from the decorrelation of the multiple occurrences of one variable during interval evaluation. Nevertheless, there is one situation when this problem does not happen, when all variables occur only once in the term (theorem from Moore). The next section will present a new kind of interval extension, based on the factorization of polynomial terms in order to decrease the number of multiple occurrences of the variables.

Let us consider a constraint $c : f \bowtie g$. An interval extension of ρ_c can be obtained from some extensions of f and g, and the interpretation of the relation symbol $\bowtie$. Given $I, J \in \mathbb{I}$, the relation symbols are interpreted as follows: we have $I = J$ if $I \cap J \neq \varnothing$, and $I \leqslant J$ if $\underline{I} \leqslant \overline{J}$. Given an interval extension F (resp. G) of f (resp. g), we define an interval extension Γ_c of ρ_c as follows:

$$\Gamma_c = \{(I_1, \ldots, I_n) \in \mathbb{I}^n \mid F(I_1, \ldots, I_n) \bowtie G(I_1, \ldots, I_n)\}$$

2.2   The Cross Nested Form

The dependency problem of IA requires the use of interval extensions containing few multiple occurrences of variables. For this purpose, we define the *cross nested form*, a new kind of interval extension of a real function based on the factorization of some common sub-expressions of a term.

Table 1 presents Algorithm CrossNested that partially factorizes a real term $p : \sum_{i=1}^{k} e_i X_i$, where for all $i \in \{1, \ldots, k\}$, e_i is a term and X_i is either 1 or a product of factors x^d, x being a variable and d a positive integer. The result is a new term for the real function defined by the initial term. The two-step procedure iterates: 1. the choice of a variable that occurs at least in two sub-terms X_i of p (otherwise, the term cannot be factorized), and 2. the computation of the Horner form of p w.r.t. the selected variable. We then define the cross nested form as follows:

Definition 2 (Cross nested form). *Let p be a term representing a real function $f : \mathbb{R}^n \to \mathbb{R}$, and fix a choice procedure in Algorithm* CrossNested. *The cross nested form of p w.r.t. this choice procedure is the natural interval extension of the term* CrossNested(p).

Since the Horner form of an univariate polynomial is optimal for interval evaluation, we expect that the cross nested form of a multivariate pseudo-polynomial (that is also a polynomial if the expressions e_i are restricted to real numbers) is a good approximation for interval evaluation of the range of a real function (see Example 1).

Example 1. Let $p : x^3 y + x^2 yz + xw - xwy + 2yz - yw$ be a term. If the variables y and z are seen as constants, then the Horner form of p (w.r.t. x) is $2yz - yw + x(w - wy + x(yz + xy))$. Given the choice procedure that selects the variable that occurs the most in the term, the resulting term from CrossNested(p) is $y(2z - w) + x(w(1 - y) + xy(z + x))$. The number of occurrences of variables is 10 for CrossNested(p), and 14 for p.

Table1. Computation of the cross nested form of a real function.

```
function CrossNested(term p : ∑ᵏᵢ₌₁ eᵢXᵢ) : term
begin
    X := {x | ∃(1 ⩽ i < j ⩽ k), x ∈ VXᵢ, x ∈ VXⱼ}
    if X = ∅
    then  return p
    else  Choose x in X
          return Horner(p, x)
    fi
end


function Horner(term p : ∑ᵏᵢ₌₁ eᵢXᵢ, variable x) : term
begin
    d := min({dᵢ ∈ ℕ⁺ | ∃i ∈ {1, …, k}, dᵢ is the degree of x in Xᵢ})
        % d is possibly equal to 0 if ∃i ∈ {1, …, k}, x ∉ VXᵢ
    J := {i ∈ ℝ | ∃i ∈ {1, …, k}, x occurs in Xᵢ with degree d}
    Let p be x^d(∑ᵢ∈J eᵢ Xᵢ/x^d + r)   % Factorization of p by x^d
    s := CrossNested(∑ᵢ∈J eᵢ Xᵢ/x^d)   % Let us remark that x ∉ Vₛ
    if r ≠ 0
    then  q := Horner(r, x)
          return x^d(CrossNested(q + s))
    else  return x^d s
    fi
end
```

The cross nested form should be compared with the nested form proposed by V. Stahl [14]. Let us define the total degree of a product $x_1^{d_1} \times \cdots \times x_n^{d_n}$ as the sum $d_1 + \cdots + d_n$, and the common total degree of two products $x_1^{d_1} \times \cdots \times x_n^{d_n}$ and $y_1^{d'_1} \times \cdots \times y_m^{d'_m}$ as the sum $\sum_{x_i = y_j} \min(d_i, d'_j)$, *i.e.*, the total degree of their common sub-terms. The nested form of a term p corresponds to the factorization at a time of two sub-terms $e_i X_i$ and $e_j X_j$ of p such that the common total degree of X_i and X_j is maximum. This process is iterated until reaching a fixed-point. In general, there is no guarantee that both forms can be compared, while the total degree of the nested form is often smaller than the one of the cross nested form. Nevertheless, we have the following result:

Proposition 1. *Given the choice procedure that selects the variable that occurs most often in the term, the number of products $x_1^{d_1} \times \cdots \times x_n^{d_n}$ in the term* CrossNested(p) *is less than or equal to the number of products in the nested form.*

Let us illustrate both forms of a real term in Example 2.

Example 2. In this example, the difference between the two forms concerns the multiple occurrences of Variable y in the nested forms, though it occurs only once in the cross nested forms.

Input polynomial	Nested form	Cross nested form
$x^3 y + x^2 + y$	$x^2(xy + 1) + y$	$y(x^3 + 1) + x^2$
$xyz + xy + yz + xz$	$xy(1 + z) + z(x + y)$	$y(x(1 + z) + z) + xz$

3 Constraint Inversion

The inversion of a constraint w.r.t. a variable is a symbolic procedure that computes an equivalent constraint whose left-hand term is reduced to this variable. In Section 4, each inverted constraint will be used for contracting the domain of the considered variable.

In the following, let us consider an occurrence of a variable x appearing in a constraint c.

3.1 Preconditioning

The constraints are ordered w.r.t. the numbers of operation symbols contained in their left-hand terms, as follows:

$$f \bowtie g \preccurlyeq f' \bowtie' g' \iff \mathsf{op}(f) \leqslant \mathsf{op}(f')$$

Let us remark that $\preccurlyeq$ is Noetherian.

In order to simplify the presentation of the inversion procedure, each constraint c to be inverted w.r.t. a variable x is preliminary rewritten in an equivalent equality constraint $x = f$. Two new operation symbols ge $: \mathbb{R} \rightarrow \mathbb{R}$ and le $: \mathbb{R} \rightarrow \mathbb{R}$ are introduced to remove the inequality relation symbols, such that for every constraint $c : f \geqslant g$ and $c' : f \leqslant g'$, we have $c \equiv (f = \mathrm{ge}(g))$ and $c' \equiv (f' = \mathrm{le}(g'))$. Their interpretations will be given later by means of IA. The following table sums up the possible preconditioning operations, where $f[x]$ is the term containing the selected occurrence of x:

$$
\begin{array}{lcl}
f[x] \geqslant g & \rightsquigarrow & f[x] = \mathrm{ge}(g) \\
f[x] \leqslant g & \rightsquigarrow & f[x] = \mathrm{le}(g) \\
g \geqslant f[x] & \rightsquigarrow & f[x] = \mathrm{le}(g) \\
g \leqslant f[x] & \rightsquigarrow & f[x] = \mathrm{ge}(g) \\
f[x] = g & \rightsquigarrow & f[x] = g \\
g = f[x] & \rightsquigarrow & f[x] = g
\end{array}
$$

Let $\mathrm{prec}(c, x)$ denote the constraint resulting from the preconditioning of c w.r.t. x. We have the following property:

Property 1. Constraint $\mathrm{prec}(c, x)$ is equivalent to c.

Finally, let us remark that systems of equalities can be more easily simplified than inequalities. Further research may consider the simplication of equalities obtained from preconditioning. Nevertheless, let us note that new simplication rules have to be designed for processing the operation symbols ge and le.

3.2 Inversion

Let $c : f = g$ be an n-ary constraint. Table 2 describes the elementary operations to invert c. Indeed, either the inverted constraint can be expressed w.r.t. the existing symbols from the real-based structure, or these symbols are no longer enough. As a consequence, a set of new operation symbols is introduced, and defined so as to guarantee the constraints' equivalence.

Table2. Elementary operations for inverting a constraint.

Rule	Constraint c	Inverted constraint $\mathrm{inv}(c, x)$
1	$f[x] + g = h$	$f[x] = h - g$
2	$f[x] - g = h$	$f[x] = h + g$
3	$g + f[x] = h$	$f[x] = h - g$
4	$g - f[x] = h$	$f[x] = g - h$
5	$f[x] \times g = h$	$f[x] = h \div g$
6	$f[x]/g = h$	$f[x] = h * g$
7	$g \times f[x] = h$	$f[x] = h \div g$
8	$g/f[x] = h$	$f[x] = g \div h$
9	$\exp(f[x]) = h$	$f[x] = \mathrm{Log}(h)$
10	$\log(f[x]) = h$	$f[x] = \exp(h)$
11	$f[x]^n = h,\ n$ even	$f[x] = r_n(h)$
12	$f[x]^n = h,\ n$ odd	$f[x] = \sqrt[n]{h}$
13	$x = h$	$x = h$

To sum up, these new symbols are: $\div$ to invert the real multiplication (Rules 5 and 7), $*$ for the division (Rules 6 and 8), Log for the logarithm (since log : $\mathbb{R}_*^+ \to \mathbb{R}$ is only defined on $\mathbb{R}_*^+$), and r_n for the power of n (in order to avoid the disjunction induced by the n-th root with n even).

Proposition 2. *For every constraint c and every variable x occurring in c, we have either* $\mathsf{inv}(c, x) \preccurlyeq c$ *or* $\mathsf{inv}(c, x)$ *is syntactically equivalent to c. Furthermore,* $\mathsf{inv}(c, x)$ *is equivalent to c.*

Proof. Since $c : f \bowtie g \preccurlyeq c' : f' \bowtie' g'$ means that $\mathsf{op}(f) \leqslant \mathsf{op}(f')$, it is clear that for all rules, except for the thirteenth one, we have $\mathsf{inv}(c, x) \preccurlyeq c$. We then remark that Rule 13 generates an equivalent constraint. Moreover, by simply rewriting ρ_c, we conclude that Rules 1, 2, 3, 4, 10, 12 preserve the equivalence of c and $\mathsf{inv}(c, x)$. For instance, Rule 1 is rewritten as follows:

$$\rho_c = \{(x_1, \ldots, x_n) \mid f[x](x_1, \ldots, x_n) + g(x_1, \ldots, x_n) = h(x_1, \ldots, x_n)\},$$

where, with a little misuse of notations, $f(x_1, \ldots, x_n)$ stands for the function of expression f with parameters $x_1, \ldots, x_n$. This implies that

$$\rho_c = \{(x_1, \ldots, x_n) \mid f[x](x_1, \ldots, x_n) = h(x_1, \ldots, x_n) - g(x_1, \ldots, x_n)\}.$$

The relation ρ_c then corresponds to a new relation $\rho_{c'}$ where c' is $f[x] = h - g$. In addition, since the symbols $\div$, r_n, $*$ and Log are defined so as to keep the equivalence property, it follows that for every constraint c, Rules 5, 6, 7, 8, 9, and 11 generate an equivalent constraint $\mathsf{inv}(c, x)$. This ends the proof. $\square$

We then define the inversion of a constraint c as the computation of a sequence of equivalent constraints by iteratively applying one elementary inversion operation, until generating two consecutive constraints that are syntactically equivalent. The procedure always terminates due to the Noetherian property of ordering $\preccurlyeq$. Let $\mathsf{Inverse}(c, x)$ denote the last constraint from the sequence. By Property 1 and Proposition 2, we have the following result:

Proposition 3. *For every constraint c and every variable x occurring in c,* $\mathsf{Inverse}(c, x)$ *is equivalent to c.*

The following example illustrates the computation of an inverted constraint:

Example 3. Let $c : 2xy \geqslant (x + 1)^2 - 1$ be a constraint. The inverted constraint with respect to y is:
$$y = \mathsf{ge}((x + 1)^2 - 1) \div (2x)$$
and is unique. By contrast, there are two different inverted constraints with respect to the two occurrences of x:

$$\begin{cases} x = \mathsf{ge}((x + 1)^2 - 1) \div (2y) \\ x = r_2(\mathsf{le}(2xy) + 1) - 1 \end{cases}$$

In practice, it is a challenge to choose correctly the tighter of the two for interval evaluation.

3.3 Interval Extension and Tightening

Let us consider a constraint $c(x_1, \ldots, x_n)$, an integer $k \in \{1, \ldots, n\}$, and an occurrence of x_k in c. Let $\pi_k(\rho_c)$ denote the projection of ρ_c over Variable x_k, i.e. the set $\{a_k \in \mathbb{R} \mid \forall i \in \{1, \ldots, k-1, k+1, \ldots, n\}, \exists a_i \in \mathbb{R}, (a_1, \ldots, a_n) \in \rho_c\}$.

Let $I = (I_1, \ldots, I_n)$ be the variables' domains, and $x_k = f(x_1, \ldots, x_n)$ the constraint obtained from the inversion of c w.r.t. x_k. If one is able to compute an interval extension F of f, the domain of x_k can be contracted by the following operation:

$$I_k := \mathsf{Hull}(I_k \cap F(I_1, \ldots, I_n))$$

while preserving all the elements of ρ_c included in the variables' domains. This completeness property is guaranteed by Proposition 4.

Proposition 4. *We have* $\pi_k(\rho_c \cap I) \subseteq \mathsf{Hull}(I_k \cap F(I_1, \ldots, I_n))$.

Proof. Since c and $\mathsf{Inverse}(c, x_k)$ are equivalent, we have:
$\rho_c \cap I = \{(a_1, \ldots, a_n) \in I \mid a_k = f(a_1, \ldots, a_n)\}$. By inclusion property of F, it follows: $\rho_c \cap I \subseteq \{(a_1, \ldots, a_n) \in I \mid a_k \in F(I_1, \ldots, I_n)\}$. Since the projection operation preserves an inclusion relation, we have: $\pi_k(\rho_c \cap I) \subseteq \{a_k \in I_k \mid a_k \in F(I_1, \ldots, I_n)\}$. Since $\{a_k \in I_k \mid a_k \in F(I_1, \ldots, I_n)\} = I_k \cap F(I_1, \ldots, I_n)$ and Hull is extensive, we have: $\pi_k(\rho_c \cap I) \subseteq \mathsf{Hull}(I_k \cap F(I_1, \ldots, I_n))$, that ends the proof. □

The tightening operation assumes the computation of an interval extension of the right-hand term of each inverted constraint. We propose to use the natural interval extension as defined in Section 2, given the following interval operations associated with the new symbols introduced for the needs for constraint inversion.

The symbols $\div$ and $*$ are respectively interpreted as the extended division and the usual multiplication of IA. We then define:

$$
\begin{cases}
r_n([a, b]) & = [-\sqrt[n]{b}, -\sqrt[n]{a}] \cup [\sqrt[n]{a}, \sqrt[n]{b}] \text{ if } a > 0 \text{ and } n \text{ is even} \\
& = [0, \sqrt[n]{b}] \qquad\qquad\quad\ \text{ if } a \leqslant 0 \leqslant b \text{ and } n \text{ is even} \\
& = \varnothing \qquad\qquad\qquad\qquad\ \text{ otherwise} \\[4pt]
\mathsf{Log}([a, b]) & = \log([a, b] \cap]0, +\infty[ ) \quad \text{ if } b > 0 \\
& = \varnothing \qquad\qquad\qquad\qquad\ \text{ otherwise} \\[4pt]
\mathsf{le}([a, b]) & =]-\infty, b] \\[4pt]
\mathsf{ge}([a, b]) & = [a, +\infty[
\end{cases}
$$

The inclusion property of these interval extensions can be easily verified. To end the section, let us illustrate the whole process from the inversion to the tightening operation by an example.

Example 4. Let $c : (x + 1)^3 = y$ be a constraint and $[-10, 10]$, $[1, 8]$ the domains
of x and y respectively. The constraint obtained from the inversion of c w.r.t. x
is $x = \sqrt[3]{y} - 1$. The tightening operation computes:

$$\mathsf{Hull}([-10, 10] \cap ([1, 2] - 1)) = [0, 1],$$

that is the new domain of x.

4   Constraint Solving

The constraint solving algorithm is presented in Table 3. The cross nested forms
of all the constraints' expressions are first computed, resulting in the set of
constraints C'. The rest of the process is a classical branch-and-prune iteration [9,
17]. The computation step consists of four operations:

1. A vector of domains J is extracted from the list of vectors S to be processed;
2. A first-order Taylor expansion of all the equations from C' (*i.e.* constraints
   that have been factorized) is generated, followed by a preconditioning of the
   new linear system, as proposed in [8, 17]. The new set of constraints C''
   is obtained from the inversion of all the constraints from C' and the linear
   system w.r.t. all the variables occurring in it.
3. The chosen domains J are contracted w.r.t. the set of inverted constraints
   C'', by Algorithm Prune, a classical AC3-like constraint propagation algorithm [11, 3]. The tightening operation defined in Section 3 is enforced over
   the inverted constraints taken as input. The output corresponds to a fixed-
   point, when no further contraction of domains happens.
4. If the new vector K is precise enough (the width of each component interval
   is less than ε), then it is added in the list of output vectors S_f; otherwise, it
   is split (generally, in 2 or 3 parts in one direction) and each new sub-vector
   is added to the list S.

The result is a set of domains' vectors S_f such that every solution of the initial
problem appears at least in one vector from this set (completeness property).
This property follows directly from Proposition 4. Moreover, the algorithm terminates in finite time (since every step is contracting and the set of intervals is
finite).

5   Experimental Results

In this section, two experiences are reported: the comparison of both nested and
cross nested forms w.r.t. the natural form, and the comparison of the branch-and-
prune algorithm with two state-of-the-art systems for solving nonlinear systems:
Numerica [17, 18] implementing interval analysis and constraint satisfaction techniques, and PHC [20] based on homotopy continuation (let us remark that PHC is
restricted to systems of polynomial equations).

Table3. The constraint solving algorithm.

BranchAndPrune(C: set of constraints; $I : \mathbb{I}^n$; $\varepsilon : \mathbb{R}_*^+$) : set of $\mathbb{I}^n$
begin
  $C' := \varnothing$
  for each constraint $f \bowtie g \in C$ do
    $C' := C' \cup \{\text{CrossNested}(f) \bowtie \text{CrossNested}(g)\}$
  od
  $S := \{I\}$ $S_f := \varnothing$
  while $S \neq \varnothing$ do
    Get J from S
    $C'' := \text{InverseAll}(C' \cup \text{Taylor}(C', J))$
    $K := \text{Prune}(C'', J)$ % *Contraction of* J
    if K $\neq \varnothing$ then
      if the precision of K is greater than ε
      then $S_f := S_f \cup \{K\}$
      else $S := S \cup \text{Bisect}(K)$ % *Partitioning of* K
      fi
    fi
  od
  return S_f
end

InverseAll(C: set of constraints) : set of constraints
begin
  $C' := \varnothing$
  for each constraint $c \in C$ do
    $c' := \text{Preprocess}(c)$ % *Preconditioning step*
    for each variable occurring in c' do
      Select in c' an occurrence x of the chosen variable
      $C' := C' \cup \{\text{Inverse}(c', x)\}$ % *Inversion of* c' *w.r.t.* x
    od
  od
  return C'
end

Prune(C: set of constraints; $I : \mathbb{I}^n$) : $\mathbb{I}^n$
begin
  $S := C$ % *S is the propagation list*
  while $S \neq \varnothing$ and $I \neq \varnothing$ do
    Get $c : x_k = f(x_1, \ldots, x_n)$ from S
    $J := \text{Hull}(I_k \cap F(I))$ % *Tightening operation*
    if $J \neq I_k$ then
      $S := S \cup \{c' \in C \mid x_k \in V_{c'}\}$ % *Constraint propagation*
      $I_k := J$
    else $S := S \setminus \{c\}$
    fi
  od
  return I
end

Table4. Expression of Problem Seyfert-filter.

$$
\begin{cases}
m_2 m_4 m_6 = 0.01 \\
a b m_4 = 7/500 \\
a^2 + m_1^2 = 2/25 \\
b^2 + m_7^2 = 37/50 \\
m_2^2 + m_3^2 + m_4^2 + m_5^2 + m_6^2 = 0.9401 \\
m_2^2 m_4^2 + m_2^2 m_5^2 + m_2^2 m_6^2 + m_3^2 m_5^2 + m_3^2 m_6^2 + m_4^2 m_6^2 = 0.038589 \\
m_1 m_3 m_5 m_7 - b m_1 m_3 m_6 + a b m_2 m_6 - a m_2 m_5 m_7 = -0.00081 \\
b m_1 m_2 m_3 m_4 + a m_4 m_5 m_6 m_7 - a b m_2^2 m_4 - a b m_4 m_6^2 = 0.39/250 \\
m_4^2 m_7^2 + m_3^2 m_7^2 - 2 b m_5 m_6 m_7 + m_2^2 m_7^2 + m_5^2 m_7^2 + b^2 m_3^2 + b^2 m_6^2 + b^2 m_2^2 = 2.7173/4 \\
a, b \in [-10^8, +10^8] \qquad m_1, \ldots, m_7 \in [-1, +1]
\end{cases}
$$

The results for `Numerica` have been extracted from the book [18]. All other results have been obtained on the same machine, namely a Sun Sparc UltraI (166MHz). The set of benchmarks originate from both computer algebra and interval analysis communities [4, 17, 19]. The expression of Problem Seyfert-filter, modelling a filter design problem, typically illustrates the kind of system to be tackled, and is given in Table 4.

The comparison of the natural, nested and cross nested forms is done by enforcing the branch-and-prune algorithm on four benchmarks, *i.e.*, the problems from our database for which the differences between these three forms are significant. The results are reported in Table 5. Each column corresponds to the use of the corresponding interval form in Algorithm `Preprocess`. The computation times for the first solution have been collected and the reported figures are the ratios w.r.t. the solving time when the natural form is used.

Table5. Comparison of interval forms.

Benchmark	Natural form	Nested form	Cross nested form
Seyfert-filter	100%	56%	36%
Rouiller-robot	100%	18%	17%
Caprasse	100%	50%	50%
Czappor-Geddes	100%	80%	60%

The conclusions are the following: the nested form is more accurate than the natural form, as shown in [14]; as a consequence, the computation is faster since less terms are evaluated and less bisections are performed; the cross nested form seems to be as accurate as the nested form for all problems, and more efficient for Seyfert-filter, whereas we remark that only two constraints out of nine have different nested and cross nested forms.

The computation times from the aforementionned constraint solvers are given in Table 6: BaP 1 (resp. BaP) is the computation time for finding the first solution (resp. all solutions) using Algorithm `BranchAndPrune`. The results from `Numerica` and `PHC` are also reported. A blank cell indicates that a result is not available. A brief description of the problems is also given (respectively, name, number of variables, and number of real solutions).

These results show the efficiency of the new algorithm, combining a very simple interval tightening procedure with fast (quadratic in the number of variables) symbolic transformations of polynomials. With respect to continuation homotopy, and more generally to symbolic algorithms, such a symbolic-numeric algorithm may compute a solution quickly, since no generation of a starting solution/system is required before the search for the solutions. This is of particular importance since in practice, a solution satisfying some requirements is often demanded(this is the case for Problem Seyfert-filter, where the solution is required to have a physical meaning). In addition, let us note that the cross nested preprocessing time is negligible and is kept in reasonable bounds w.r.t. the numeric solving time when the number of variables increases.

Table6. Comparison of constraint solving algorithms (times in seconds).

Benchmark	Var	Sol	BaP 1	BaP	Numerica	PHC
Nbody	6	12	120.10	1185.00		508.60
Dessin-d'enfant 1	8	6	111.20	1047.50		271.65
Seyfert-filter	9	128	1.75	328.85		22824.20
Noonburg-network	5	11	0.05	80.55		261.25
Bellido-kinematics	9	8	3.50	81.55		689.30
Rouiller-robot	9	24	7.20	53.50		6640.55
Dessin-d'enfant 2	10	3	8.00	57.50		181.40
Neurophysiology	8	8	6.25	46.15	108.00	68.85
Kinematics 2	8	10	0.40	19.20	243.30	624.80
Rose	3	3	7.80	15.70		126.25
Katsura-magnetism	6	12	0.20	7.20		14.50
Ku	10	2	0.75	5.70		3.80
Caprasse	4	18	0.15	5.00	21.80	31.90
Trinks	6	2	0.45	3.00		6.25
Wood-function	4	3	0.60	2.10		2.00
Sendra	2	6	0.10	1.25		17.65
Brown	5	2	0.10	0.85	2.90	0.70
Kinematics 1	12	16	0.05	0.75	7.20	6929.55
Czapor-Geddes	3	2	0.30	0.75		1.10
Cyclohexane	3	16	0.10	0.75	1.60	3.65
Cox-Little-O'Shea	3	2	0.05	0.15		51.15

6   Conclusion

A symbolic-numeric algorithm for solving general nonlinear systems has been devised. It combines a preprocessing step of the constraint' expressions to simplify it w.r.t. interval arithmetic, with a numeric branch-and-prune iterative process to derive a set of precise interval vectors of domains enclosing the real solutions. The contraction of domains is based on the inversion procedure of a constraint to generate a new equivalent constraint with respect to which an efficient, fast tightening operation over intervals can be enforced. A set of experimental results from a prototype are reported, as well as comparisons with other systems.

Some directions for further research are sketched. The nested and cross nested forms must be clearly compared. In particular, what are the situations where one can choose the tighter according to *ad hoc* criteria? This is also closely connected with the use of particular orderings of variables. The combination of other interval extensions, such as the Bernstein form and the Taylor form of order $k \geqslant 2$ [12], is a promising approach to design more precise pruning algorithms. In this framework, some heuristics have to be developed in order to prevent unnecessary/redundant computations.

Acknowledgements. We are grateful to Frédéric Benhamou and Éric Monfroy for interesting discussions on these topics.

References

[1] Götz Alefeld and Jürgen Herzberger. *Introduction to Interval Computations*. Academic Press, 1983.

[2] Philippe Aubry, Daniel Lazard, and Marc Moreno Maza. On the Theories of Triangular Sets. *Journal of Symbolic Computation*, 28:105–124, 1998.

[3] Frédéric Benhamou and Laurent Granvilliers. Combining Local Consistency, Symbolic Rewriting and Interval Methods. In J. Calmet, J.A. Campbell, and J. Pfalzgraf, editors, *Proceedings of the 3rd International Conference on Artificial Intelligence and Symbolic Mathematical Computation*, volume 1138 of *LNCS*, pages 144–159, Steyr, Austria, 1996. Springer-Verlag.

[4] Dario Bini and Bernard Mourrain. Handbook of Polynomial Systems. http://www-sop.inria.fr/saga/POL/, 1996.

[5] Bruno Buchberger. Gröbner Bases: an Algorithmic Method in Polynomial Ideal Theory. In *Multidimensional Systems Theory*, pages 184–232. D. Reidel Publishing Company, 1985.

[6] John G. Cleary. Logical Arithmetic. *Future Computing Systems*, 2(2):125–149, 1987.

[7] Georges Edwin Collins and Hoon Hong. Partial Cylindrical Algebraic Decomposition for Quantifier Elimination. *Journal of Symbolic Computation*, 12(3):299–328, 1991.

[8] Eldon Robert Hansen. *Global Optimization using Interval Analysis*. Marcel Dekker, 1992.

[9] Ralph Baker Kearfott. Some Tests of Generalized Bisection. *ACM Transactions on Mathematical Software*, 13(3):197–220, 1987.

[10] Shankar Krishnan and Dinesh Manocha. Numeric-symbolic Algorithms for Evaluating One-dimensional Algebraic Sets. In *Proceedings of International Symposium on Symbolic and Algebraic Computation*, pages 59–67. ACM Press, 1995.

[11] Alan K. Mackworth. Consistency in Networks of Relations. *Artificial Intelligence*, 8(1):99–118, 1977.

[12] Kyoko Makino and Martin Berz. Efficient Control of the Dependency Problem based on Taylor Model Method. *Reliable Computing*, 5:3–12, 1999.

[13] Ramon Edgar Moore. *Interval Analysis*. Prentice-Hall, Englewood Cliffs, NJ, 1966.

[14] Volker Stahl. *Interval Methods for Bounding the Range of Polynomials and Solving Systems of Nonlinear Equations*. PhD thesis, University of Linz, Austria, 1995.

[15] Josef Stoer and Roland Bulirsch. *Introduction to Numerical Analysis*. Texts in Applied Mathematics. Springer, 2nd edition, 1993.

[16] Maarten Hermann Van Emden. Canonical extensions as common basis for interval constraints and interval arithmetic. In F. Benhamou, editor, *Proceedings of the 6th French Conference on Logic Programming and Constraint Programming*, pages 71–83. Hermès, 1997.

[17] Pascal Van Hentenryck, David McAllester, and Deepak Kapur. Solving Polynomial Systems Using a Branch and Prune Approach. *SIAM Journal on Numerical Analysis*, 34(2):797–827, 1997.

[18] Pascal Van Hentenryck, Laurent Michel, and Yves Deville. *Numerica: a Modeling Language for Global Optimization*. MIT Press, 1997.

[19] Jan Verschelde. Database of Polynomial Systems. Michigan State University, USA, 1999. http://www.math.msu.edu/~jan/demo.html.

[20] Jan Verschelde. PHCpack: A General-purpose Solver for Polynomial Systems by Homotopy Continuation. *ACM Transactions on Mathematical Software*, 25(2):251–276, 1999.

Basic Operators for Solving Constraints via Collaboration of Solvers

Carlos Castro[1] and Eric Monfroy[2]

[1] Departamento de Informática, Universidad Técnica Federico Santa María
Avenida España 1680, Casilla 110-V, Valparaíso, Chile
ccastro@inf.utfsm.cl
[2] Centrum voor Wiskunde en Informatica, CWI
P.O. Box 94079, NL-1090 GB Amsterdam, The Netherlands
Eric.Monfroy@cwi.nl

Abstract. In this paper, we propose a strategy language for designing schemes of constraint solver collaborations: a set of strategy operators enables one to design several kinds of collaborations. We exemplify the use of this language by describing some well known techniques for solving constraints over finite domains and non-linear constraints over real numbers via collaboration of solvers.

1 Introduction

In constraint programming, the programming process consists of formulating problems with constraints. Solutions of these so called Constraint Satisfaction Problems (CSPs) are generated by solvers. Numerous algorithms have been developed for solving CSPs and the resulting technology has been successfully applied for solving real-life problems. The design and implementation of these constraint solvers is generally an expensive and tedious task. Thus, the idea of reusing existing solvers is very interesting, but it also implies that we must have some tools to integrate them. Even more important, considering that some problems cannot be tackled or efficiently solved with a single solver, we definitively realize the interest of integrating and making cooperate several solvers [19, 4, 13, 20, 18]. This is called collaboration of solvers [15]. In order to make solvers collaborate, the need of powerful strategy languages to control their integration and application has been well recognized [16, 17, 1].

The existing approaches are generally not generic: they consider fixed domains (linear constraints [4], non-linear constraints over real numbers [18, 13, 3]), fixed strategies, or fixed scheme of collaboration (sequential [18, 3], asynchronous [13]). In the language **BALI**, collaborations are specified using control primitives and the constraint system is a parameter. Although **BALI** is more generic and flexible, the control capabilities for specifying strategies are not always fine enough [17]. In the system COLETTE [7, 8], a solver is viewed as a strategy that specifies the order of application of elementary operations expressed by transformation rules.

J.A. Campbell and E. Roanes-Lozano (Eds.): AISC 2000, LNAI 1930, pp. 142–156, 2001.

Extending ideas of **BALI** and COLETTE, we consider collaborations of solvers as strategies that specify the order of application of component solvers. In [9], we propose a strategy language for designing component or elementary constraint solvers and we exemplify its use by specifying several solvers (such as solvers for constraints over finite domains and real numbers). In this paper, we present the application of our language for prototyping constraint solving schemes via collaboration of solvers.

The main motivation for this work is to propose a general framework in which one can design component constraint solvers as well as solver collaborations. This approach makes sense since the design of constraint solvers and the design of collaborations require similar methods (strategies are often the same: *don't-care, fixed point, iteration, parallel, concurrent, ...*). In other words, we propose a language for writing component solvers and designing collaborations of several solvers at the same level.

This paper is organized as follows: Section 2 presents basic definitions and notations. In Section 3, we present an overview of our strategy language whereas in Section 4 we detail its basic operators. In Section 5, we use our language for solving constraints over finite domains and real numbers via the collaboration of several solvers. Finally, we conclude in Section 6.

2   Definitions

Definition 1 (Constraint Systems and Constraint Solvers). *A constraint system is a 4-tuple $(\Sigma, \mathcal{D}, \mathcal{V}, \mathcal{L})$ where Σ is a first-order signature given by a set of function symbols $\mathcal{F}_\Sigma$ and a set of predicate symbols $\mathcal{P}_\Sigma$, $\mathcal{D}$ is a Σ-structure (its domain being denoted by $|\mathcal{D}|$), $\mathcal{V}$ is an infinite denumerable set of variables, and $\mathcal{L}$ is a set of constraints: a non-empty set of $(\Sigma, \mathcal{V})$-atomic formulae, called* atomic constraints, *closed under conjunction and disjunction.*

We denote by $\perp$ the *unsatisfiable* constraint and the *true* constraint by $\top$. The set of atomic constraints is denoted by $\mathcal{L}_{At}$. An assignment is a mapping $\alpha : \mathcal{V} \to |\mathcal{D}|$. The set of all assignments is denoted by $ASS_\mathcal{D}^\mathcal{V}$. An assignment α extends uniquely to an homomorphism $\underline{\alpha} : T(\Sigma, \mathcal{V}) \to |\mathcal{D}|$. The set of solutions of a constraint $c \in \mathcal{L}$ is the set $Sol_\mathcal{D}(c)$ of assignments $\alpha \in ASS_\mathcal{D}^\mathcal{V}$ such that $\underline{\alpha}(c)$ holds. A constraint c is valid in $\mathcal{D}$ (denoted by $\mathcal{D} \models c$) if $Sol_\mathcal{D}(c) = ASS_\mathcal{D}^\mathcal{V}$. We use $Var(c)$ to denote the set of variables from $\mathcal{V}$ occurring in the constraint c.

Given a constraint system $(\Sigma, \mathcal{D}, \mathcal{V}, \mathcal{L})$, a *solver* is a computable function $S : \mathcal{L} \to \mathcal{L}$ satisfying the correctness and completeness properties, i.e., $\forall C \in \mathcal{L}$, $Sol_\mathcal{D}(S(C)) \subseteq Sol_\mathcal{D}(C)$ and $Sol_\mathcal{D}(C) \subseteq Sol_\mathcal{D}(S(C))$. We extend S to a constraint system $(\Sigma, \mathcal{D}, \mathcal{V}, \mathcal{L}')$, where $\mathcal{L} \subseteq \mathcal{L}'$, in the following way: $\forall C \in \mathcal{L}' \setminus \mathcal{L}$, $S(C) = C$. We say that a constraint C is in solved form with respect to S if $S(C) = C$.

In order to be able to manipulate specific parts of a constraint, we introduce the notions of *syntactical form* and *sub-constraint*.

Definition 2 (Syntactical Forms and Sub-constraints). *We say that C' is a* syntactical form *of C, denoted by $C' \approx C$, if $C' = C$ modulo the associativity and commutativity of $\wedge$ and $\vee$, and the distributivity of $\wedge$ on $\vee$ and of $\vee$ on $\wedge$* [1]. *We say that $C' \in \mathcal{L}$ is a* sub-constraint *of C, denoted by $C_{[C']}$, if:*

- $C = C'$
- *or* $\exists C_1 \in \mathcal{L}, \omega \in \{\wedge, \vee\}, C = C_1 \omega C'$
- *or* $\exists C_1 \in \mathcal{L}, \omega \in \{\wedge, \vee\}, C = C' \omega C_1$
- *or* $\exists C_1, C_2 \in \mathcal{L}, \omega \in \{\wedge, \vee\}, C = C_1 \omega C_2$ *and* $(C_{1[C']}$ *or* $C_{2[C']})$

A couple (C'', C') such that C'' is a sub-constraint of C' and $C' \approx C$ is called an *applicant* of C. We denote by $\mathcal{SF}(C)$ the finite set of all the syntactical forms of a constraint C: $\mathcal{SF}(C) = \{C' | C' \approx C\}$ [2]. We denote by $\mathcal{LA}$ the set of all the lists of applicants, and by $\mathcal{LC}$ the set of all the lists of constraints. Generally, we will use LA (respectively LC) to denote a list of applicants (respectively constraints). We denote by $\mathcal{P}(\mathcal{L} \times \mathcal{L})$ the power-set of all the sets of couples of constraints. $\mathcal{A}tom(C)$ denotes the set of atomic constraints that occur in C: $\{c | c \in \mathcal{L}_{At}$ and $C_{[c]}\}$.

Finally, in order to explicitly handle sub-parts of a constraint, we define the notions of *filter* to select specific parts of a constraint, and *sorter* to classify the elements of a list w.r.t. a given order [3].

Definition 3 (Filters and Sorters). *Given a constraint system $(\Sigma, \mathcal{D}, \mathcal{V}, \mathcal{L})$, a* filter *$\phi$ is a computable function $\phi : \mathcal{L} \to \mathcal{P}(\mathcal{L} \times \mathcal{L})$ such that $\phi(C) = \{(Cf_i, C_i), \ldots, (Cf_n, C_n)\}$ for all $C \in \mathcal{L}$, where each C_i is a syntactical form of C and Cf_i is a sub-constraint of C_i.*

A sorter $Sorter$, *w.r.t. a partial order $\preceq$, is a computable function $Sorter : \preceq \times \mathcal{P}(\mathcal{L} \times \mathcal{L}) \to \mathcal{LA}$ such that $\forall \{(Cf_{i_1}, C_{i_1}), \ldots, (Cf_{i_n}, C_{i_n})\} \in \mathcal{P}(\mathcal{L} \times \mathcal{L})$:*

1. $Sorter(\preceq, \{(Cf_{i_1}, C_{i_1}), \ldots, (Cf_{i_n}, C_{i_n})\}) = [(Cf_1, C_1), \ldots, (Cf_n, C_n)]$
2. $\forall k \in [1, \ldots, n], \exists j \in [1, \ldots, n], Cf_{i_j} = Cf_k$ *and* $C_{i_j} = C_k$
3. $\forall j \in [1, \ldots, n-1], Cf_j \preceq Cf_{j+1}$

The elements of $\phi(C)$ are called *candidates*. We define the filter Id which returns the initial set of constraints and the order $None$ which returns the initial list of candidates. Considering the filters ϕ_1 and ϕ_2 on $(\Sigma, \mathcal{D}, \mathcal{V}, \mathcal{L})$, then $\phi_1; \phi_2$ defined by $\phi_1(C) \cap \phi_2(C)$ is also a filter on $(\Sigma, \mathcal{D}, \mathcal{V}, \mathcal{L})$ for all $C \in \mathcal{L}$.

3   An Overview of the Strategy Language

Most of the application mechanisms that we use in our strategy language are based on the same technique when applied to a constraint C:

---

[1] We consider that "=" is purely syntactic.

[2] The ACD theory defines a finite set of quotient classes that we can effectively filter.

[3] These transformations are normally hidden in existing solvers. In [9], we detail examples of the definition of filters and sorters.

1. A set SC of candidates is built using the filter ϕ on C.
2. The set SC is sorted using the partial order $\preceq$. We obtain LC, a sorted list of candidates.
3. The solver S is applied to one (e.g., the "best" w.r.t. $\preceq$) or several elements of LC.
4. Each occurrence of the sub-constraint(s) modified by S are replaced in their corresponding (w.r.t. candidates) syntactical form of C.

The idea behind this scheme can be better understood in the following example. Suppose we are given the CSP over finite domains:

$$x \in [1, \ldots, 10] \ \wedge \ y \in [1, \ldots, 5] \ \wedge \ x \geq y$$

In order to find a solution we can carry out enumeration as follows:

– We first filter domain constraints in order to obtain a set of candidates:

$$\{(x \in [1, \ldots, 10], \quad x \in [1, \ldots, 10] \ \wedge \ y \in [1, \ldots, 5] \ \wedge \ x \geq y),$$
$$(y \in [1, \ldots, 5], \quad x \in [1, \ldots, 10] \ \wedge \ y \in [1, \ldots, 5] \ \wedge \ x \geq y)\}$$

– If we want to use the minimum domain criterion, a sorter will return the following sorted list of candidates:

$$[(y \in [1, \ldots, 5], \quad x \in [1, \ldots, 10] \ \wedge \ y \in [1, \ldots, 5] \ \wedge \ x \geq y),$$
$$(x \in [1, \ldots, 10], \quad x \in [1, \ldots, 10] \ \wedge \ y \in [1, \ldots, 5] \ \wedge \ x \geq y)]$$

– Applying a solver to split the "best" domain constraint we obtain:

$$y \in [1, \ldots, 2] \ \vee \ y \in [3, \ldots, 5], \quad x \in [1, \ldots, 10] \ \wedge \ y \in [1, \ldots, 5] \ \wedge \ x \geq y$$

– After replacing the original constraint in the corresponding syntactical form we finally obtain:

$$x \in [1, \ldots, 10] \ \wedge \ (y \in [1, \ldots, 2] \ \vee \ y \in [3, \ldots, 5]) \ \wedge \ x \geq y$$

This syntactical form is equivalent to the original set of constraints and once we activate operators properties we could continue the solving process.

4   The Strategy Language

Now we briefly present several *application mechanisms* to apply solvers to constraints. We assume that a solver is applied only once to a given set of constraints. In the following, we consider given a constraint system $CS = (\Sigma, \mathcal{D}, \mathcal{V}, \mathcal{L})$, solvers $S_1, \ldots, S_n$, filters $\phi_1, \ldots, \phi_n$, and partial orders $\preceq_1, \ldots, \preceq_n$.

We also use the notion of separators that are mainly defined to manipulate elements of conjunctions and disjunctions of constraints as elements of lists. A $\wedge\_separator$ $\delta_\wedge$ is a function $\delta_\wedge : \mathcal{L} \rightarrow \mathcal{LC}$ s.t.: $\forall \ C \in \mathcal{L}, \exists n \in \mathbb{N}, \delta_\wedge(C) = [C_1, \ldots, C_n]$ where $C \approx C_1 \wedge \ldots \wedge C_n$. Similarly, a $\vee\_separator$ $\delta_\vee$ is a function

$\delta_\vee : \mathcal{L} \to \mathcal{LC}$ such that: $\forall\ C \in \mathcal{L}, \exists n \in \mathbb{N}, \delta_\vee(C) = [C_1, \ldots, C_n]$ where $C \approx C_1 \vee \ldots \vee C_n$.

Finally, we use the notion of a constraint property p on a constraint system $(\Sigma, \mathcal{D}, \mathcal{V}, \mathcal{L})$ which is a function from constraints to Booleans (i.e., $p : \mathcal{L} \to \mathcal{B}oolean$).

We use five basic operators that are analogous to function compositions and that allow to design solvers by combining "basic" functions (non decomposable solvers), or to create solver collaborations by combining component solvers. Consider two solvers S_i and S_j. Then, for all $C \in \mathcal{L}$:

- $S_i^0(C) = C$ (*Identity*)
- $S_i; S_j(C) = S_j(S_i(C))$ (*solver concatenation*)
- $S_i^n(C) = S_i^{n-1}; S_i(C)$ if $n > 0$ (*solver iteration*)
- $S_i^\star(C) = S_i^n(C)$ such that $S_i^{n+1}(C) = S_i^n(C)$ (*solver fixed-point*)
- $(S_i, S_j)(C) = S_i(C)$ or $S_j(C)$ (*solver don't-care*)

Property 1. Let S_i and S_j be two solvers. Then, $S_i; S_j$, S_i^n, $S_i^\star$, and (S_i, S_j) are solvers.

We also use high level operators: two operators to apply a solver to specific components of a constraint, two operators to apply several solvers on a constraint, and two operators to apply a solver on each component of a conjunction or disjunction of constraints. Note that in the following, substitutions apply to every occurrence of sub-constraints.

dc$(S_i, \phi)(C)$: this operator restricts the use of the solver S_i to one randomly chosen sub-constraint of a syntactical form of C (obtained using the filter ϕ). For all $C \in \mathcal{L}$, **dc**$(S_i, \phi)(C) = C'$, where:

- $[(Cf_1, C_1), \ldots, (Cf_n, C_n)] = \phi(C)$
- if there exists $i \in [1, \ldots, n]$ such that $S_i(Cf_i) \neq Cf_i$, then $C' = C_i\{Cf_i \mapsto S_i(Cf_i)\}$, otherwise $C' = C$.

best$(S_i, \preceq, \phi)(C)$: this operator restricts the use of the solver S_i to the best (w.r.t. the partial order $\preceq$) sub-constraint of a syntactical form of C (obtained using the filter ϕ) that S_i is able to modify. For all $C \in \mathcal{L}$, **best**$(S_i, \preceq, \phi)(C) = C'$, where:

- $[(Cf_1, C_1), \ldots, (Cf_n, C_n)] = Sorter(\preceq, \phi(C))$
- if there exists $i \in [1, \ldots, n]$, such that $S_i(Cf_i) \neq Cf_i$, and $\forall j \in [1, \ldots, n]$ $(S_i(Cf_j) \neq Cf_j \Rightarrow i \leq j)$ then $C' = C_i\{Cf_i \mapsto S_i(Cf_i)\}$, otherwise $C' = C$.

pcc$(p, (S_1, \preceq_1, \phi_1), \ldots, (S_n, \preceq_n, \phi_n))(C)$: this operator applies once one of the solvers S_i and returns a constraint that verifies the property p. For all $C \in \mathcal{L}$, **pcc**$(p, [S_1, \preceq_1, \phi_1], \ldots, [S_n, \preceq_n, \phi_n])(C) = C'$, where:

- for all $i \in [1, \ldots, n]$ $[(Cf_{i,1}, C_{i,1}), \ldots, (Cf_{i,m_i}, C_{i,m_i})] = Sorter(\preceq_i, \phi_i(C))$
- if there exists $(i,j) \in [1, \ldots, n] \times [1, \ldots, m_i]$ such that $p(S_i(Cf_{i,j}))$, and $S_i(Cf_{i,j}) \neq Cf_{i,j}$ then $C' = C_{i,j}\{Cf_{i,j} \mapsto S_i(Cf_{i,j})\}$, otherwise $C' = C$.

$\mathbf{bp}((S_1, \preceq_1, \phi_1), \ldots, (S_n, \preceq_n, \phi_n))(C)$: this operator applies n solvers $S_1, \ldots, S_n$ on n sub-constraints of one syntactical form of the constraint. For all $C \in \mathcal{L}$, $\mathbf{bp}([S_1, \preceq_1, \phi_1], \ldots, [S_n, \preceq_n, \phi_n])(C) = C'$, where[4]:

- for all $i \in [1, \ldots, n]$ $[(Cf_{i,1}, C''), \ldots, (Cf_{i,m_i}, C'')] = Sorter(\preceq_i, \phi_i(C))$
- for all $i \in [1, \ldots, n]$, if there exists $j \in [1, \ldots, m_i]$, s.t. $S_i(Cf_{i_j}) \neq Cf_{i_j}$, and for all $k < j$, $S_i(Cf_{i_k}) = Cf_{i_k}$, then $\sigma_i = \{Cf_{i,i_j} \mapsto S_i(Cf_{i,i_j})\}$, else $\sigma_i = \emptyset$.
- $C' = C''\sigma$ where $\sigma = \bigcup_{i \in [1, \ldots, n]} \sigma_i$.

$\wedge\_\mathbf{p}(S_i, \delta_\wedge)(C)$: this operator applies (in parallel) the solver S_i to several conjuncts (determined by $\delta_\wedge$) of the constraint C and the final result is obtained by conjunction of the results computed in parallel. For all $C \in \mathcal{L}$, $\wedge\_\mathbf{p}(S_i, \delta_\wedge)(C) = C'$, where:

- $[C_1, \ldots, C_n] = \delta_\wedge(C)$
- $C' = S_i(C_1) \wedge \ldots \wedge S_i(C_n)$

$\vee\_\mathbf{p}(S_i, \delta_\vee)(C)$: this operator is analogous to $\wedge\_\mathbf{p}$ but $\delta_\vee$ determines disjuncts, and the final result is the disjunction of the results computed in parallel. For all $C \in \mathcal{L}$, $\vee\_\mathbf{p}(S_i, \delta_\vee)(C) = C'$, where:

- $[C_1, \ldots, C_n] = \delta_\vee(C)$
- $C' = S_i(C_1) \vee \ldots \vee S_i(C_n)$

In spite of its simplicity, the following property is essential because it allows us to manipulate component functions and solvers at the same level, and thus to create solver collaboration with the same strategy language.

Property 2. Consider n solvers $S_1, \ldots, S_n$, n filters $\phi_1, \ldots, \phi_n$, n partial orders $\preceq_1, \ldots, \preceq_n$, a constraint property p, separators $\delta_\wedge$ and $\delta_\vee$. Then, $\mathbf{dc}(S_i, \phi)$, $\mathbf{best}(S_i, \preceq, \phi)$, $\mathbf{pcc}(p, (S_1, \preceq_1, \phi_1), \ldots, (S_n, \preceq_n, \phi_n))$, $\mathbf{bp}((S_1, \preceq_1, \phi_1), \ldots, (S_n, \preceq_n, \phi_n))$, $\wedge\_\mathbf{p}(S_i, \delta_\wedge)$, and $\vee\_\mathbf{p}(S_i, \delta_\vee)$ are solvers.

5   Some Examples of Solver Collaborations

In this section we exemplify the use of our strategy language specifying solvers for constraints over finite domains and real numbers.

---

[4] Here we need the list of filters $[\phi_1, \ldots, \phi_n]$ to be stable and pairwise disjoint.

5.1   Solving Constraints over Finite Domains

A CSP P over finite domains is any conjunction of formulae of the form:

$$\bigwedge_{x_i \in \mathcal{X}} (x_i \in D_{xi}) \wedge C$$

where a domain constraint $x_i \in D_{xi}$ is created for each variable x_i occurring in the constraint C, D_{xi} being a finite set of values.

Solving this kind of problem can be seen as an interleaving process between local consistency verification and enumeration. The most widely used level of consistency verification, *Arc-Consistency*, can be expressed as the repeated application of the following transformation rule that reduces the set of possible values the variables can take.

$$x_i \in D_{x_i} \wedge c \wedge C \implies x_i \in RD(x_i \in D_{x_i}, c) \wedge c \wedge C \quad \textbf{if} \ \ RD(x_i \in D_{x_i}, c) \neq D_{x_i}$$

where $RD(x_i \in D_{x_i}, c) = \{v_i \in D_{x_i} \mid (\exists v_1 \in D_{x_1}, \dots, v_{i-1} \in D_{x_{i-1}}, v_{i+1} \in D_{x_{i+1}}, \dots, v_n \in D_{x_n}) : c(v_1, \dots, v_i, \dots, v_n)\}$.

Then, we define the solver $LocalConsistency$ which applies this rule. In order to carry out enumeration, we consider the solver $SplitDomain$ which transforms a domain constraint into a disjunction of two domain constraints if the width of the original domain is greater than or equal to a "minimal" width ϵ. For finite domains, ϵ is generally set to 1. For all $c = X \in D_X$ from $\mathcal{L}$:

- if $c \in \mathcal{L}_{Dom}$ such that $width(c) \geq \epsilon$, then

$$SplitDomain(c) = X \in D'_X \vee X \in D''_X$$

  where $D_X = D'_X \cup D''_X$ [5],
- otherwise, $SplitDomain(c) = c$.

In order to select domain constraints, we define the filter ϕ_D that returns all domain constraints of the form $X \in D_X$, where D_X specifies the values that the variable X can take.

We also define the filter $\phi_{D \wedge c \wedge Ds}$ that returns sub-constraints which are the conjunction of a domain constraint, an atomic constraint, and a conjunction of domain constraints, i.e. , an atomic constraint with all the domain constraints of the variables occurring in it.

Finally, we define the sorter $\preceq_{Dom}$ that returns the candidate whose domain constraint is the one with the minimum set of values.

Then, the solver $FullLookahead_{MinDom}$, which returns all solutions to a CSP over finite domains, is defined in the following way:

$$\begin{aligned}
FullLookahead_{MinDom} = \ &\mathbf{dc}(LocalConsistency, \phi_{D \wedge c \wedge Ds})^\star; \\
&(\mathbf{best}(SplitDomain, \preceq_{Dom}, \phi_D); \\
&\mathbf{dc}(LocalConsistency, \phi_{D \wedge c \wedge Ds})^\star)^*
\end{aligned}$$

---

[5] We generally also enforce that $D'_X \cap D''_X = \emptyset$.

This heuristic firstly enforces local consistency. Then, it carries out an enumeration step on the variable with the minimum set of remaining values, followed again by local consistency verification. Local consistency verification is always carried out on the whole set of constraints.

Using δ_{Var}, a $\wedge$-**separator** which splits a set of constraints into n variable-disjoint subsets of constraints, the application of $FullLookahead_{MinDom}$ can be improved when solving CSPs that can be decomposed:

$$\wedge\text{\_}\mathbf{p}(FullLookahead_{MinDom}, \delta_{Var})$$

In this way, we are solving several CSPs in parallel. The obvious advantage is to deal with simpler problems. The solution to the original problem will be in the union of the solutions to all subproblems.

5.2   Optimization Problems over Finite Domains

Here, we concentrate on an extension of a CSP called Constraint Satisfaction Optimization Problem (CSOP). CSOP consists in finding an optimal (i.e., maximal or minimal) value for a given function, such that the set of constraints is satisfied [21]. The work of Bockmayr and Kasper [5] seems to be the best currently available reference that explains the approach generally used by the constraint solving community to deal with this problem. In this section, we first explain two approaches for solving CSOPs, and then, we show how they can be combined, all of that using our strategy language.

A CSOP can be described by a tuple $\langle P, f, lb, ub \rangle$ representing a CSP, an optimization function, and the lower and upper bounds of this function. Without loss of generality, we consider the case of minimization of a function f over integers. To deal with this problem, we consider two approaches, both of them requiring an initial step verifying that $Sol(C \wedge f \leq^? ub) \neq \emptyset$, i.e., there exists a solution to the constraint C satisfying the additional constraint $f \leq^? ub$.

The first approach consists in applying the following rule until it cannot be applied any more:

$$\langle P, f, lb, ub \rangle \quad \rightarrow \quad \langle P, f, lb, \alpha(f) \rangle \text{ if } \alpha \in Sol(C \wedge f <^? ub)$$

Each iteration of this rule tries to decrease the upper bound ub by at least one unit until an unsatisfiable problem is obtained. That is why we call this technique *satisfiability to unsatisfiability*. The minimum value of the function f represents the upper bound of the last successful application of this rule. Thus, we define the solver $MinSatToUnsat$ implementing this approach. We do not detail here this definition, but it is obvious that for solving the CSPs, as it is needed by this approach, we could use the already defined solver $FullLookahead_{MinDom}$.

The second approach applies the following rules until they cannot be applied any more:

$$\langle P, f, lb, ub \rangle \rightarrow \langle P, f, lb, \alpha(f) \rangle \text{ if } \alpha \in Sol(C \wedge f <^? \tfrac{(lb+ub)}{2})$$
$$\langle P, f, lb, ub \rangle \rightarrow \langle P, f, \tfrac{(lb+ub)}{2}, ub \rangle \text{ if } lb \neq ub \text{ and } Sol(C \wedge f <^? \tfrac{(lb+ub)}{2}) = \emptyset$$

The first rule tries to find a new value for the upper bound ub and reduces, by at least one-half, the range of possible values of the function f each time a new solution is obtained[6]. The second rule similarly updates the lower bound lb in the opposite situation. We call this approach *binary splitting* and we define the solver $MinSplitting$ implementing it.

Concerning the behavior of these strategies, we can note that the strategy $MinSatToUnsat$ takes a lot of time for reaching the minimal value of f, when it is located too far from the initial upper bound. On the other hand, applying the strategy $MinSplitting$, the same situation happens when the minimal value of f is close to the initial upper bound. Since it is not evident to know where the optimal solution is located, an *a priori* choice between these approaches is not possible in the general case. In order to improve the performances of these two basic solvers, we could make them collaborate in order to profit from the advantages of both of them, and to avoid their drawbacks.

A first scheme of cooperation between the solvers $MinSatToUnsat$ and $MinSplitting$ is expressed by the strategy $SeqOpt$:

$$SeqOpt = (MinSatToUnsat; MinSplitting)^\star$$

Using the strategy $SeqOpt$ both solvers are executed sequentially. Its obvious disadvantage is leaving a solver inactive, while the other one is working. Moreover, due to the exponential complexity of the problem under consideration, the whole process could be blocked if one solver cannot find a solution. To avoid this situation, we can think of running them concurrently, updating the current solution as soon as a new one is available, and stopping the other solver.

$$ParOpt = (\mathbf{pcc}(first, [MinSatToUnsat, None, Id], [MinSplitting, None, Id]))^\star$$

We do not filter the initial set of constraints and so we do not have any sorter. In this case, we are interested in the solver that will be the faster, that is why we use the *first* property[7]. Using this strategy, a solver never waits for a solution coming from the other one. In the extreme case that all solutions are read from the same elementary solver until the final solution is obtained, the performance of this new solver, $ParOpt$, is the same as if one of the elementary solvers runs independently.

5.3   Combining Symbolic Rewriting and Interval Methods

Here, we consider systems of non-linear equations, and two solvers. Gröbner bases computation [6] (i.e., the gb solver) transforms a set of multivariate polynomial equalities into a normal form from which solutions can be derived more

---

[6] Of course, we can think of different ratios. Thus, the first approach can be seen as a particular case of the second one.

[7] Here, since we consider parallel computation, we extend properties of constraints to properties of constraints and computations.

easily than from the initial set. The second solver, int, is a propagation-based numerical solver over the real numbers. We assume that every constraint of the CSPs we consider can be processed by int.

It is generally very efficient to pre-process a CSP with symbolic rewriting techniques before applying a propagation-based solver. In fact, the pre-processing may add redundant constraints (in order to speed-up propagation), simplify constraints, deduce some univariate constraints (whose solutions can easily be extracted by propagation), and reduce the variable dependency problem.

Thus, we consider sc, a simple collaboration where Gröbner bases computation pre-processes equality constraints before the interval solver is applied on the whole CSP:

$$sc = \mathbf{dc}(gb, \phi_=); int$$

where the filter $\phi_=$ selects equalities of polynomials.

Consider, for example, the following problem:

$$x^3 - x * y^2 + 2 = 0 \ \wedge \ x^2 - y^2 + 2 = 0 \ \wedge \ y > 0$$

Most of the solvers based on propagation require splitting to isolate the solutions of this CSP. However, using gb (with a lexicographic order $x \succ y$), the problem becomes

$$y^2 - 3 = 0 \ \wedge \ -1 + x = 0 \ \wedge \ y > 0$$

and int can easily isolate solutions.

However, as stressed in [3], Gröbner bases computation may require too much memory and be very time-consuming compared to the speed-up they introduce. Thus, in [3] the authors propose a trade-off between pruning and computation time: gb is applied on subsets of the initial CSP, and the union of the resulting bases and of the constraints that are not rewritten (such as inequalities, and equalities of non-polynomial expressions) forms the input of the propagation-based solver. We can describe this collaboration as follows:

$$\wedge\_\mathbf{p}(\mathbf{dc}(gb, \phi_=), \delta_{part}); int$$

where δ_{part} is the $\wedge$_**separator** corresponding to the partitioning of the initial system introduced in [3].

5.4   The Solvers of CoSAc

CoSAc [18] is a constraint logic programming system for non-linear polynomial equalities and inequalities. The solving mechanism of **CoSAc** consists of five heterogeneous solvers working in a distributed environment, and cooperating through a client/server architecture:

- chr_lin [11], implemented with CHRs, for solving linear constraints (equalities and inequalities),

- gb [10] for computing Gröbner bases (note that this solver is itself based on a client/server architecture),
- $maple\_uni$ for computing roots of a univariate polynomial equality, i.e., $maple\_uni$ extracts solutions from one equation, not from a set of equations,
- $maple\_exp$ for simplifying and transforming constraints (both this solver and the previous one are Maple [12] programs), and
- ecl for testing closed inequalities using ECL^{i}PS^e [14] features.

Since **CoSAc** uses several solving strategies, these solvers cooperate in three collaborations: S_{inc}, S_{fin} and S'_{fin}. We now focus on how these collaborations could be described in a simple way using our control language. The collaborations are thus clarified: 1) every constraint cannot be treated by all the solvers, and using filters, we can make it clear and formalized; 2) distributed applications are implicit and part of the primitive semantics; 3) it becomes clear where improvements/strategies can be integrated.

S_{inc} is the incremental (in the sense of **CoSAc**) collaboration, i.e., it is applied as soon as a new constraint is added to the store. $maple\_exp$ transforms (e.g., expands polynomials, and simplify arithmetic expressions) all constraints so eq_lin can propagate information and simplify the set of linear equations (equalities and inequalities) filtered by $\phi_{=,<,lin}$:

$$S_{inc} = maple\_exp \; ; \; \mathbf{dc}(eq\_lin, \phi_{=,<,lin})$$

S_{fin} is one of the final solvers of **CoSAc**. It is applied once to the remaining constraints. First, constraints are simplified again by $(maple\_exp)$ since S_{inc} may transform constraints into a syntax gb cannot understand. After computing Gröbner bases of the set of non-linear polynomial equalities (filtered by $\phi_{=}$), variables are eliminated (by $maple\_uni$) one by one from univariate polynomials (filtered by $\phi_{=,uni}$), solutions are propagated, and linearized equations are solved (eq_lin). This process terminates when each variable has been eliminated or when there is no more univariate polynomial:

$$\begin{aligned} S_{fin} = \;& maple\_exp \; ; \\ & \mathbf{dc}(gb, \phi_{=}) \; ; \\ & (\mathbf{dc}(maple\_uni, \phi_{=,uni}); \mathbf{dc}(eq\_lin, \phi_{=,<,lin}))^\star \end{aligned}$$

Here, we can see the flexibility and the simplicity of our control language. In **CoSAc**, the S_{fin} collaboration is fixed. From its description in our language, we can notice that $maple\_uni$ is applied by a *don't care* primitive. Some strategies can easily be introduced to improve the collaboration. In fact, $maple\_uni$ could be applied with a "best" primitive, ordering possible candidates with respect to the increasing degree of univariate polynomial equations (with a $\preceq_{degree}$ sorter). Using $\mathbf{best}(maple\_uni, \preceq_{degree}, \phi_{=,uni})$ variables could be eliminated from lower degree equations first, and thus less arithmetic errors/roundings could be propagated to the system (and that is a weak point of **CoSAc**). Concerning gb and eq_lin, a "best" primitive would not help since these solvers consider the "maximal" set of filtered constraints.

S'_{fin} is an alternative to S_{fin} which is more efficient when eliminations of non-linear variables do not linearize any other constraint and only ground inequalities have to be checked by ecl. We can write it as:

$$S'_{fin} = maple\_exp \ ;$$
$$\mathbf{dc}(gb, \phi_=) \ ;$$
$$(\mathbf{dc}(maple\_uni, \phi_{=,uni}))^\star \ ;$$
$$(\mathbf{dc}(ecl, \phi_{<,ground}))^\star$$

Again, strategies can be introduced since ground inequalities can be checked simultaneously. Using δ_{one}, a $\wedge$-**separator** that splits a set of n constraints into n singletons of atomic constraints, the application of ecl is improved:

$$\wedge\_\mathbf{p}(\mathbf{dc}(ecl, \phi_{<,ground}), \delta_{one})$$

We remark that we still need a filter for ecl since δ_{one} does not perform any filtering.

As mentioned in [17], the first solvers of S_{fin} and S'_{fin} can be "factorized":

$$S''_{fin} = maple\_exp \ ;$$
$$\mathbf{dc}(gb, \phi_=) \ ;$$
$$\mathbf{pcc}(first, \ [(\mathbf{dc}(maple\_uni, \phi_{=,uni}); \mathbf{dc}(eq\_lin, \phi_{=,<,lin}))^\star \ , None, Id],$$
$$[(\mathbf{dc}(maple\_uni, \phi_{=,uni}))^\star \ ; (\mathbf{dc}(ecl, \phi_{<,ground}))^\star, None, Id])$$

The remaining parts of the collaborations are executed concurrently. No filtering is needed (Id for both sub-collaborations), and thus we do not have any sorter ($None$) since there is only one candidate after filtering, i.e., the initial set of constraints. We do not impose any property on the result, and we are interested in the sub-collaboration that will be the faster ($first$ property). Note that improvements for applying ecl and $maple\_uni$ still hold in S''_{fin}.

5.5   Combining Consistencies

Box consistency [2] is a local consistency notion for interval constraints that relies on bounds of domains of variables: it is generally implemented as a (local) splitting of domains combined with the interval Newton method for determining consistent bounds of intervals. Hull consistency is another notion of consistency, stronger than box consistency. However, it can only be applied on primitive constraints that are either part of the original CSP, or are obtained by decomposing the constraints of the CSP. Then, the reduction of the "decomposed" CSP is weaker, but also faster. The idea of [2] is to combine these two consistencies in order to reduce the computation time for enforcing box consistency.

Let us consider $Hull$ and Box, two solvers that respectively enforce hull and box consistency of a CSP. Then, the combination of [2] can be described by:

$$(Hull \ ; \ Box)^\star$$

Since we can define both solvers and collaboration in our language, we now specify the *Hull* and *Box* solvers:

$$Box = (\mathbf{dc}(box, \phi_{\neg p}))^{\star} \quad \text{and} \quad Hull = (\mathbf{dc}(hull, \phi_{p}))^{\star}$$

where ϕ_p (respectively $\phi_{\neg p}$) filters one primitive (respectively non-primitive) constraint together with the domain constraints (e.g., $x \in [a, b]$) associated with each of its variables, *box* (respectively *hull*) is a component solver that given a constraint c enforces box (respectively hull) consistency of c w.r.t. each of its variables.

We can also consider some inner strategies, such as reducing the variable with the largest domain. Then, *Hull* and *Box* are defined as follows:

$$Box = (\mathbf{best}(box, \gg, \phi_{\neg p}))^{\star} \quad \text{and} \quad Hull = (\mathbf{best}(hull, \gg, \phi_{p}))^{\star}$$

where "$\gg$" selects the candidate with the largest domain.

Note that we could once again decompose these solvers into solvers that enforce box (or hull) consistency of one constraint with respect to one variable. Note also that $(Hull \; ; \; Box)^{\star}$ can represent the solver *int* considered in Section 5.3. We could also think about some other description of *Hull* and *Box* (e.g., using parallel application of solvers), but then we would not respect anymore the original combination of [2].

6   Conclusions

We have presented a strategy language for solving CSPs via collaboration of solvers. A key point in this work is the introduction of basic strategy operators that allow the design of solvers by combining basic functions as well as the collaboration of solvers by combining component solvers. We have exemplified the use of this language by the simulation of well-known techniques for solving CSPs over finite domains and non-linear constraints over real domains. To show the broad scope of our control language's potential applications, we have designed several solvers that are considered of different nature (such as propagation based solvers, optimization over finite domain, and Gröbner bases computation). We are currently working on the implementation of this language in order to evaluate the real applicability of this framework. From a more theoretical point of view, we are considering as further work the verification of the termination properties of the strategy operators.

7   Acknowledgments

We are grateful to the anonymous referees who pointed out very accurate issues that allowed us to improve our work and the quality of this paper.

References

[1] F. Arbab and E. Monfroy. Heterogeneous distributed cooperative constraint solving using coordination. *ACM Applied Computing Review*, 6:4–17, 1999.

[2] F. Benhamou, F. Goualard, L. Granvilliers, and J. Puget. Revising Hull and Box Consistency. In *Proc. of International Conference on Logic Programming*, pages 230–244, Las Cruces, USA, 1999. The MIT Press.

[3] F. Benhamou and L. Granvilliers. Combining Local Consistency, Symbolic Rewriting, and Interval Methods. In *Proc. of AISMC3*, volume 1138 of *LNCS*, pages 144–159, Steyr, Austria, 1996. Springer-Verlag.

[4] H. Beringer and B. DeBacker. Combinatorial Problem Solving in Constraint Logic Programming with Cooperative Solvers. In C. Beierle and L. Plümer, editors, *Logic Programming: Formal Methods and Practical Applications*, Studies in Computer Science and Artificial Intelligence. North Holland, 1995.

[5] A. Bockmayr and T. Kasper. A unifying framework for integer and finite domain constraint programming. Research Report MPI-I-97-2-008, Max Planck Institut für Informatik, Saarbrücken, Germany, Aug. 1997.

[6] B. Buchberger. Gröbner Bases: an Algorithmic Method in Polynomial Ideal Theory. In N. K. Bose, editor, *Multidimensional Systems Theory*, pages 184–232. D. Reidel Publishing Company, Dordrecht - Boston - Lancaster, 1985.

[7] C. Castro. Building Constraint Satisfaction Problem Solvers Using Rewrite Rules and Strategies. *Fundamenta Informaticae*, 34(3):263–293, June 1998.

[8] C. Castro. COLETTE, Prototyping CSP Solvers Using a Rule-Based Language. In J. Calmet and J. Plaza, editors, *Proc. of The Fourth International Conference on Artificial Intelligence and Symbolic Computation, AISC'98*, volume 1476 of *LNCS*, pages 107–119, Plattsburgh, NY, USA, Sept. 1998. Springer-Verlag.

[9] C. Castro and E. Monfroy. A Control Language for Designing Constraint Solvers. In *Proc. of Third International Conference Perspective of System Informatics, PSI'99*, volume 1755 of *LNCS*, pages 402–415, Novosibirsk, Russia, 2000. Springer-Verlag.

[10] J.-C. Faugere. *Résolution des systèmes d'équations algébriques*. PhD thesis, Université Paris 6, France, 1994.

[11] T. Frühwirth. Constraint handling rules. In A. Podelski, editor, *Constraint Programming: Basics and Trends*, volume 910 of *LNCS*. Springer-Verlag, 1995.

[12] K. Geddes, G. Gonnet, and B. Leong. *Maple V: Language reference manual*. Springer Verlag, New York, Berlin, Paris, 1991.

[13] P. Marti and M. Rueher. A Distribuited Cooperating Constraints Solving System. *International Journal of Artificial Intelligence Tools*, 4(1-2):93–113, 1995.

[14] M. Meier and J. Schimpf. ECLiPSe User Manual. Technical Report ECRC-93-6, ECRC (European Computer-industry Research Centre), Munich, Germany, 1993.

[15] E. Monfroy. *Collaboration de solveurs pour la programmation logique à contraintes*. Phd thesis, Université Henri Poincaré - Nancy 1, France, Nov. 1996. Also available in english. On-line at: http://www.cwi.nl/~eric/Private/Publications/index.html.

[16] E. Monfroy. An environment for designing/executing constraint solver collaborations. *ENTCS (Electronic Notes in Theoretical Computer Science)*, 16(1), 1998.

[17] E. Monfroy. The Constraint Solver Collaboration Language of BALI. In D. Gabbay and M. de Rijke, editors, *Frontiers of Combining Systems 2*, volume 7 of *Studies in Logic and Computation*, pages 211–230. Research Studies Press/Wiley, 2000.

[18] E. Monfroy, M. Rusinowitch, and R. Schott. Implementing Non-Linear Constraints with Cooperative Solvers. In K. M. George, J. H. Carroll, D. Oppenheim, and J. Hightower, editors, *Proc. of ACM Symposium on Applied Computing (SAC'96), Philadelphia, PA, USA*, pages 63–72. ACM Press, February 1996.

[19] G. Nelson and D. C. Oppen. Simplifications by Cooperating Decision Procedures. *ACM Trans. on Programming Languages and Systems*, 1(2):245–257, Oct. 1979.

[20] C. Ringeissen. Cooperation of decision procedures for the satisfiability problem. In F. Baader and K. Schulz, editors, *Proc. of First Int. Workshop Frontiers of Combining Systems, FroCoS'96*, pages 121–139. Kluwer Academic Publishers, 1996.

[21] E. Tsang. *Foundations of Constraint Satisfaction*. Academic Press, 1993.

Automatic Determination of Geometric Loci. 3D-Extension of Simson-Steiner Theorem[*]

Eugenio Roanes-Macías and Eugenio Roanes-Lozano

Universidad Complutense de Madrid, Dept. Algebra
c/ Rector Royo Villanova s/n, Edificio "La Almudena", 28040-Madrid, Spain
{roanes,eroanes}@eucmos.sim.ucm.es

Abstract. A method for determining loci without using a deep algebraic background is presented. It uses pseudodivision techniques (Wu's algorithm). The key idea is to make the hypothesis conditions depend on an indeterminate point, X. When forcing the thesis condition to be a consequence of hypothesis conditions, a new condition involving X appears. That condition leads to the locus. The method is applied to prove a new theorem: the generalization of Simson-Steiner Theorem to 3D.

Keywords. *Automatic theorem proving. Geometric loci. Pseudodivisions.*

1 Introduction

Applying Wu's techniques to discover geometric theorems was already suggested in [2,10]. The key idea is to add new additional hypotheses to a set of original hypotheses (from which the thesis can't be deduced) in order for the thesis to become a consequence of the extended set of hypotheses. Different authors have treated the problem from different points of view.

Kapur and Mundy [5] apply this to perspective viewing. In the first step they use the Ritt-Wu characteristic method to obtain the characteristic set of the hypotheses' ideal. The conclusion is pseudo-divided by a polynomial ideal. If the pseudo-remainder is 0, the thesis is a consequence of the hypotheses. Otherwise it is factored and some of its factors are possible candidates to become additional hypotheses. This approach is summarised, together with other different methods, in the excellent paper [4].

T. Recio and M. P. Vélez developed [7] a method for automatic discovery of theorems based on Gröbner-base computation. It makes use of Hilbert's Nullstellensatz and clearly details the mathematics in the background (ideal/variety duality). It consists of finding complementary hypotheses until a statement that becomes true is obtained.

A specific use related to triangles and measure also based on Gröbner-base computation can be found in [6].

Reading these ingenious articles suggested to us the idea to try to specifically determine geometric loci automatically. In our first article in this line [9] we

---

[*] Partially supported by project DGES PB96-0098-C04-03 (Spain).

J.A. Campbell and E. Roanes-Lozano (Eds.): AISC 2000, LNAI 1930, pp. 157–173, 2001.
© Springer-Verlag Berlin Heidelberg 2001

reproved a theorem about geometric loci recently discovered [3]. This method is generalized here.

Wu's algorithm is used for operating on polynomials, but without explicitly using the ideal/variety duality. Therefore the method can be understood and justified without a deep algebraic background (in fact only pseudodivisions and linear algebraic combinations of polynomials are used). Moreover, as Hilbert's Nullstellensatz is not required at any step, the base fields do not need to be algebraically closed. Let us observe that the method uses sufficient conditions at different steps, so if it is not able to provide any possible additional hypotheses, that doesn't mean that the result can't be reached with the help of other methods.

This paper begins describing an adaptation of Wu's method for mechanical geometry theorem proving (not Wu's complete method -in Chou's terminology [2]) to our goal. The process for determining loci is described in detail afterwards. Finally, it is used to determine a locus, which can be considered a generalization to 3D of Simson-Steiner's theorem for 2D reproved in [9].

2   Basic Algebraic Tools. Adaptation of Wu's Algorithm

In this section, we briefly summarize concepts relative to the basic algebraic ideas used in Wu's algorithm and adapt them to introduce the algebraic tools used in this paper to determine geometric loci automatically. Introductory books for those concepts are [1,2,10].

Let $K[v, w, ..., z]$ be a polynomial ring in the indeterminates $v, w, ..., z$ over the field K of characteristic 0. In the usual polynomial division of polynomials belonging to this ring, the quotient and the remainder obtained are rational expressions, which can be non-integer expressions (i.e. variables can appear in denominators). To avoid this inconvenience, division can be substituted by pseudodivision.

Given the polynomials $f, g \in K[v, w, ..., z]$, the *pseudodivision* of f by g with respect to the variable v consists of the usual polynomial division, after substituting f by its product by the *multiplier*

$$m = lcoeff(g, v)^{1 + deg(f,v) - deg(g,v)}$$

where $lcoeff(g, v)$ and $deg(g, v)$ are the leading coefficient and the degree of g with respect to the variable v, respectively. The quotient and remainder obtained this way are called *pseudoquotient* and *pseudoremainder*. This pseudoremainder will be briefly denoted $prem(f, g, v)$ and the corresponding multiplier $mulf(f, g, v)$. It can be proved that the pseudoremainder and pseudoquotient are integer expressions (variables do not appear in their denominators). As in usual polynomial division, the pseudoremainder (r) and the pseudoquotient (q) verify $m \cdot f = g \cdot q + r$ and hence r is in the ideal $< f, g >$ of the polynomial ring

$$r \in< f, g > K[v, w, ..., z] \qquad (1)$$

Besides, $deg(r, v) < deg(g, v)$. So, if $deg(g, v) = 1$, then v does not appear in r.

The pseudodivision operation can be applied to reduce a finite family of multivariate polynomials to *triangular form*. Starting with a list of s indeterminates or variables $[v_1, v_2, ..., v_s]$ over the field K, and a list of s polynomials, $[h_1, h_2, ..., h_s]$, belonging to the ring $K[v_1, v_2, ..., v_s]$, a process similar to Gaussian-elimination, but substituting linear operations by pseudodivisions, can be applied until a triangular system of polynomials

$$g_1 = g_1(v_1, v_2, v_3, ..., v_s), g_2 = g_2(v_2, v_3, ..., v_s), g_3 = g_3(v_3, ..., v_s), ..., g_s = g_s(v_s)$$

is obtained, by applying a constructive algorithm (described for example in [1] and [2]), which will be denoted hereafter by *trian_*,

$$trian\_([h_1, h_2, ..., h_s], [v_1, v_2, ..., v_s]) = [g_1, g_2, ..., g_s] \qquad (2)$$

These polynomials g_i satisfy two conditions that will be essential later: for each $i = 1, ..., s$, $deg(g_i, v_i) > 0$ and, as a consequence of (1),

$$g_1, g_2, ..., g_s \in < h_1, h_2, ..., h_s > K[v_1, v_2, ..., v_s] \qquad (3)$$

Now, starting from the lists $[v_1, ..., v_s]$ and $[g_1, ..., g_s]$ mentioned above and a new polynomial, $t_h \in K[v_1, ..., v_s]$, the following sequence of pseudoremainders can be considered

$$r_1 = prem(t_h, g_1, v_1), \ r_2 = prem(r_1, g_2, v_2), ..., r_s = prem(r_{s-1}, g_s, v_s) \qquad (4)$$

The last pseudoremainder obtained this way, r_s, will be called *final pseudoremainder* and the process to compute it will be denoted *fin_prem*

$$fin\_prem(t_h, [g_1, g_2, ..., g_s], [v_1, v_2, ..., v_s]) = r_s$$

Now, as a consequence of (2), r_s can be directly obtained from $[h_1, h_2, ..., h_s]$ and the process to compute it directly will be denoted *final_prem*

$$r\_s = final\_prem(t_h, [h_1, ..., h_s], [v_1, ..., v_s]) =$$
$$= fin\_prem(t_h, trian\_([h_1, ..., h_s], [v_1, ..., v_s]), [v_1, ..., v_s])$$

The sequence of multipliers used in the pseudodivisions (4)

$$m_1 = mulf(t_h, g_1, v_1), m_2 = mulf(r_1, g_2, v_2), ..., m_s = mulf(r_{s-1}, g_s, v_s)]$$

can be joined in a list and the process to compute it will be denoted *m_list*

$$m\_list(t_h, [g_1, g_2, ..., g_s], [v_1, v_2, ..., v_s]) = [m_1, m_2, ..., m_s]$$

In the same way, as a consequence of (2), this list can be directly obtained from $[h_1, h_2, ..., h_s]$ and the process to compute it directly will be denoted *mulf_list*

$$[m_1, m_2, ..., m_s] = mulf\_list(t_h, [h_1, ..., h_s], [v_1, ..., v_s]) =$$
$$= m\_list(t_h, trian\_([h_1, ..., h_s], [v_1, ..., v_s]), [v_1, ..., v_s])$$

In the classic Wu's method, if $r_s = 0$, then the relation $t_h = 0$ (considered as the thesis condition) follows from the set of relations $h_1 = 0, ..., h_s = 0$ (considered as hypothesis conditions) and from $mulf(r_{i-1}, g_i, v_i) \neq 0$ the non-degenerate conditions are obtained.

In the method of determination of geometric loci that will be presented hereafter the same algebraic tools used in Wu's method will be used, but they are used in a different way in order to reach a different target. The following lemma will be essential in this process.

Lemma 1. *In accordance with the notation mentioned above, the final pseudoremainder can be expressed in the form:*

$$r_s = m_s \cdot m_{s-1} \cdot ... \cdot m_1 \cdot t_h + \sum_{i=1}^{s} \gamma_i h_i \quad ; \quad \gamma_i \in K[v_1, v_2, ..., v_s] \tag{5}$$

Proof. The s pseudoremainders (4) can be written in the form

$$r_1 = m_1 t_h - g_1 q_1, \ r_2 = m_2 r_1 - g_2 q_2, \ ..., \ r_s = m_s r_{s-1} - g_s q_s$$

where $q_1, q_2, ..., q_s$ and $m_1, m_2, ..., m_s$ are the pseudoquotients and the multipliers, respectively. Substituting the value of each one of these pseudoremainders in the following equality, r_s can be expressed in the form

$$r_s = m_s \cdot m_{s-1} \cdot ... \cdot m_1 \cdot t_h + \sum_{j=1}^{s} \beta_j g_j \ ; \ \beta_j \in K[v_1, v_2, ..., v_n]$$

As, in accordance with (3), the g_j can be expressed in the form

$$g_j = \sum_{i=1}^{s} \delta_{ji} h_i \ ; \ \delta_{ji} \in K[v_1, v_2, ..., v_n]$$

consequently

$$r_s = m_s \cdot m_{s-1} \cdot ... \cdot m_1 \cdot t_h + \sum_{j=1}^{s} \beta_j \left(\sum_{i=1}^{s} \delta_{ji} h_i \right)$$

Now, the lemma equality follows immediately by denoting $\sum_{j=1}^{s} \beta_j \delta_{ji} = \gamma_i$.

Overview of the algorithms used: In accordance with the preceding explanations, the polynomials and list of polynomials involved in the calculations are:

$V = [v_1, v_2, ..., v_s]$ (list of variables or indeterminates over K)

$H = [h_1, h_2, ..., h_s]$ (list of polynomials in $K[v_1, v_2, ..., v_s]$)

$G = [g_1, g_2, ..., s_s]$ (list of triangulated polynomials)

t_h (polynomial in $K[v_1, v_2, ..., v_s]$)

r_s (final pseudoremainder)

$[m_1, m_2, ..., m_s]$ (list of multipliers)

and the algorithms used are the following:

$trian_-(H, V) = G$

$fin\_prem(t_h, G, V) = r_s$

$final\_prem(t_h, H, V) = fin\_prem(t_h, trian_-(H, V), V) = r_s$

$m\_list(t_h, G, V) = [m_1, m_2, ..., m_s]$

$mulf\_list(t_h, H, V) = m\_list(t_h, trian_-(H, V), V) = [m_1, m_2, ..., m_s]$

As the calculations mentioned here are very laborious, they must be automated. This can be implemented on a CAS containing a command that calculates pseudodivisions. We have developed an implementation of these algorithms in Maple, using techniques that are described in [8]. The code is omitted for the sake of brevity, but anyone interested in obtaining it is welcome (it occupies about 3K in its readable form).

3   A Previous Example to Illustrate the Method

In order to illustrate the method, we shall begin by showing the ideas over an easy well known example of locus, the Simson-Wallace Theorem generalized by Jakob Steiner. We shall state it as a problem and reprove this theorem using our own method.

Problem: *Let X be a point in the plane of triangle ABC and let M, N, P be the orthogonal projections of X in the side-lines AB, BC, CA, respectively. Let us move X in the plane of ABC in such a way that the area of triangle MNP is kept unchanged (as a constant, a). What is the locus of points X?* (Fig. 1).

In the preceding problem one can distinguish three types of points. Points A, B, C are freely chosen in the plane (except for exceptional positions in which they are collinear) and they are consequently called *free points*. The indeterminate point X, which gives the geometric locus, is called *locus point*. Finally, points M, N, P, determined from locus and free points by geometric conditions, are called *linked points* or *dependent points*.

In order to clarify the description of the method, it is convenient to distinguish several steps in its execution.

STEP 1. Select the coordinates

For the sake of simplicity of calculations, it is convenient to select a coordinate system such that most of the free points have coordinates as simple as possible.

Free points: $A(0, 0), B(b, 0), C(c, e)$.

Locus point: $X(x, y)$.

Linked or dependent points: $M(m, 0), N(n, u), P(p, q)$

STEP 2. Convert hypothesis and thesis conditions into polynomial equations

Conditions that determine linked points (starting from free points and locus point) are called *hypothesis conditions* and conditions that determine the locus point are called the *thesis conditions*.

Hypothesis conditions:

$$m - x = 0 \qquad\qquad (XM \perp AB)$$
$$(n - b) \cdot e - u \cdot (c - b) = 0 \qquad\qquad (N \in BC)$$
$$(n - x) \cdot (c - b) + e \cdot (u - y) = 0 \qquad (XN \perp BC)$$
$$(p - x) \cdot c + e \cdot (q - y) = 0 \qquad\qquad (XP \perp CA)$$
$$p \cdot e - q \cdot c = 0 \qquad\qquad (XN \perp BC)$$

Thesis condition:

$$\det \begin{pmatrix} f & d & 1 \\ r & s & 1 \\ t & z & 1 \end{pmatrix} = 2a \qquad\qquad (\text{area}(FTR) = a)$$

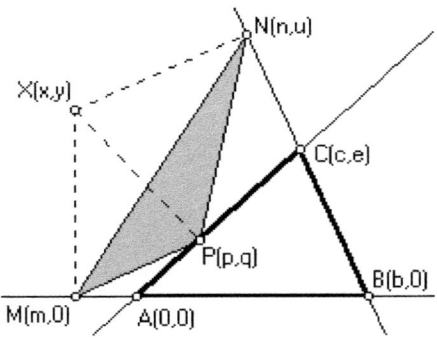

Fig. 1. The locus of point X

STEP 3. Establish parameters and coordinates

  Free coordinates: b, c, e (coordinates of free points)
  Locus coordinates: x, y (coordinates of locus point)
  Linked or dependent coordinates: m, n, u, p, q (coordinates of linked points)
  Parameters: b, c, e, a (freely chosen variables, including free coordinates)
  Parameter conditions: $b \neq 0 \neq e$ (A, B, C are non-collinear points)

STEP 4. Input the hypotheses and thesis polynomials

Starting from the left hand side of hypotheses and thesis equalities of STEP 2 by substituting dependent coordinates by independent variables over $\mathbb{R}$, we obtain the *hypothesis polynomials* (denoted $h1, h2, h3, h4, h5$) and the *thesis polynomial* (denoted th). Let us write the Maple code:

```
> h1 := m - x:
> h2 := (n - b)*e - u*(c - b):
> h3 := (n - x)*(c - b) + e*(u - y):
> h4 := p*e - q*c:
> h5 := (p - x)*c + e*(q - y):
> th: = det(matrix([[f,g,1],[r,s,1],[t,z,1]])) - 2*a:
```

STEP 5. Compute the final pseudoremainder and factorize it

Starting from the lists of ordered variables and hypothesis polynomials, which are denoted V and H, respectively, the final pseudoremainder ($r5$) can be computed:

```
> V := [q,p,u,n,m]:
> H: = [h5,h4,h3,h2,h1]:
> r5:=final_prem(th,H,V):
```

After factoring, collecting a and sorting it, r5 can be written:

$$r5 = 2 \cdot e^2 \cdot (2 \cdot b \cdot e^3 \cdot x^2 - 2 \cdot b^2 \cdot e^3 \cdot x + 2 \cdot b \cdot e^3 \cdot y^2 + 2 \cdot b \cdot e^2 \cdot y \cdot (b \cdot c - c^2 - e^2) \\ -a \cdot (b^2 - 2 \cdot b \cdot c + c^2 + e^2) \cdot (c^2 + e^2))$$

Let us denote by *rho* the product of factors not depending on locus coordinates (x, y) and let us denote by *phi* the only factor depending on them:

```
> rho:=2*e^2:
> phi:=simplify(r5/rho);
```

$$\phi = 2 \cdot b \cdot e^3 \cdot x^2 - 2 \cdot b^2 \cdot e^3 \cdot x + 2 \cdot b \cdot e^3 \cdot y^2 + 2 \cdot b \cdot e^2 \cdot y \cdot (b \cdot c - c^2 - e^2) \\ -a \cdot (b^2 - 2 \cdot b \cdot c + c^2 + e^2) \cdot (c^2 + e^2)$$

In accordance with the lemma, r_5 can be expressed in the form:

$$r_5 = m_5 m_4 m_3 m_2 m_1 t_h + w_1 h_1 + w_2 h_2 + \ldots\ldots + w_5 h_5$$

where $w_1, w_2, \ldots\ldots, w_5$ are polynomials in the variables. Now, substituting variables by dependent coordinates, we have $h_1 = 0 \wedge \ldots \wedge h_5 = 0$. Hence, for the thesis condition, $th = 0$, to be verified, r_5 must be zero. As $2 \cdot e^2 \neq 0$ (under parameter condition $e \neq 0$), for the thesis condition be verified, we must have $\phi = 0$. This is the equation (with respect to x, y) of a sheaf of concentric circles whose radio depends on area a.

We have seen that points X such that $area(FRT) = a$ verify $\phi = 0$. But the reciprocal question arises: does every point in $\phi = 0$ verify the thesis condition? To answer it, $\phi = 0$ must be input as a new hypothesis condition.

STEP 6. Input the new hypothesis polynomial

Consequently, the left hand side of $\phi = 0$ must be input as new hypothesis polynomial, which will be denoted h_6:

```
> h6 := phi:
```

STEP 7. Establish new parameters and coordinates

As point $X(x, y)$ must verify $\phi = 0$, one of its coordinates (abscissa x, for example) can be freely chosen, and so it will be a new parameter, and the other, a new dependent coordinate.

New parameters: b, c, e, a, x.

New variables: m, n, u, p, q, y.

STEP 8. Compute the new final pseudoremainder

Starting from the new lists of variables and hypothesis polynomials, which are denoted VV and HH, respectively, the final pseudoremainder ($r6$) can be computed by applying the operator *final_prem* of section 2.

```
> VV := [q,p,u,n,m,y]:
> HH: = [h5,h4,h3,h2,h1,h6]:
> r6:=final_prem(th,HH,VV):
```

$$r6 = 0$$

STEP 9. Compute the value of the multipliers

The list of multipliers, denoted by M, can be computed by applying the operator *mulf_list* of section 2.

```
> M := mulf_list(th,HH,VV);
```

$$M = [e, e^2 + c^2, e, e^2 + c^2 - 2cb + b^2, 1, e^3 b]$$

Therefore, each multiplier is nonzero, under parameter conditions ($b \neq 0 \neq e$).

In accordance with the lemma, r_6 can be expressed in the form:

$$r_6 = m_6 m_5 \ldots\ldots m_2 m_1 t_h + \gamma_1 h_1 + \gamma_2 h_2 + \ldots\ldots + \gamma_6 h_6$$

where $\gamma_1, \gamma_2, \ldots\ldots, \gamma_6$ are polynomials in the variables. As $r_6 = 0$, substituting the variables by the dependent coordinates ($h_1 = 0 \wedge \ldots \wedge h_6 = 0$), we have

$$0 = m_6 m_5 \ldots\ldots m_2 m_1 t_h$$

Finally, as multipliers are all non-zero, the thesis condition $t_h = 0$ follows.

Summary: *the locus of X such that the area of the triangle MNP is the constant a, is the circle of equation $\phi = 0$. It can be easily verified that this circle is centered in the circumcenter of triangle ABC. In particular, for $a = 0$ (i.e., M, N, P being collinear) the circle passes through point A and therefore it is the circumcircle of ABC (Simson-Steiner Theorem).*

4   Loci: General Method of Determination

The concepts introduced through the preceding example will be generalized here. In order to state the ideas precisely, we shall begin by defining the concept of geometric loci.

Definition 1. *Let us consider the euclidean space K^n, over a field K, not necessarily algebraically closed, containing a base field, k, of characteristic 0. Let us consider, in this euclidean space:*

- *a subset of points, called* free points, *freely chosen, except for exceptional situations (such as non-collinear A,B,C in the preceding section)*

- *an indeterminate point, X, called* locus point

- *a subset of points, called* linked points *or* dependent points, *determined from the locus point and some free points by geometric conditions, called* hypothesis conditions

- *another condition, called the* thesis *condition, involving some of the linked, locus and free points and any parameter c_0 (a in the example of section 3)*

Then, the subset of points X in K^n that satisfy the thesis condition, under hypotheses conditions and parameters conditions, is called the geometric locus *or, briefly, the* locus.

Note 1. It will be supposed that all the conditions mentioned can be expressed by polynomial equations (otherwise the method described hereafter is not operative).

A computer algebraic method to determine geometric loci, based on pseudo-divisions, is described hereafter. The key idea is to make the hypothesis conditions depend on an indeterminate point, X (locus point). When trying to obtain the thesis condition from the hypothesis conditions, a new condition involving X appears. And this condition leads to the determination of the locus. In order to clarify the description of the method, it is convenient to distinguish several steps in its execution.

STEP 1. Select the coordinate system

Coordinates of the free points (concatenated coordinates of all the considered free points) will be denoted $a_1, a_2, ..., a_r$. Coordinates of the locus point, X, will be $(x_1, ..., x_n)$. Coordinates of all the linked points (concatenated coordinates of all the considered linked points) will be denoted $d_1, d_2, ..., d_s$. For the sake of simplicity of calculations, it is convenient to select a coordinate system such that most of the free points have coordinates as simple as possible.

STEP 2. Convert hypothesis and theses conditions into polynomial equations

Geometric conditions that determine linked points (starting from free points and locus point), which we have called hypothesis conditions, are to be expressed by polynomial equations among coordinates:

$$h_i(a_1, a_2, ..., a_r; x_1, ..., x_n; d_1, d_2, ..., d_s) = 0 \quad ; \quad i = 1, 2, ..., s \tag{6}$$

and the same has to be done with the geometric condition to be satisfied, which we have called the thesis condition:

$$t_h(c_0, a_1, a_2, ..., a_r; x_1, ..., x_n; d_1, d_2, ..., d_s) = 0 \tag{7}$$

STEP 3. Establish parameters and coordinates

Coordinates of free points $(a_1, a_2, ..., a_r)$ are called *free coordinates*. Coordinates of the locus points $(x_1, ..., x_n)$ are called *locus coordinates*. Coordinates of linked or dependent points $(d_1, ..., x_d)$ are called *linked coordinates* or *dependent coordinates*. Variables that can be freely chosen (including free coordinates and c_0) are called *parameters*. Conditions of free coordinates that consider the exceptional positions of free points that must be excluded are called *parameter conditions*. They are usually inequality conditions of the form $a_i \neq 0$ or $a_i < a_j$.

STEP 4. Input hypothesis and thesis polynomials

From the left hand side of each equations in (6) and (7), we define a polynomial by substituting linked coordinates, $d_1, ..., d_s$, for independent variables over K, denoted $v_1, ..., v_s$, respectively. These polynomials will be denoted, respectively

$$h_i(a_1, ..., a_r; x_1, ..., x_n; v_1, v_2, ..., v_s) \quad ; \quad i = 1, 2, ..., s \tag{8}$$

$$t_h(c_0, a_1, ..., a_r; x_1, ..., x_n; v_1, v_2, ..., v_s) \tag{9}$$

where $c_0, a_1, a_2, ..., a_r \in K$ are considered as parameters and $v_1, v_2, ..., v_n$ as independent indeterminates over the field K. Polynomials (8) and (9) will be called *hypothesis polynomials* and *thesis polynomial*, respectively.

As a consequence, substituting every variable in (8) and (9) by its corresponding linked coordinate, expressions (6) and (7) are obtained, respectively.

For the sake of simplicity of calculations, it is convenient to express (8) and (9) as polynomials with integer coefficients (in $\mathbb{Z}[c_0, a_1, ..., a_r, x_1, ..., x_n][v_1, ..., v_s]$).

STEP 5. Compute the final pseudoremainder and factorize it

Starting from the lists of ordered variables and hypothesis polynomials, which are denoted V and H, respectively, the final pseudoremainder (r_s) can be computed by applying the operator *final_prem* of section 2. (An appropriate selection of the order of variables and hypothesis polynomials can shorten its calculation).

After factoring this final pseudoremainder polynomial, r_s, it can be written

$$r_s = \rho \prod_j \phi_j(c_0, a_1, ..., a_r, x_1, ..., x_n, v_1, ..., v_s) \tag{10}$$

where ρ is the product of the factors not containing locus coordinates and the ϕ_j are the factors containing them.

Therefore, the ϕ_j are the candidates to define equations of component of the geometric locus and we shall refer to them as *locus polynomial factors*.

STEP 6. Add a new hypothesis polynomial

From now on, it will be supposed that point X is in the algebraic variety of K^n defined by ϕ_j. Hence, the new condition $\phi_j(a_1, ..., a_r, x_1, ..., x_n, d_1, ..., d_s) = 0$ is to be added to the list of the s hypotheses conditions considered in STEP 2. As a consequence, the locus polynomial factor $\phi_j(a_1, ..., a_r, x_1, ..., x_n, v_1, ..., v_s)$ is to be added to the list of the s hypotheses polynomials considered in STEP 3, to form the new list of $s + 1$ hypotheses polynomials $[h_1, h_2, ..., h_s, \phi_j]$.

Consequently, one of the loci coordinates $x_1, ..., x_n$ can no longer be freely chosen and therefore it must be considered as a new linked coordinate. For instance, if $degree(\phi_j, x_n) = 1$, then x_n can be selected as new linked coordinate to be added to the old ones (such degree is to be chosen positive, but as small as possible). Therefore, the variable x_n is to be added to the list of s variables considered in STEP 3, to form the new list of $s + 1$ variables $[v_1, v_2, ..., v_s, x_n]$.

STEP 8. Compute the new final pseudoremainder

Starting from the new variables list, $VV = [v_1, ..., v_s, x_n]$, the new hypothesis polynomial list, $HH = [h_1, h_2, ..., h_s, \phi_j]$ and the thesis polynomial t_h, the new final pseudoremainder can be computed by applying the operator *final_prem*:

$$r'_{s+1} = final\_prem(t_h, HH, VV)$$

In order for the process to succeed, r'_{s+1} must be zero.

STEP 9. Compute the value of the multipliers

Finally, the list of multipliers, denoted by M, can be computed by applying operator *mulf_list* of section 2.

$$mulf\_list(t_h, HH, VV) = [m'_1, ..., m'_s, m'_{s+1}]$$

In order the process to succeed, it is expected that all these multipliers must be nonzero (after substituting variables, $v_1, ..., v_s$, by linked coordinates, $d_1, ..., d_s$, and under parameters conditions considered in STEP 3), as is established by the following theorem.

Note 2. Steps 6 to 9 must be repeated for each one of the locus polynomial factors obtained in STEP 5.

Theorem 1. *In accordance with the notation above, if $\rho \neq 0$ (under parameter conditions) and for $j = 1, 2, ...$, the locus polynomial factor ϕ_j satisfies the two following conditions:*

  1) $r'_{s+1} = 0$ (the final pseudoremainder is zero)

  2) $m'_1 \neq 0, ..., m'_{s+1} \neq 0$ (under parameter conditions) *then the geometric locus that satisfies the thesis condition (7), under hypothesis conditions (6) and parameters conditions, is the union of the subvarieties in K^n of equations*

$$\phi_j(c_0, a_1, ..., a_r, x_1, ..., x_n, d_1, ..., d_s) = 0 \; ; \; j = 1, ..., z$$

Proof. Let us suppose that $(x_1, ..., x_n)$ is a point of the locus defined by conditions (6), that verifies (7). Let us prove that this point is the union of the mentioned subvarieties.

As the final remainder, r_s, obtained in STEP 6, verifies lemma 1 and it can be factored as in (10), it follows that

$$\rho \prod_{j=1}^{e} \phi_j = m_s \cdot m_{s-1} \cdot ... \cdot m_1 \cdot t_h + \sum_{j=1}^{s} \gamma_{ij} h_j$$

As ρ has been supposed nonzero, substituting every indeterminate, v_i by its corresponding linked coordinate, d_i, from (6) and (7) it follows that

$$\prod_{j=1}^{e} \phi_j(a_1, a_2, ..., a_r, x_1, ..., x_n, d_1, d_2, ..., d_s) = 0$$

Hence, one, at least, of these z factors must be null. If, for instance, this occurs for $j = u$, then $(x_1, ..., x_n)$ belongs to the subvariety defined by ϕ_u.

Reciprocally, suppose now that $(x_1, ..., x_n)$ is a point of the union of the mentioned subvarieties and let us prove that this point satisfies (7). If this point is in the subvariety defined by $\phi_u = 0$, then

$$\phi_u(a_1, ..., a_r, x_1, ..., x_n, d_1, ..., d_s) = 0. \tag{11}$$

As a consequence, this equation can be added to the hypothesis conditions and hence $\phi_u(a_1, ..., a_r, x_1, ..., x_n, v_1, ..., v_s)$ can be added as a new hypothesis polynomial. Now, by applying lemma 1 to the new system of hypothesis polynomials,

$$r'_{s+1} = m'_{s+1} \cdot m'_s \cdot m'_{s-1} \cdot ... \cdot m'_1 \cdot t_h + \sum_{i=1}^{s} \gamma'_i h_i + \gamma^*_u \phi_u$$

As r'_{s+1} has been supposed zero (condition 1 of hypothesis), it follows that

$$m'_{s+1} \cdot m'_s \cdot m'_{s-1} \cdot ... \cdot m'_1 \cdot t_h = -\sum_{i=1}^{s} \gamma'_i h_i - \gamma^*_u \phi_u$$

and substituting variables $v_1, ..., v_s$ by dependent coordinates $d_1, ..., d_s$, from (6) and (11), it follows that

$$m'_{s+1} \cdot m'_s \cdot m'_{s-1} \cdot ... \cdot m'_1 \cdot t_h(a_1, ..., a_r, x_1, ..., x_n, d_1, ..., d_s) = 0$$

As these multipliers have been supposed to be nonzero (condition 2 of hypothesis), condition (7) is therefore satisfied.

5   Extension of the Simson-Steiner Theorem to $3D$

As an application of the method of determination of loci explained in the preceding section, we shall try to extend Simson-Steiner Theorem considered in section 3 to $3D$. This, as far as we know, is a new theorem. As in section 3, we shall state it as a problem.

Problem: *Let us consider in the real euclidean space $\mathbb{R}^3$ a tetrahedron (OABC), an arbitrary point (X) and the orthogonal projections (M, N, P, Q) of X on face-planes of OABC. What is the locus of X, such that $vol(MNPQ) = v$?*

Let us adapt the problem to apply the method described in the preceding section. Vertices (O, A, B, C) of the tetrahedron are the free points and X is the locus point. The orthogonal projections (M, N, P, Q) are the linked or dependent points and the geometric conditions that determine these points as projections of X give the hypothesis conditions. Finally, the condition that the volume of tetrahedron $OABC$ is kept unchanged (equal to the constant v) is the thesis condition. As in the preceding section, we shall distinguish nine steps.

STEP 1. Select the coordinates
Vertices of the tetrahedron: $O(0,0,0)$, $A(1,0,0)$, $B(0,b,0)$, $C(1,c,e)$.
Locus point: $X(x, y, z)$.
Projections of X: $M(m1, m2, m3)$, $N(n1, n2, n3)$, $P(p1, p2, p3)$, $Q(q1, q2, q3)$

STEP 2. Convert hypothesis and thesis conditions into polynomial equations

Hypothesis conditions:

$[\overrightarrow{OM}, \overrightarrow{OA}, \overrightarrow{OB}] = 0$	O, A, B, M coplanar (triple product $= 0$)
$\overrightarrow{XM} \cdot \overrightarrow{OA} = 0$	$XM \perp OA$ (inner product $= 0$)
$\overrightarrow{XM} \cdot \overrightarrow{OB} = 0$	$XM \perp OB$
$[\overrightarrow{ON}, \overrightarrow{OB}, \overrightarrow{OC}] = 0$	O, B, C, N coplanar
$\overrightarrow{XN} \cdot \overrightarrow{OB} = 0$	$XN \perp OB$
$\overrightarrow{XN} \cdot \overrightarrow{OC} = 0$	$XN \perp OC$
$[\overrightarrow{OP}, \overrightarrow{OC}, \overrightarrow{OA}] = 0$	O, C, A, P coplanar
$\overrightarrow{XP} \cdot \overrightarrow{OC} = 0$	$XP \perp OC$
$\overrightarrow{XP} \cdot \overrightarrow{OA} = 0$	$XP \perp OA$
$[\overrightarrow{AQ}, \overrightarrow{AB}, \overrightarrow{AC}] = 0$	A, B, C, Q coplanar
$\overrightarrow{XQ} \cdot \overrightarrow{AB} = 0$	$XQ \perp AB$
$\overrightarrow{XQ} \cdot \overrightarrow{AC} = 0$	$XQ \perp AC$

Thesis condition:

$[\overrightarrow{MN}, \overrightarrow{MP}, \overrightarrow{MQ}] - 6v = 0$	$vol(MNPQ) = v$

STEP 3. Establish parameter and coordinates

Free coordinates: b, c, e
Locus coordinates: x, y, z
Linked or dependent coordinates: $m1, m2, m3, n1, n2, n3, p1, p2, p3, q1, q2, q3$
Parameters (including free coordinates): b, c, e, v
Parameter conditions: $b \neq 0 \neq e$

As O, A, B, C must be non-coplanar points, parameter conditions follow from
$vol(OABC) = \frac{1}{6}[\overrightarrow{OM}, \overrightarrow{OA}, \overrightarrow{OB}] = \frac{1}{6}be \neq 0.$

STEP 4. Input the hypothesis and thesis polynomials

To make it easier, three procedures have been implemented:

$escl(R,S,T,U)$ for inner product $\overrightarrow{RS} \cdot \overrightarrow{TU}$
$vect(R,S,T,U)$ for cross product $\overrightarrow{RS} \times \overrightarrow{TU}$
$tripl(R,S,T,U)$ for triple product $[\overrightarrow{RS}, \overrightarrow{RT}, \overrightarrow{RU}]$

where R, S, T, U are points in $\mathbb{R}^3$. Thus, we have (in Maple):

```
> h1  := tripl(O,M,A,B):
> h2  := escl(X,M,O,A):
> h3  := escl(X,M,O,B):
> h4  := tripl(O,N,B,C):
> h5  := escl(X,N,O,B):
> h6  := escl(X,N,O,C):
> h7  := tripl(O,P,C,A):
> h8  := escl(X,P,O,C):
> h9  := escl(X,P,O,A):
> h10 := tripl(A,Q,B,C):
> h11 := escl(X,Q,A,B):
> h12 := escl(X,Q,A,C):
> th: = tripl(M,N,P,Q) - 6*v:
```

STEP 5. Compute the final pseudoremainder and factorize it

Lists of variables and hypothesis polynomials are denoted V and H, respectively:

```
> V := [m1,m2,m3,n1,n2,n3,p1,p2,p3,q1,q2,q3]:
> H := [h2,h3,h1,h6,h5,h4,h9,h7,h8,h11,h10,h12]:
> r12:=final_prem(th,G,V):
> factor(r12);
```

$$b^5 e^2 (b^2 + 1)(-6e^4 b^2 v - 6e^6 v - e^5 bz^2 - 12e^2 c^2 v - 6c^4 v - e^4 bc^2 x^2 z$$
$$+e^4 b^2 x^2 zc - e^5 b^2 x^2 y - e^6 bx^2 z - 6e^2 b^2 c^2 v - e^5 b^2 z^2 y + e^4 bc^2 zx$$
$$-e^4 b^2 zcx + 2e^5 bz^2 cy + 2e^4 bzcxy + e^6 bzx - e^3 bc^2 z^2 + e^5 bxz^2$$
$$-e^5 by^2 x - e^6 by^2 z + e^4 b^2 z^3 c - e^4 bz^3 c^2 - 6c^4 e^2 v + e^5 b^2 xy + e^6 b^2 zy$$
$$-e^5 b^2 z^2 c - 6b^2 e^6 v - 6e^4 v - 12c^2 e^4 v - 6b^2 e^4 c^2 v)$$

Denote by ρ the product of factors not depending on x, y, z and denote by ϕ the only factor depending on x, y, z:

```
> rho:=b^5e^2(b^2+1):
> phi:=simplify(r12/rho);
```

For the thesis condition, $th = 0$, to be verified, r_{12} must be zero. Note that $\rho = b^5 e^2(b^2 + 1) \neq 0$, under parameter conditions. Hence, for the thesis condition, $th = 0$, to be verified, ϕ must be zero. But, does every point in $\phi = 0$ verify the thesis condition? To answer it, ϕ must be added as a new hypothesis condition.

We have seen that points X such that $vol(MNPQ) = v$ verify $\phi = 0$. But the reciprocal question arises: does every point in $\phi = 0$ verify the thesis condition? To answer it, $\phi = 0$ must be added as a new hypothesis condition.

STEP 6. Input the new hypothesis polynomial

Consequently, ϕ must be added as new hypothesis polynomial, which will be denoted $h13$:

```
> h13 := phi:
```

STEP 7. Precise new parameters and coordinates

As point $X(x, y, z)$ must verify $h_{13} = 0$, one of its coordinates (z, for example) can be freely chosen, and so it will be a new parameter, and the other ones will be new dependent coordinates.

STEP 8. Compute the new final pseudoremainder

New lists of variables and hypothesis polynomials are denoted VV and HH:

```
> VV := [m1,m2,m3,n1,n2,n3,p1,p2,p3,q1,q2,q3,z]:
> HH := [h2,h3,h1,h6,h5,h4,h9,h7,h8,h11,h10,h12,h13]:
> r13 := final_prem(th,HH,VV);
```

$$r13 = 0$$

STEP 9. Compute the value of the multipliers

```
> M := mulf_list(th,HH,VV);
```

$$M = [1, b, b, 1, b, -b^2 - b^2 e^2, 1, e, e^2 + c^2, -1, -e - b^2 e, -e^2 - b^2 e^2 - c^2, -e^4 b^2 c + e^4 b c^2]$$

Under parameter conditions ($b \neq 0 \neq e$), every element in M is non-zero except the last one, which must also satisfy: $0 \neq c \neq b$ (this condition can be avoided, treating it as a particular case or selecting another coordinate system).

In accordance with theorem 5, we have obtained the following extension of Simson-Steiner Theorem to $3D$:

Let us consider in the real euclidean space $\mathbb{R}^3$: a tetrahedron $(OABC)$, an arbitrary point (X) and its orthogonal projections (M, N, P, Q) on the face-planes of $OABC$. Then the locus of point X, such that $vol(MNPQ) = v$ (v constant) is the cubic surface $\phi = 0$.

In particular, for $b = 1, c = -1, e = 1$, this cubic surface equation is

$$z^2 + 3z^2y - 3zx - xz^2 + 2zxy + y^2z - xy - zy + y^2x + 3x^2z + x^2y + 2z^3 + 72v = 0$$

For $v = 0$ (M, N, P, Q coplanar points), vertices of $OABC$ are singular points of the surface and the border-lines of $OABC$ are contained in the surface. This surface is visualized in Fig. 2 using package DPGraph2000.

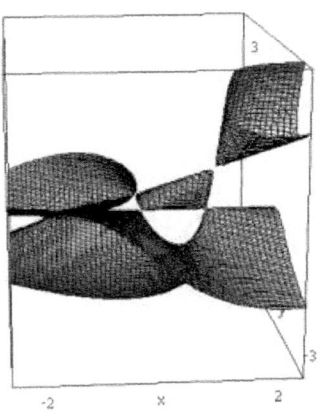

Fig. 2. Cubic surface loci $\phi = 0$ for $v = 0$

6   Conclusions

The method to determine automatically geometric loci we have developed has been justified without using a deep algebraic background, and the base field does not need to be algebraically closed.

It is useful when some of the points that are directly involved in the (thesis) condition that defines the locus are determined by a non empty set of (hypotheses) conditions. Both of them (hypothesis and thesis conditions) must be converted into polynomial equations.

As we have shown in the preceding section, the method can be applied to detect geometric loci in an n-dimensional euclidean space (for a given n, fixed in advance).

Finally, we would like to remark that we do not intend to retire classical methods to find geometric loci, that use techniques of synthetic geometry. But many times it is not easy to find the key idea that solves the problem with classic techniques. Then the standard technique shown here can be very useful.

References

1. D. Cox, J. Little, D. O'Shea: Ideals, Varieties, and Algorithms. Springer, 1991.
2. S. C. Chou: Mechanical Geometry Theorem Proving. Reidel, 1988.
3. M. de Guzmán: An Extension of the Wallace-Simson Theorem: Projecting in Arbitrary Directions. Mathematical Monthly, 106, 6, June 1999 (pages 574-580).
4. D. Kapur: Automated Geometric Reasoning: Dixon Resultants, Groebner Bases and Characteristic Sets. In: D. Wang (ed.): Proceedings International Workshop on Automated Deduction in Geometry. Springer-Verlag, LNAI 1360.
5. D. Kapur, J.L. Mundy: Wu's method and its application to perspective viewing. In: D. Kapur and J.L. Mundy (eds.): Geometric Reasoning. MIT Press, 1989.
6. W. Koepf: Gröbner bases and triangles. Int. J. of Comput. Algebra in Math. Education, 4, 4, 1998 (pages 371-386).
7. T. Recio, M. P. Vélez: Automatic Discovery of Theorems in Elementary Geometry. Journal of Automated Reasoning, 23, 1999 (pages 63-82).
8. E . Roanes M., E. Roanes L.: Cálculos automáticos por ordenador con Maple V.5. Rubiños, 1999.
9. E . Roanes M., E. Roanes L.: Búsqueda Automática de Lugares Geométricos. Bol. Soc. Puig Adam de Profs. de Matemáticas, 53, 1999 (pages 67-77).
10. W. T. Wu: Mechanical Theorem Proving in Geometries. Text and Monographs in Symbolic Computation. Springer-Verlag, 1994.

Numerical Implicitization of Parametric Hypersurfaces with Linear Algebra*

Robert M. Corless, Mark W. Giesbrecht, Ilias S. Kotsireas, and
Stephen M. Watt

University of Western Ontario, London, Ontario, N6A 5B9 Canada
Ontario Research Centre for Computer Algebra
{Rob.Corless, Mark.Giesbrecht, Ilias.Kotsireas, Stephen.Watt}@orcca.on.ca
http://www.orcca.on.ca

Abstract. We present a new method for implicitization of parametric curves, surfaces and hypersurfaces using essentially numerical linear algebra. The method is applicable for polynomial, rational as well as trigonometric parametric representations. The method can also handle monoparametric families of parametric curves, surfaces and hypersurfaces with a small additional amount of human interaction. We illustrate the method with a number of examples. The efficiency of the method compares well with the other available methods for implicitization.

1 Introduction

The problem of implicitization for curves, surfaces and hypersurfaces is an important problem in Algebraic Geometry with immediate practical applications in such areas as Geometric Modeling, Graphics, Computer Aided Geometric Design (see [Hof89]). The implicitization problem has been addressed using a variety of mathematical methodologies and techniques including Gröbner bases, (see [Buc88], [Kal90], [GC92], [LM94], [FHL96]) Characteristic sets, (see [Li89], [Gao00]) Resultants, (see [SAG84], [CG92]) Perturbation, (see [Hob91], [MC92a], [MC92b], [SSQK94], [Hon97], [SGD97]), Multidimensional Newton formulae (see [GV97]), Elimination theories, (see [SAG85], [Wan95]) and Symmetric functions (see [GVT95]).

We note that the inverse problem of parameterization, which is an equally important problem in Algebraic Geometry with direct practical applications, has also been investigated by many authors (see for example [AB88], [AGR95], [HS98], [Sch98], [SW91], [SW98]).

Some of the above methods work for special categories of curves, surfaces and hypersurfaces. Moreover, some methods handle only special kinds of parametric representations, like polynomial, rational or trigonometric ones.

* Work supported by the Ontario Research Centre for Computer Algebra the Ontario Research and Development Challenge Fund and the Natural Sciences and Engineering Research Council of Canada.

J.A. Campbell and E. Roanes-Lozano (Eds.): AISC 2000, LNAI 1930, pp. 174–183, 2001.

It is important to have efficient algorithms to solve the implicitization and parameterization problems. This is mainly because in many practical applications and depending on the particular circumstances we want to use the parametric equations or the implicit equation.

In this paper we present a new implicitization method for curves, surfaces and hypersurfaces that works for polynomial, rational and trigonometric parameterizations. The method uses an alternative interpretation of the implicitization problem as an eigenvalue problem, inspired by theoretical considerations coming from the area of the Calculus of Variations (see [Tro83]). The method ultimately uses numerical linear algebra to recover the implicit (cartesian) equation from the parametric equations.

2   Description of the Problem

In what follows the term, geometric object, will be used to describe a curve, a surface, or a general hypersurface.

A parameterization of a geometric object in a space of dimension n can be described by the following set of parametric equations:

$$x_1 = f_1(t_1, \ldots, t_k), \ldots, x_n = f_n(t_1, \ldots, t_k) \tag{1}$$

where the $t_1, \ldots, t_k$ are parameters and the functions $f_1, \ldots, f_n$ can be polynomial, rational or trigonometric functions. The case $n = 2$ corresponds to curves, the case $n = 3$ corresponds to surfaces and the case $n \geq 4$ corresponds to hypersurfaces in general. The implicitization problem consists of computing the polynomial cartesian (implicit) equation

$$p(x_1, \ldots, x_n) = 0, \tag{2}$$

of the geometric object described by the parametric equations (1), which satisfies

$$p(f_1(t_1, \ldots, t_k), \ldots, f_n(t_1, \ldots, t_k)) = 0,$$

for all values of the parameters $t_1, \ldots, t_k$.

3   Implicitization as an Eigenvalue Problem

Suppose that $g(x, y) = M(x, y)\mathbf{a}$ where $\mathbf{a}$ is the vector of (unknown) coefficients of the polynomial $g(x, y)$ and $M(x, y) = [1, x, y, \ldots, y^m]$ where m is the total degree of $g(x, y)$. Given a parameterization $(x(s), y(s))$ of $g(x, y) = 0$, numerical or exactly-known, we can ask for the vector $\mathbf{a}$ that minimizes

$$J(g) = \int_{s_0}^{s_1} w(s)g^* g \, ds$$

$$= \int_{s_0}^{s_1} w(s)\mathbf{a}^* M^*(x(s), y(s)) M(x(s), y(s))\mathbf{a} \, ds$$

(for a specified positive weight function $w(s)$) subject to the constraint $\|\mathbf{a}\|^2 = 1$. Forming the Lagrange multiplier we get the standard Rayleigh-Ritz problem of minimizing $K(\mathbf{a}) = J + \lambda(1 - \mathbf{a}^*\mathbf{a})$.

Consider $K(\mathbf{a} + \Delta\mathbf{a}) - K(\mathbf{a}) =$

$$2\Delta\mathbf{a}^* (G - \lambda I)\,\mathbf{a} + \Delta\mathbf{a}^* (G - \lambda I)\,\Delta\mathbf{a} \tag{3}$$

where G is the (Hermitian, positive semi-definite) structured matrix

$$G = \int_{s_0}^{s_1} w(s) M^* M \, ds \,, \tag{4}$$

and we therefore see that if

$$[G - \lambda I]\,\mathbf{a} = 0 \,, \tag{5}$$

then λ is an eigenvalue of G with eigenvector $\mathbf{a}$; therefore $\lambda \geq 0$ because G is positive semidefinite. This gives $K(\mathbf{a} + \Delta\mathbf{a}) - K(\mathbf{a}) = \Delta\mathbf{a}^*(G - \lambda I)\Delta\mathbf{a}$. Now if λ is the *smallest* eigenvalue of G, the eigenvalues of $(G - \lambda I)$ are all non-negative, and hence

$$K(\mathbf{a} + \Delta\mathbf{a}) - K(\mathbf{a}) = \Delta\mathbf{a}^*(G - \lambda I)\Delta\mathbf{a} \geq 0 \,. \tag{6}$$

Thus our minimum will occur at an eigenvector of G corresponding to its smallest eigenvalue.

Moreover, the standard theory [Tro83, p. 343] shows that the *eigenvalue* λ is exactly $J(M(x, y)\mathbf{a})$ for the corresponding eigenvector $\mathbf{a}$.

Finally, equality in (6) occurs if and only if $\Delta\mathbf{a}$ is also an eigenvector corresponding to λ. This is possible only if the smallest eigenvalue is multiple.

More generally, the conditioning of these eigenvectors depends on the distances to the nearest other small eigenvalues [GVL95]. Errors are amplified by a factor of essentially $1/(\lambda_k - \lambda_m)$.

If an implicitization exists with the support $M(x, y)$, then finding a vector in the null space of G finds this implicitization.

4   Description and Implementation of the Method

In this section we describe the algorithmic steps of the method in pseudo-code. We have implemented the method in MAPLE and tested our implementation with all the examples given in the next section.

Input: Parametric equations of the form (1) for specific n, k.
Output: The cartesian (implicit) equation for the geometric object
        represented by these equations.

Step 1:   choose m (total degree of the implicit equation).
Step 2:   construct the line matrix v of all power products of total degree
          up to m in the variables $x_1, \ldots, x_n$.
Step 3:   compute the matrix $M = v^t \cdot v$.
Step 4:   substitute $x_1, \ldots, x_n$ by their parametric representations (1),
          in the matrix M.
Step 5:   integrate the elements of the matrix M successively over each
          parameter $t_1, \ldots, t_k$.
Step 6:   compute a null-vector nv of the matrix resulting from Step 5.
Step 7:   recover the implicit equation as the product $M \cdot nv$.

Several comments are in order to clarify certain points in the above description of our implicitization method.

1. During the integration step, care should be taken so as to avoid integrals with infinite values or divergent integrals. Such degenerate cases may occur when for example the parametric equations contain denominators or trigonometric functions. Usually it is an easy matter to choose suitable intervals of integration. In the case of rational parametric equations this is the problem of base points (see for example [CG92], [MC92a], [MC92b]).
2. Another issue related to the integration step, is that sometimes it is inevitable to perform the integrations numerically, simply because the analytic expression is either too complicated to be of any use, or is not elementary. When numerical integration is employed, the resulting matrix will have floating point elements and one should be very careful about how to compute correctly the nullspace. Indeed, it may happen that according to the precision used for the computation, one obtains one or more vectors as a basis for the nullspace.
3. In the last step of the algorithm, we obtain the cartesian equation in the variables $x_1, \ldots, x_n$, but this will not always be a polynomial with integer coefficients. Some more processing is necessary to discover the integer relations among the coefficients and finally multiply by the appropriate number to unveil a polynomial with integer coefficients. This can be done using integer relation-finding algorithms as they are implemented in Maple.

5   Application of the Method

In this section we give some examples to illustrate the use of the implicitization method for curves and surfaces. The method works for rational as well as for trigonometric parameterizations and can also be used to recover cartesian equations for monoparametric families of curves or surfaces.

Example 1. (The Descartes Folium)
Consider the following parametric equations for the plane algebraic curve known
as the Descartes Folium:

$$x = \frac{3t^2}{t^3 + 1}, \quad y = \frac{3t}{t^3 + 1}. \tag{7}$$

Choose $m = 3$ and define the line matrix $v = [1, x, y, x^2, xy, y^2, x^3, x^2y, xy^2, y^3]$
and form the associated 10×10 matrix $M = v^t \cdot v$:

$$M = \begin{bmatrix}
1 & x & y & x^2 & xy & y^2 & x^3 & x^2y & xy^2 & y^3 \\
x & x^2 & xy & x^3 & x^2y & xy^2 & x^4 & x^3y & x^2y^2 & xy^3 \\
y & xy & y^2 & x^2y & xy^2 & y^3 & x^3y & x^2y^2 & xy^3 & y^4 \\
x^2 & x^3 & x^2y & x^4 & x^3y & x^2y^2 & x^5 & x^4y & x^3y^2 & x^2y^3 \\
xy & x^2y & xy^2 & x^3y & x^2y^2 & xy^3 & x^4y & x^3y^2 & x^2y^3 & xy^4 \\
y^2 & xy^2 & y^3 & x^2y^2 & xy^3 & y^4 & x^3y^2 & x^2y^3 & xy^4 & y^5 \\
x^3 & x^4 & x^3y & x^5 & x^4y & x^3y^2 & x^6 & x^5y & x^4y^2 & x^3y^3 \\
x^2y & x^3y & x^2y^2 & x^4y & x^3y^2 & x^2y^3 & x^5y & x^4y^2 & x^3y^3 & x^2y^4 \\
xy^2 & x^2y^2 & xy^3 & x^3y^2 & x^2y^3 & xy^4 & x^4y^2 & x^3y^3 & x^2y^4 & xy^5 \\
y^3 & xy^3 & y^4 & x^2y^3 & xy^4 & y^5 & x^3y^3 & x^2y^4 & xy^5 & y^6
\end{bmatrix}.$$

We substitute equations (7) into the matrix M to obtain a new matrix M'.
Integrate all the elements of the matrix M' with respect to t over the interval
$[0, 2]$. Since the denominators in equations (7) have a singularity at $t = -1$, we
choose an interval which does not contain that point. The integrations can be
performed symbolically or numerically. We prefer the numerical evaluation in
this example, because the analytical expressions for the integrals yield a fairly
complicated matrix. The difference in the computing times between calculating
the nullvector for the analytical and the numerical matrix, is dramatic. The
numerical rank of the resulting matrix is 9, which means that its nullspace is
of dimension 1 and thus generated by one nullvector. The Maple environment
variable *Digits* is set to 15 in order to achieve a better accuracy. We compute
the nullvector and multiply it by v, to obtain the equation

$$-0.9045\, xy + 0.3015\, x^3 + 0.3015\, y^3 = 0,$$

which shows that the implicit equation of the Descartes Folium is:

$$x^3 + y^3 - 3\, x\, y = 0.$$

The most time-consuming part of the computation (2.5 sec) is the numerical
evaluation of the 100 definite integrals involved.

Example 2.
Consider the following trigonometric parametric equations of the unit sphere in three-dimensional space:

$$x = \cos\theta \, \sin\phi, \; y = \cos\theta \, \cos\phi, \; z = \sin\theta.$$

We form the 10×10 matrix with first row $[1, x, y, z, x^2, xy, xz, y^2, yz, z^2]$ and integrate from 0 to $\pi/3$ for θ and ϕ successively. The resulting matrix has rank 9 and its nullspace is spanned by the vector $[-1, 0, 0, 0, 1, 0, 0, 1, 0, 1]$. This gives directly the cartesian equation of the unit sphere:

$$-1 + x^2 + y^2 + z^2 = 0.$$

The method works also for rational parameterizations of the unit sphere.
Example 3.
Let a be a parameter and consider the family of curves defined by the following rational parametric equations:

$$x = \frac{t\left(a - t^2\right)}{\left(1 + t^2\right)^2} \; y = \frac{t^2\left(a - t^2\right)}{\left(1 + t^2\right)^2}$$

We compute the cartesian equation for some values of a and by extrapolation we have that the general monoparametric cartesian equation for the family of curves is:

$$x^4 - a\,yx^2 + 2\,x^2y^2 + y^3 + y^4 = 0.$$

Now an easy computation shows that this equation is indeed valid for arbitrary a.
Example 4.
Unfortunately other monoparametric families present bigger difficulties. Consider the family of curves given by the polynomial parametric equations:

$$x = t + t^2, \; y = t + t^n \tag{8}$$

where n is a parameter. We compute the cartesian equation for some values of n and we see that we have to distinguish two cases according to the parity of n, for the general monoparametric cartesian equation of the family of the curves. For n even the cartesian equation has $n + 1$ terms, and is of the form:

$$x^n + \sum_{i=2}^{k}(a_i + b_iy)x^i + y^2 - nxy = 0, \quad k = \frac{n}{2}, \tag{9}$$

where the a_i, b_i are constants. For n odd the cartesian equation has $n + 3$ terms, and is of the form:

$$x^n + 2x^{k+1} + \sum_{i=1}^{k}(c_i + d_iy)x^i - 2y - y^2 = 0, \quad k = \frac{n-1}{2}, \tag{10}$$

where the c_i, d_i are constants. We have not solved the problem completely, unlike the situation in the previous example. But now for an arbitrary positive integer n we can substitute the parametric equations (8) into (9) for even n (resp. to (10) for odd n) and determine the unknown coefficients a_i, b_i (resp c_i, d_i) by solving a highly structured linear system of $2n - 1$ equations in $n - 2$ (resp. $n - 1$ unknowns).

Example 5.
The following example is taken from [GC92]. Consider the parametric equations for a Bézier curve:

$$x(t) = \frac{8\,t^6 - 12\,t^5 + 32\,t^3 + 24\,t^2 + 12\,t}{t^6 - 3\,t^5 + 3\,t^4 + 3\,t^2 + 3\,t + 1}$$

and

$$y(t) = \frac{24\,t^5 + 54\,t^4 - 54\,t^3 - 54\,t^2 + 30\,t}{t^6 - 3\,t^5 + 3\,t^4 + 3\,t^2 + 3\,t + 1}.$$

We form the 10×10 matrix with first row $[1, x, y, x^2, xy, y^2, x^3, x^2y, xy^2, y^3]$ and perform the integration from $t = 1$ to $t = 2$. The resulting equation to an accuracy of 7 decimal digits is:

$$0.0001644979\,y^3 + (0.005604679 - 0.001665542\,x)\,y^2$$

$$+ \left(-0.001110361\,x - 0.3527776 - 0.00003965575\,x^2\right)y$$

$$+0.8819439\,x + 0.02516157\,x^3 - 0.3115356\,x^2 = 0.$$

If we normalize this equation by dividing with the smallest coefficient in absolute value, then we get the following equation, in which some of the integer relations between the coefficients of the final equation appear already:

$$4.148147\,y^3 + (141.3333 - 42.00001\,x)\,y^2$$

$$+ \left(-1.0\,x^2 - 8896.001 - 28.0\,x\right)y$$

$$-7856.001\,x^2 + 22240.0\,x + 634.4999\,x^3 = 0.$$

The final step is provided by either Maple (using the `convert/rational` command in conjunction with a small value of the `Digits` environment variable, say 5) or RevEng, the newer version of the Inverse Symbolic Calculator available on-line from the CECM (http://www.cecm.sfu.ca/MRG/INTERFACES.html). We disregard the integer coefficients in the above equation and after processing the remaining three non-integer coefficients we discover that:

$$4.148147 \simeq \frac{112}{27}, \quad 141.3333 \simeq \frac{424}{3} \quad 634.4999 \simeq \frac{1269}{2}$$

Multiplying the equation with the lcm of the denominators which is 54, we get the final cartesian equation:

$$224\,y^3 + (7632 - 2268\,x)\,y^2 + \left(-1512\,x - 480384 - 54\,x^2\right)y$$

$$+34263\,x^3 - 424224\,x^2 + 1200960\,x = 0,$$

which can easily be verified with Maple.

Example 6.

The following example is taken from [SAG85]. Consider a rational cubic Bézier curve with control points

$$P_0 = (4,1), P_1 = (5,6), P_2 = (5,0), P_3 = (6,4),$$

with respective weights $w_0 = 1, w_1 = 2, w_2 = 2, w_3 = 1$. The parametric equations of the curve are given by:

$$x = 2\,t^3 - 18\,t^2 s + 18\,ts^2 + 4\,s^3$$
$$y = 39\,t^3 - 69\,t^2 s + 33\,ts^2 + s^3$$
$$z = -3\,t^2 s + 3\,ts^2 + s^3$$

We choose to work with a 20×20 matrix (total degree 3) and integrate from 0 to 1 for t and s successively. The resulting matrix has rank 19 and its nullspace is spanned by a vector with small rational coefficients. Multiplying by the lcm of the denominators we get the following cartesian equation:

$$224\,y^3 - 7056\,y^2 x + 33168\,y^2 z + 60426\,yx^2 - 562500\,yxz + 1322088\,yz^2 - 156195\,x^3$$

$$+2188998\,x^2 z - 10175796\,xz^2 + 15631624\,z^3 = 0.$$

6 Conclusion

We present a new method for doing implicitization of curves, surfaces and hypersurfaces, using essentially linear algebra. The method works for polynomial, rational and trigonometric parametric equations. The method also applies to monoparametric families of parametric curves, surfaces and hypersurfaces, with a small amount of extra work. The method is quite efficient due to the fact that it does not use Gröbner bases or multivariate factorization computations. The efficiency of the method can be improved by taking into account the special structure of the matrices involved in the computation.

References

[AB88]   Shreeram S. Abhyankar and Chanderjit L. Bajaj. Automatic parameterization of rational curves and surfaces iii: Algebraic plane curves. *Computer Aided Geometric Design*, 5:309–321, 1988.

[AGR95]   Cesar Alonso, Jaime Gutierrez, and Tomas Recio. An implicitization algorithm with fewer variables. *Computer Aided Geometric Design*, 12(3):251–258, 1995.

[Buc88]   Bruno Buchberger. Applications of gröbner bases in non-linear computa-
            tional geometry. In J. R. Rice, editor, *Mathematical Aspects of Scientific
            Software*, volume 14 of *IMA Volumes in Mathematics and its applications*,
            pages 59–87. Springer-Verlag, 1988.
[CG92]   Eng-Wee Chionh and Ronald N. Goldman. Degree, multiplicity, and in-
            version formulas for rational surfaces using u-resultants. *Computer Aided
            Geometric Design*, 9:93–108, 1992.
[FHL96]   George Fix, Chih-Ping Hsu, and Tie Luo. Implicitization of rational para-
            metric surfaces. *Journal of Symbolic Computation*, 21:329–336, 1996.
[Gao00]   Xiao-Shan Gao. Conversion between implicit and parametric represen-
            tations of algebraic varieties. In *Mathematics Mechanization and Appli-
            cations*, Academic Press, pages 1–17, 2000. personal communication, to
            appear.
[GC92]   Xiao-Shan Gao and Shang-Ching Chou. Implicitization of rational para-
            metric equations. *Journal of Symbolic Computation*, 14:459–470, 1992.
[GV97]   Laureano González-Vega. Implicitization of parametric curves and surfaces
            by using multidimensional newton formulae. *Journal of Symbolic Compu-
            tation*, 23:137–151, 1997.
[GVL95]   Gene H. Golub and Charles Van Loan. *Matrix Computations*. Johns Hop-
            kins, 3rd edition, 1995.
[GVT95]   L. González-Vega and G. Trujillo. Implicitization of parametric curves and
            surfaces by using symmetric functions. In A.H.M. Levelt, editor, *Proc.
            ISSAC-95*, ACM Press, pages 180–186, Montreal, Canada, july 1995.
[Hob91]   John D. Hobby. Numerically stable implicitization of cubic curves. *ACM
            Transactions on Graphics*, 10(3):255–296, 1991.
[Hof89]   Christoph M. Hoffmann. *Geometric and Solid Modeling: An Introduction*.
            Morgan Kaufmann Publishers, Inc. California, 1989.
[Hon97]   Hoon Hong. Implicitization of nested circular curves. *Journal of Symbolic
            Computation*, 23:177–189, 1997.
[HS98]   Hoon Hong and Josef Schicho. Algorithms for trigonometric curves (sim-
            plification, implicitization, parameterization). *Journal of Symbolic Compu-
            tation*, 26:279–300, 1998.
[Kal90]   Michael Kalkbrener. Implicitization of rational parametric curves and sur-
            faces. In S. Sakata, editor, *Proc. AAECC-8*, volume 508 of *Lecture Notes
            in Computer Science*, pages 249–259, Tokyo, Japan, august 1990.
[Li89]   Ziming Li. Automatic implicitization of parametric objects. *Mathematics-
            Mechanization Research Preprints*, 4:54–62, 1989.
[LM94]   Sandra Licciardi and Teo Mora. Implicitization of hypersurfaces and curves
            by the primbasissatz and basis conversion. In Mark Giesbrecht and Joachim
            von zur Gathen, editors, *Proc. ISSAC-94*, ACM Press, pages 191–196, Ox-
            ford, England, april 1994.
[MC92a]   Dinesh Manocha and John F. Canny. Algorithm for implicitizing rational
            parametric surfaces. *Computer Aided Geometric Design*, 9:25–50, 1992.
[MC92b]   Dinesh Manocha and John F. Canny. Implicit representation of rational
            parametric surfaces. *Journal of Symbolic Computation*, 13:485–510, 1992.
[SAG84]   T. W. Sederberg, D. C. Anderson, and R. N. Goldman. Implicit represen-
            tation of parametric curves amd surfaces. *Computer Vision, Graphics, and
            Image Processing*, 28:72–84, 1984.
[SAG85]   T. W. Sederberg, D. C. Anderson, and R. N. Goldman. Implicitization, in-
            version, and intersection of planar rational cubic curves. *Computer Vision,
            Graphics, and Image Processing*, 31:89–102, 1985.

[Sch98]   Josef Schicho. Rational parametrization of surfaces. *Journal of Symbolic Computation*, 26(1):1–30, 1998.

[SGD97]   Thomas Sederberg, Ron Goldman, and Hang Du. Implicitizing rational curves by the method of moving algebraic curves. *Journal of Symbolic Computation*, 23(2-3):153–175, 1997.

[SSQK94]   Thomas W. Sederberg, Takafumi Saito, Dongxu Qi, and Krzysztof S. Klimaszewski. Curve implicitization using moving lines. *Computer Aided Geometric Design*, 11(6):687–706, 1994.

[SW91]   J.R. Sendra and F. Winkler. Symbolic parametrization of curves. *Journal of Symbolic Computation*, 12(6):607–631, 1991.

[SW98]   J.R. Sendra and F. Winkler. Real parametrization of algebraic curves. In J. Calmet and J. Plaza, editors, *Proc. AISC'98*, volume 1476 of *Lecture Notes in Artificial Intelligence*, pages 284–295, Plattsburgh, New York, USA, september 1998.

[Tro83]   John L. Troutman. *Variational Calculus with Elementary Convexity*. Undergraduate Texts in Mathematics. Springer-Verlag, New York, 1983.

[Wan95]   Dongming Wang. Reasoning about geometric problems using an elimination method. In Jochen Pfalzgraf and Dongming Wang, editors, *Automated Practical Reasoning Algebraic Approaches*, Lecture Notes in Computer Science, pages 147–185, Tokyo, Japan, 1995.

A Note on Modeling Connectionist Network Structures: Geometric and Categorical Aspects

Jochen Pfalzgraf

Department of Computer Science
University of Salzburg
Jakob Haringer Str.2
A-5020 Salzburg, Austria
jpfalz@cosy.sbg.ac.at

Abstract. This position paper proposes a mathematical modeling approach for a certain class of connectionist network structures. Investigation of the structure of an artificial neural network (ANN) in that class (paradigm) suggested the use of geometric and categorical modeling methods in the following sense. A (noncommutative) geometric space can be interpreted as a so-called geometric net. To a given ANN a corresponding geometric net can be associated. Geometric spaces form a category. Consequently, one obtains a category of geometric nets with a suitable notion of morphism. It is natural to interpret a learning step of an ANN as a morphism, thus learning corresponds to a finite sequence of morphisms (the associated networks are the objects). An associated ("local") geometric net is less complex than the original ANN, but it contains all necessary information about the network structure. The association process together with learning (expressed by morphisms) leads to a commutative diagram corresponding to a suitable natural transformation. Commutativity can be exploited to make learning "cheaper". The simplified mathematical network model was used in ANN simulation applied in an industrial project on quality control. The "economy" of the model could be observed in a considerable increase of performance and decrease of production costs.

1   Introduction

In this note we give a brief overview of work (in progress) dealing with a mathematical approach for modeling the network stucture of an artificial neural network (ANN). The work originates in a fruitful cooperation with H.Geiger who introduced his own network paradigms with neuron types developed on basis of neuro-physiolocial insight into biological information processing of cells. The architecture of these networks is very flexible, allowing a modularized design of network stuctures, integrating cascades of various neuron layers having different functionality (like feature extraction). Feedback loops can be modeled, too. The main neuron models are CCM (Conductivity Coupled Model), RCM (Rate Coded Model), SSM (Single Spike Model). The latter one is a rather complex neuron type (with many parameters) used to model time-dependent behavior

J.A. Campbell and E. Roanes-Lozano (Eds.): AISC 2000, LNAI 1930, pp. 184–199, 2001.
© Springer-Verlag Berlin Heidelberg 2001

in ANNs. In principle, all kinds of learning can be applied. The main learning rules used in Geiger's paradigms are forced learning (processing activities of neurons), delta rule learning, Hebbian learning (both processing synaptic weights, respectively). The first version of a simulator for these network types was implemented by H.Geiger (late 1970s) in terms of a Network Command Interpreter (NCI). Under his guidance this tool was developed further (by several diploma and doctoral students at TU Munich and colleagues in his company), resulting in the simulator "NeuroTools" which, since then, has been successfully applied in various industrial projects by Geiger and his coworkers. Later, in our group, a new command language was developed (in close cooperation with Geiger and his group) improving the older tool. This new version, "NeuroTools 6.0", is now being tested.

In this article our interest focuses on the *net structure* of an ANN. Consequently, neurons are abstractly modeled as *nodes* of a directed (and colored) graph, independend of their specific neuron types, respectively. Therefore the previously mentioned neuron models and rules are not of relevance in this contribution. A detailed presentation will be included in [GP]. Analyzing how the previously mentioned networks are designed by its inventor, we can observe that their architecture is amenable to mathematical modeling in a natural way. It has turned out in the course of our cooperation that methods from geometry (so-called "noncommutative geometric spaces") and category theory can be applied to model the network structures. The rough idea can be described as follows.

It can be observed that the corresponding ANNs are regularly structured. That means, considered from the "view point" of each node, the "local" structure (including connections) of the network is the same. Mathematically, one speaks of a "pointed space" when selecting a distinguished point in a space and describing the structure "locally". In this sense, the whole ("global") network structure can be homogeneously described by its pointed spaces, i.e. the essential information for structuring the net is given locally. Which roles do geometry and category theory play - how do they arise ?

The regularly structured networks can be interpreted as *directed colored graphs*. Input and output layers do not have to be distinguished. Concentrating on the local and global network structure, it turns out that it is reasonable to associate an (abstract) "geometric net" to a given ANN, globally, and to introduce a smaller associated net that reflects the local network structure (containing the essential information). A (noncommutative) *geometric space* has a natural interpretation as a directed colored graph - we call it *geometric net*. This suggests interpretation of an ANN structure as a geometric net. Noncommutative geometric spaces form a *category* (NCG) with geometric spaces as objects and structure preserving maps as morphisms. This, in turn, leads to the category of geometric nets and to the interpretation of an ANN structure as the object of a corresponding category. Now, from this point of view, we would like to model a *learning step as a morphism* between networks (objects) and *learning* as a finite *sequence of morphisms* in such a category.

Subsequently, we introduce the basic mathematical notions and apply them in the previously mentioned way. For details we refer to the literature. A detailed exposition with much more material on the ANN paradigms and the mathematical modeling aspects is in preparation ([GP]). A very short version containing the first basic ideas is [GP95]. At the end of this contribution we briefly present an industrial application on optical quality control carried out by H.Geiger (cf. [Gei94]). Deployment of the mathematical model on a macro level for purposes of implementation in the simulation ("NeuroTools") and application of the networks leads to a considerable improvement of performance. Shortening of program code and run time speed-up can be achieved and less storage space is needed. Summarizing, the production costs in that project could be enormously reduced. This is how the simplifying mathematical model has an "economic" impact on applications.

A final remark on the use of categorical notions. A priori, there is no need at all for using categorical modeling. But in fact, it was "categorical thinking" that opened our eyes for simplifications which, together with the geometric viewpoint, led to the interpretation of a network as an "object" and a learning step as a "morphism".

Concluding the introduction, we point to the interesting article by R. Eckmiller ([Eck90]) where he suggests the use of various mathematical methods, especially from geometry, in the field of connectionist network modeling. In some respect our work is in this spirit.

2   Remark on Geiger's Network Paradigm

We recall that this contribution concentrates on modeling network structures. We do not consider specific neuron types and learning rules. Neurons are just abstractly modeled as nodes of an associated network and learning will be interpreted in terms of certain mappings. Thus, our approach has more of a qualitative nature.

As already indicated in the introduction, the ANN structures developed and (industrially) used by H.Geiger (and implemented for simulation in "Neuro-Tools") are regularly structured in the following sense. As illustrated in the figure the cascaded network structure consists of an *input retina* FL_0 followed by several (problem dependend) *feature layers* $FL_1, \ldots, FL_n$ - *receptive fields (feature detectors)*.

Of basic importance is the observation that the networks are locally prestructured in a way which is amenable to a geometric interpretation in the sense of local *configurations*. A local description of the structure means selecting a fixed (but arbitrary) node (neuron), say x_0, of the network and specifying the connections (directed edges) with other nodes in a well defined neighborhood. We call this a *pointed network* (pointed space, in general mathematical modeling). We can take a special feature detector as in the figure ("local view"), for example.

Considering a pointed network in a node x_0, the *local structure* consists of *configurations*. In a geometric sense this means triangles (3-tuples), tetrahedrons

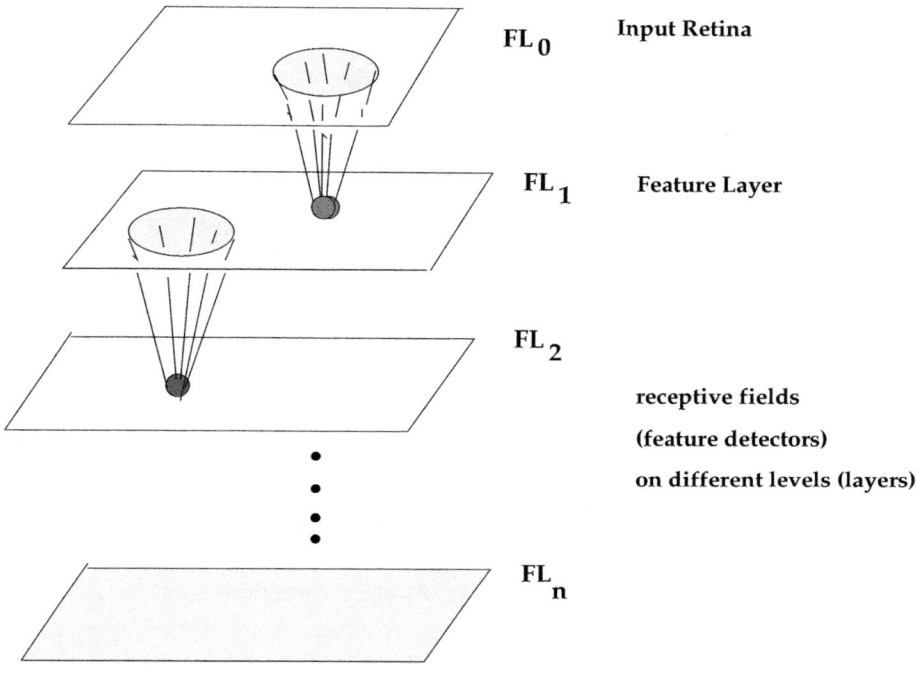

FL$_0$ **Input Retina**

FL$_1$ **Feature Layer**

FL$_2$

receptive fields
(feature detectors)
on different levels (layers)

FL$_n$

Global Multi-Layer Network

Fig. 1. a multilayer network (cascades)

(4-tuples), more generally *higher dimensional abstract simplices* (n-tuples) consisting of neighboring nodes of x_0 and corresponding connecting synaptic (colored) edges. This yields a very rich structure from a geometric standpoint.

For illustration we refer to the following figures representing a local feature detector and its associated (abstract) local net (the weights of the directed edges are not depicted). This special picture corresponds to a local edge detector that will be discussed later in section 4.

As we have mentioned above, in each neuron (node) of a layer the corresponding pointed network (local view) is the same. This leads to a *homogeneous structure* from a global point of view. Actually, this is where we turn to our geometric model: to every pointed network a geometric net will be associated encoding all the relevant information of the ANN locally, but since this situation is the same in each node, the global network can be reconstructed ("shifting" the pointed local network over all other nodes). Geometrically, this amounts to shifting all the geometric configurations in x_0 to all other nodes.

Now, a crucial aspect is that *learning* can be done locally, replacing the learning procedure of the entire (global), possibly large, neural network. This

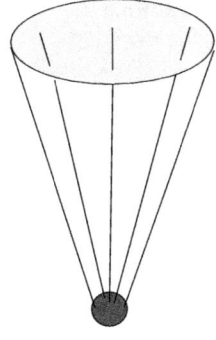
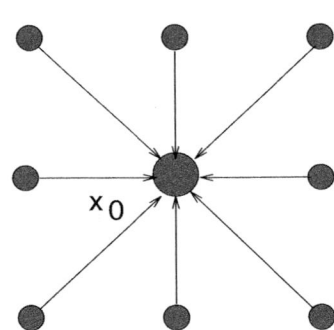

local feature detector   associated local net (abstraction)

the "local view" of the network

Fig. 2. "local view" of a network (feature detectors)

reduces the complexity of learning considerably. Later we will come back to that aspect. By the way, simulation of ANNs consisting of about one million neurons or more is no problem for "NeuroTools".

In practice (and in the simulator NeuroTools) nodes of a particular ANN are placed in a regular grid. As a 2-dimensional grid we can use, for example, $\mathbb{Z} \times \mathbb{Z}$. Furthermore, Geiger's modeling approach uses an *"embedding principle"*, i.e. a real (finite) ANN, designed for a particular application, is embedded in a larger network. This allows us to verify the previously mentioned homogeneously distributed local regular structure of a corresponding ANN - in each node (pointed net, "local view") there is the same net structure. We point out here, that for our mathematical model of a network structure we do not resort to an underlying grid. The associated nets will be abstract mathematical models, but they contain all the relevant information of the corresponding network (set of nodes and synaptic connection structure).

3   Some Basic Mathematical Notions

In this section we develop the basic mathematical notions, especially from geometry, and show how the concept of a noncommutative geometric space naturally arises in our study of networks. Accordingly, we introduce the notion of a *geometric net*. Such geometric nets can be associated to connectionist networks . As already indicated we will use the language of *category theory* as "linguistic basis". It is an experience that "categorical modeling" can lead to the effect of a "formal economy". (We note here that our notion of a category has to be

distinguished from the categories discussed in the cognitive sciences. We refer
to the remarks in [NKK94], section 12.3; this has nothing in common with our
work below). Thus, we are interested in a *category of associated geometric nets*
that incorporates all the necessary information which is needed to represent a
corresponding connectionist network mathematically. Additionally, we aim at
modeling *learning* in terms of the categorical notion of *morphisms*.

In [NKK94] a mathematical definition of an ANN is introduced which fits
the common "classical" neural network paradigms. For our purposes here and
for future use we propose the following general (preliminary) definition (more
details will be discussed in [GP]). A network (ANN) consists of an underlying
set of nodes ("points", in terms of elements of a corresponding space) $\mathbf{X}$ and a
bivariate map $<,>: \mathbf{X} \times \mathbf{X} \longrightarrow R$, where R is a set. The map $<,>$ assigns to
each pair of nodes (points) x, y a "weight" ("color") $< x, y > \in R$. In classical
ANNs this is a numerical value, but we want to be able to model more general
data for an ordered pair ("edge") (x, y) – like predicates, states, activities, etc..
$\mathbf{X}$ is partitioned into disjoint subsets $\mathbf{X_1}, ..., \mathbf{X_n}$, called "layers". There are two
distinguished layers, say $\mathbf{X_1}$ ("input layer") and $\mathbf{X_n}$ ("output layer"), being of
relevance for real applications. The other layers are often called "hidden layers".
A layer $\mathbf{X_i}$ can be interpreted as a (sub-)network of $\mathbf{X}$ given by
$<,>_i: \mathbf{X_i} \times \mathbf{X_i} \longrightarrow R_i$, ($<,>_i$ is the corresponding restriction of $<,>$ and R_i
the corresponding subset of R). It is clear, that the "local" and "global" network
structure is determined by $<,>$. In practical applications a node (point) of $\mathbf{X}$
will be a "neuron" ("processing element"); it has to be specified by corresponding
parameters and information processing capabilities (e.g. threshold unit, tranfer
functions). In this contribution we do not deal with these aspects, we are only
interested in the network structure ("topology"). As discussed below, $\mathbf{X}$ will be
interpreted as an object of a suitable category and learning as (a sequence of)
morphisms.

3.1   Some Basic Notions from Category Theory

For convenience of reading we recall the basic notions from category theory, like
category, functor, natural transformation, as we need them for later use.

Definition 1. *A category $\mathbf{A}$ consists of a class of objects, denoted by A, B,
C, ... $\in Obj(\mathsf{A})$ (the objects of $\mathbf{A}$), and for each pair of objects A, B a set of
morphisms, $Mor(A, B)$, also denoted by $\mathbf{A}(A, B)$ and a composition relation on
morphisms such that if $f : A \to B$ and $g : B \to C$ are morphisms, then there is
a morphism $g \circ f : A \to C$, the composition of f and g. For these notions the
following two axioms are required for a category.*
   (i) The composition of morphisms is associative, that is $h \circ (g \circ f)$
       $= (h \circ g) \circ f$.
   *(ii) For every object A there is the identity morphism id_A with the
       properties $f \circ id_A = f$ and $id_B \circ f = f$ for all $f : A \to B$.*
(Note that $Mor(A, B)$ can be empty).

We briefly emphasize here that the *arrow notation* for morphisms is of basic importance. We shall use $f : A \to B$ as well as $A \xrightarrow{f} B$ to denote morphisms. The arrow notation is well suited to illustrate (to "visualize") a broad spectrum of modeling problems in a categorical sense (e.g. everything dealing with relational structures).

Some typical examples of categories in mathematics are, among others: the category of *sets, groups, monoids, topological spaces, vector spaces* over a field, etc.. General relational structures can be interpreted in categorical terms (cf. [Pfa94]). Summarizing, one can say that category theory discusses the basic features of "everyday work" when dealing with spaces in a certain discipline and studying structure preserving functions (the morphisms) between spaces.

For later use we introduce the notion of the *category of pointed sets*, $\mathbf{SET}_*$. A pointed set (set with base point) X_a is a set X together with a selected "base point" $a \in X$. If X_a, X_b are pointed sets, then a base-point-preserving map is a map $f : X_a \longrightarrow X_b$, s.th. $f(a) = b$. With pointed sets as objects, base-point-preserving maps as morphisms, and ordinary composition of maps we obtain the category $\mathbf{SET}_*$. The notion of a pointed space (space with base point) is basic in algebraic topology (homotopy theory).

Definition 2. *The notion of a functor constitutes a concept of "function" between categories. Let* $\mathbf{X}$ *and* $\mathbf{Y}$ *denote two categories. Then a functor* $F : \mathbf{X} \longrightarrow \mathbf{Y}$ *assigns to every object* $A \in Obj(\mathbf{X})$ *an object* $F(A)$ *in the category* $\mathbf{Y}$ *and to every morphism* $f : A \to B$ *in* $\mathbf{X}$ *a morphism* $F(f) : F(A) \to F(B)$ *in* $\mathbf{Y}$ *such that the following holds for morphisms* $f : A \to B$, $g : B \to C$ *and* id_A *in* $\mathbf{X}$

(1) $F(g \circ f) = F(g) \circ F(f)$
(2) $F(id_A) = id_{F(A)}$

More specifically, such a functor is called covariant; it is called contravariant, if it reverses arrows and thus reverses the order of the arrows of a composition of morphisms (i.e. $F(g \circ f) = F(f) \circ F(g)$*).*

On the next higher level of abstraction the notion of a *natural transformation* is settled. It is a kind of a function between functors and is defined as follows.

Definition 3. *Let* $F : \mathbf{X} \longrightarrow \mathbf{Y}$ *and* $G : \mathbf{X} \longrightarrow \mathbf{Y}$ *be two functors. A natural transformation* $\alpha : F \longrightarrow G$ *is given by the following data.*

For every object A *in* $\mathbf{X}$ *there is a morphism* $\alpha_A : F(A) \to G(A)$ *in* $\mathbf{Y}$ *such that for every morphism* $f : A \to B$ *in* $\mathbf{X}$ *the following diagram (square) is commutative.*

$$
\begin{array}{ccc}
F(A) & \xrightarrow{\;\alpha_A\;} & G(A) \\
{\scriptstyle F(f)}\downarrow & & \downarrow{\scriptstyle G(f)} \\
F(B) & \xrightarrow[\;\alpha_B\;]{} & G(B)
\end{array}
$$

Commutativity means (in terms of equations) that the following compositions of morphisms are equal: $G(f) \circ \alpha_A = \alpha_B \circ F(f)$.

The morphisms α_A, $A \in Obj(\mathbf{A})$, *are called the components of the natural transformation* α.

For more material about categories we refer to the rather extensive literature (here only citing [AHS90], [Lan98], [LS96], [Pie91]).

3.2   Colored Graphs

In our considerations below we are working with the following definition of a graph and (general) net. We keep close to the exposition of that material as presented in [Pfa95] and the literature used there.

Definition 4. *A directed graph (with orientation) Y is given by a set of vertices (nodes)* $VY \neq \emptyset$ *and a set of edges (arcs)* EY, *where* $VY \cap EY = \emptyset$, *and two incidence maps* $\iota : EY \to VY$, $\tau : EY \to VY$.
For $e \in EY$ *we say that* ιe *is the initial point and* τe *is the terminal point of the edge e, respectively. Through the maps* ι *and* τ *each edge e obtains an orientation and in that way the graph is directed with orientation. In general, we allow* $\iota e = \tau e$, *that means e is a loop.*
A (general) net is defined as a directed (and, if necessary, oriented) graph Y with a valuation or weighting or coloring $\omega : EY \to R$, $R \neq \emptyset$.

Let Γ and X denote two directed (oriented) graphs, as defined at the beginning. A *graph morphism* $\alpha : \Gamma \to X$ is defined in the obvious way as a structure preserving map from vertices to vertices and edges to edges. More explicitly, let e be an edge with vertices $\iota e = v$, $\tau e = w$, then the image $\alpha(e)$ is the edge with vertices $\alpha(v)$, $\alpha(w)$.

It is then clear what is the definition of a *graph isomorphism* and *graph automorphism*. Accordingly, for general nets a *morphism* is defined as a graph morphism which respects the coloring on the edges. That means $\omega(e_1) = \omega(e_2) \Rightarrow \omega(\alpha e_1) = \omega(\alpha e_2)$ - this is compatible with the definition of a morphism in the category of geometric spaces given below.

3.3   A Category of Geometric Spaces

We briefly recall here the definition of a *geometric space* and for the details we refer to the literature. *Noncommutative geometry* was introduced by J.André more than 20 years ago as a natural generalization of classical affine geometry. A line $x \sqcup y$ joining two points $(x, y \in X)$ of the underlying point set X is *directed* (e.g. like a ray in Euclidean spaces); thus, $x \sqcup y \neq y \sqcup x$ (noncommutativity of the join operation), in general. As usual, the notion of *parallelism* is defined as an equivalence relation on the set of all lines of a space. As a very brief selection of references of the extensive work of J.André we cite [And88, And92, And93]. In [Pfa85] a new modeling approach for noncommutative spaces was introduced leading to an algebraization of the notion of a noncommutative geometric space. It turns out that the parallelism of a space encodes already the whole geometric structure. In this approach a geometric space is defined by a map $<,>: X^2 \to R$ called *parallel map* or *parallelism*; a *line* joining two points $x, y \in X$ is defined as follows:

$x \square y := x \sqcup y \cup \{< x, y >\}$,   where $< x, y >$ is called the *ideal point* or *direction* or *color* of the line and $x \sqcup y := \{x\} \cup \{z \mid < x, z >=< x, y >\}$ is the set of *proper points* of the line – it is the solution set of the equation $< x, \zeta >=< x, y >$. R denotes the set of *directions, colors* of a space. The validity of *geometric axioms, configurations* can be expressed in terms of the solvability of *corresponding equations* in this "calculus" (cf. [Pfa85]). Concerning geometric axioms, the parallel shifting of triangles and, more generally, the shifting of simplices across a space, is of basic importance. If such configurational conditions hold, a space has a rich geometric structure (cf. e.g. [Pfa87]).

An example for illustration (*ray space*): Let $X := \mathbb{R}^n$ be the real n-space, then a non-commutative space with a rich geometric structure whose lines are the (directed) rays can be defined by ($\mathbb{R}_+$ denotes the non-negative reals): $x \sqcup y := x + \mathbb{R}_+.(y - x)$,   the ray beginning in x going through y. Two rays are parallel, if they have the same direction). In terms of a suitable parallel map $<,>$ the ray space can be defined as follows. $X := \mathbb{R}^n$, $R := S^{n-1} \cup \{0\}$, where S^{n-1} denotes the $(n-1)$- dimensional unit sphere in $\mathbb{R}^n$. Then the n-dimensional ray space is defined by the parallel map $<,>: X^2 \to R$, $< x, y >:= \frac{y-x}{\|y-x\|}$ if $x \neq y$, and $< x, x >:= 0$. Applying the definition of a line, the set of proper points of the line joining two different points x and y is given by the ray $x \sqcup y = x + \mathbb{R}_+ \cdot (y - x)$, whereas $x \sqcup x = \{x\}$.

A *morphism* between two geometric spaces $(X_1, <, >_1, R_1), (X_2, <, >_2, R_2)$ is defined as a map $f : X_1 \longrightarrow X_2$ which respects the underlying geometric structure, namely the parallelism, i.e. the following condition holds for all $x, y, u, v \in X_1$: if $< x, y >_1 =< u, v >_1$ then $< f(x), f(y) >_2 =< f(u), f(v) >_2$.

With these notions of a geometric space and corresponding morphisms we obtain the *category of geometric spaces* which we denote by **NCG** (cf. [Pfa98] for a summary). The *composition of morphisms* is the usual one (as in set theory). For illustrational purposes (and to prepare pictorially how we pass on to geometric nets later) we include Fig.3 below showing the "local view" of a geometric space in a selected point x (i.e. the "pointed space" $X_x, x \in X$).

3.4   Geometric Nets

After this brief collection of the basic introductory notions from geometric spaces we come to the natural link with net theory.

The "$<, >$-notion" for geometric spaces leads in a natural way to an associated net which we will call *geometric net*. Let $(X, <, >, R)$ be a given geometric space. The following directed (oriented) colored graph Y (geometric net) can be associated with it, naturally:

$VY := X$, $EY := X^2$, $\iota e := x$, $\tau e := y$,   for $e = (x, y) \in EY$.

The coloring $\omega : EY \to R$ is defined by $\omega(e) :=< x, y >$, for $e = (x, y) \in EY$. This leads to the associated *geometric net*. In all those cases where we want to

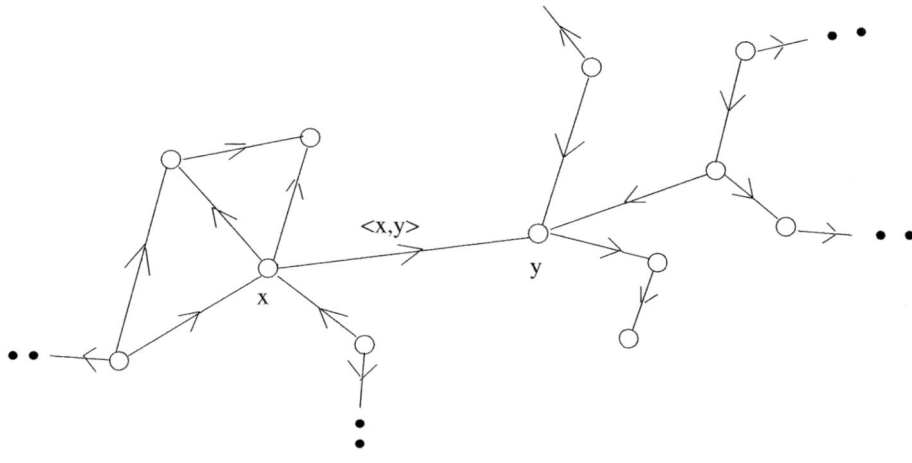

illustration of a geometric space (locally)

Fig. 3. "local view" (pointed space)

avoid loops $e = (x, x)$ we can work with $EY = X^2 \setminus i_X$ (where $i_X = \{(x, x) \mid x \in X\}$, all diagonal pairs), but in general we do not assume this, a priori. Thus, in our geometric nets it is suggestive to illustrate an edge $e = (x, y)$ with $\iota e \neq \tau e$ as a directed line segment with color (label) $< x, y >$ (cf. the edge joining x and y in the "local view" of a geometric space (Fig.3)). Conversely, given a (general) net, we obtain a geometric space having the nodes of the net as points and the corresponding parallel map induced by the coloring. In general, the underlying graph of a net is not complete (i.e. there are nodes not connected by an edge). If necessary we can make it complete: let x, y be two nodes that are not connected, then we define the direction $< x, y >= \infty$ in the induced geometric space, where ∞ is an additional, artificial direction (color). In this way we obtain the induced parallel map on all pairs of the underlying point set of the induced geometric space. Analogous to the category of geometric spaces **NCG** we obtain **GeoNET**, the category of geometric nets.

4 Associated Geometric Networks

As we already mentioned in section 2, the design of the network structures in Geiger's paradigm is amenable to mathematical modeling with methods presented in the foregoing section.

For illustration, below we briefly sketch in a concrete example how geometric features arise naturally. To this end we take a standard network structure which arises in applications. In our case we consider the following *filter* for line segment detection in pixel images; it is commonly in use (cf. for example [Roj93], section 3.4). All points which we consider are elements of a regular *integer grid* (we simply take $\mathbb{Z} \times \mathbb{Z}$). Locally, that means from the view of a selected grid point, denoted by x_0, we consider a specific neighborhood, in our case the eight nearest neighbors in the grid, denoted by $x_1, x_2, \ldots, x_8$, sitting around the center x_0. The filter we are considering is then defined by the *connection matrix*

$$\begin{pmatrix} -1 & -1 & -1 \\ -1 & +8 & -1 \\ -1 & -1 & -1 \end{pmatrix} .$$

This means that the central node receives a signal input weighted by -1 from all those neighboring pixels which are set to one (i.e. "black colored") and all these input signals are summed up and added to $+8$, the result is compared with a threshold and if this threshold is exceeded an output value 1 is produced indicating that the corresponding central point is part of a line segment. This is, roughly speaking, how the local feature extractor in a corresponding perceptron network would evaluate this filter. Now, we are associating to this situation the following network (local geometric net) which we illustrate in the following picture

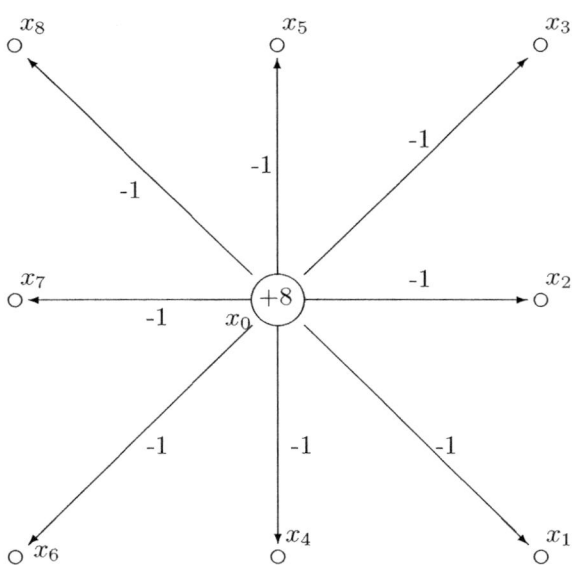

The corresponding net weights on the edges (coloring of the net) can be interpreted in terms of the following *parallel map (coloring)*

$$<,>: \{x_0, x_1, \dots, x_8\} \times \{x_0, x_1, \dots, x_8\} \longrightarrow \{-1, +8\},$$

with $< x_0, x_0 >:= +8, \quad < x_0, x_i >:= -1, \quad \text{for all} \quad i = 1, 2, \dots, 8$

defining an associated *geometric space* locally, that corresponds to our associated geometric net. "Locally" here means that we are only defining the values $< x_0, x_i >$ with respect to the "base point" x_0 from which the space is locally regarded. We can easily extend this definition of $<,>$ to the whole grid by setting $< x_0, z >:= 0$ (or $< x_0, z >:= \infty$), for all other grid points z, and repeat this definition in every point x_0, accordingly. This, finally, yields the complete parallel map (coloring). Thus, evaluating the line $x_0 \sqcup x_1$, which is equal to $x_0 \sqcup x_i$, for all $i = 1, 2, \dots, 8$, we obtain $x_0 \sqcup x_1 = \{x_0\} \cup \{x_i \mid < x_0, x_i >=< x_0, x_1 >= -1\} = \{x_0\} \cup \{x_1, x_2, x_3, x_4, x_5, x_6, x_7, x_8\}$ and this is exactly the "discrete" circle around x_0 with "radius" (color) equal to -1. Hence we have obtained a space which is related to a "circle space" (cf. [Pfa85], [Pfa98]). We point out once more, this associated net is to be seen as an abstraction of the concrete local feature extractor as used in Geiger's paradigm (and similarly in a perceptron-like network). It encodes the essential information that is needed.

Based on the previous considerations it turns out that a certain step of abstraction is convenient, in the following sense . Given a concrete "real" ANN (designed for a particular task), denoted by $\mathcal{N}$, we assign to it an "abstract associated network" This *associated net*, denoted by $\mathsf{GNET}(\mathcal{N})$, encodes all the essential network information, but is much better suited for mathematical modeling purposes. The nodes of $\mathsf{GNET}(\mathcal{N})$ represent the neurons of $\mathcal{N}$, the edges represent the synaptic connections (including weights), and the structure of $\mathsf{GNET}(\mathcal{N})$ represents the network structure ("topology") of $\mathcal{N}$ (numerically expressible in terms of a corresponding connection matrix). With respect to the category of geometric nets $\mathsf{GNET}(\mathcal{N})$ is interpreted as an object of **GeoNET**.

Now we come to the "local view" of a network as illustrated in the figures above. Mathematically, the local view of a network corresponds to a *pointed net* (analogous to pointed set), i.e. we select a node and describe the (local) network structure from the "viewpoint" of this specific node. Due to the fact that an ANN $\mathcal{N}$, and consequently $\mathsf{GNET}(\mathcal{N})$, in each node has the same local structure (cf. section 2), it is reasonable to assign to a particular $\mathsf{GNET}(\mathcal{N})$, a specific pointed net, with respect to a selected node, say x_0 as base point (cf. the category of pointed sets in section 3). This local description of the network contains all the essential information, because the entire (global) network can be reconstructed by taking all local networks together (shifting the pointed net over all other nodes). To shorten notation we set $X :=\mathsf{GNET}(\mathcal{N})$ and let X_{x_0} denote the pointed (local) network structure. It is important to note, that the (geometric, local) net structure of X_{x_0} is completely given by the "local data" (connections, colors) $< x_0, y >$, where y runs through all the points (nodes) of X. Due to local regularity (as previously mentioned), all pointed nets are isomorphic to each other. This means, in every base point (node) we have the same situation concerning net structure and synaptic connections (colors). Thus the local description is less complex, but represents the essential net information. In terms of geometric spaces this amounts to saying that in every pointed space the same geometric configurations appear. This is a very strong regularity condition and leads to a rich geometric structure.

On the basis of these abstractions we formulate the process of *learning*. The standard way to describe learning is to change (update) the synaptic weights (colors) of the directed edges of a network. Thus, if i and j are the indices of connected neurons and w_{ij} denotes the connection strength, then a *learning step* is described by $w_{ij} \leftarrow w_{ij} + \Delta w_{ij}$. Usually, w_{ij} is an integer or real number and Δw_{ij} is determined by a learning rule. In accordance with our notation in the categories **NCG** and **GeoNET**, respectively, we can express a learning step as follows. Let x_i, x_j denote the nodes in $X = \mathsf{GNET}(\mathcal{N})$ representing the corresponding neurons in the ANN, then $< x_i, x_j >= w_{ij}$. We point out that in our proposed approach the data $< x, y >$ associated with a pair of nodes is not restricted to a numerical value only, but can consist of further data like states (activity modes) of the nodes (neurons). The following argument applies to Geiger's regularly structured ANNs. If there is a pair of nodes (neurons) x_k, x_l with $< x_k, x_l >= w_{kl} = w_{ij} =< x_i, x_j >$, then a learning step causes the same change of weights, i.e. $\Delta w_{ij} = \Delta w_{kl}$. In other words, if two connection weights are equal, then after a learning step both can be changed, but their equality is preserved. This motivates us to model a *learning step* as a *morphism* in the sense of the categories **NCG**, **GeoNET**, respectively, in the following way. Let X_0 denote the initial associated network and X_1 the resulting network after the first learning step. Supposing that we do not alter the set of nodes (neurons) – this is usually the case – then a learning step can be modeled as an identity mapping on the nodes and a change of the corresponding weights or coloring. Geometrically, this means that the parallel map (coloring) $<, >_0 : X_0^2 \longrightarrow R$ after learning step 1 changes into $<, >_1 : X_1^2 \longrightarrow R$. Due to the previous remark, such a learning step corresponds to a morphism, denoted by $L : X_0 \longrightarrow X_1$, since the following holds. For points (nodes) x, y, u, v in X_0 we obtain:

$$< x, y >_0 =< u, v >_0 \quad \Longrightarrow \quad < L(x), L(y) >_1 =< L(u), L(v) >_1.$$ Thus L, being the identity on nodes (points) of X_0, has the property of a morphism. We suggest giving this the name "homomorphic learning", formulated in a categorical environment. Consequently, a *learning process* can be expressed in terms of a sequence of learning steps (composition of morphisms L), i.e.

$$X_0 \xrightarrow{L} X_1 \xrightarrow{L} X_2 \xrightarrow{L} \ldots \ldots \xrightarrow{L} X_{n-1} \xrightarrow{L} X_n.$$

If no confusion can arise we use the same symbol L for each morphism in this sequence.

Analogously, to each associated (global) geometric net X_i in the above sequence we obtain a corresponding pointed net $(X_i)_{x_0}$, locally, in base point (selected node) x_0. Learning w.r.t. these pointed nets functions in exactly the same way as described previously and the local change of synaptic connections matches the corresponding changes in X_i. Consequently, we have a corresponding sequence of pointed geometric nets representing "local" learning which is "cheaper", than "global" learning considered above:

$$(X_0)_{x_0} \xrightarrow{L} (X_1)_{x_0} \xrightarrow{L} (X_2)_{x_0} \xrightarrow{L} \ldots \ldots \xrightarrow{L} (X_{n-1})_{x_0} \xrightarrow{L} (X_n)_{x_0}.$$

Analogous to the category $\mathbf{SET}_*$, we introduce the category of pointed geometric nets $\mathbf{GeoNET}_*$, a subcategory of $\mathbf{GeoNET}$. The process of assigning a geometric net X to the pointed net X_{x_0} (with base point x_0, a node of X) is of functorial nature in the sense of a functor in category theory. Let us consider a fixed (finite) set of points (nodes) P interpreted as nodes of (associated) geometric nets. Then the collection of all geometric nets having the same underlying set of nodes (points) P is a subcategory of $\mathbf{GeoNET}$, denoted by $\mathbf{GeoNET}(P)$. Let x_0 be a node of P, then we can define the following functor $\mathsf{P}_{x_0} : \mathbf{GeoNET}(P) \longrightarrow \mathbf{GeoNET}_*$, that assigns to each geometric net X in $\mathbf{GeoNET}(P)$ the corresponding pointed net X_{x_0}. It is clear that a morphism of $\mathbf{GeoNET}(P)$ leads to a corresponding morphism of $\mathbf{GeoNET}_*$. The properties of a functor are naturally verified.

Finally, we can establish the following commutative diagram, using the previously introduced notation.

$$
\begin{array}{ccc}
X & \xrightarrow{\ \sigma_X\ } & X_{x_0} \\
{\scriptstyle L}\big\downarrow & & \big\downarrow{\scriptstyle L} \\
X' & \xrightarrow{\ \sigma_{X'}\ } & X'_{x_0}
\end{array}
$$

For the objects and morphisms on the right side of the commutative square the following relations hold $X_{x_0} = \mathsf{P}_{x_0}(X)$, $X'_{x_0} = \mathsf{P}_{x_0}(X')$ and $L = \mathsf{P}_{x_0}(L)$. This diagram corresponds to the natural transformation $\sigma : \mathsf{Id} \longrightarrow \mathsf{P}_{x_0}$, where Id is the *identity functor* on $\mathbf{GeoNET}$ (and thus on $\mathbf{GeoNET}(P)$), P_{x_0} is the previously defined functor. $\sigma_X, \sigma_{X'}$ are abitrary components of the natural transformation. Analyzing this diagram we can observe, that "Learning on the right is cheaper" (for more details we refer to the remark on the "local view" of a network in section 4). This fact can be used to reduce the complexity of training (learning) in a particular ANN. In the next section we briefly present an application where this effect was exploited.

5   An Industrial Application: Optical Quality Control

In this section we briefly describe an industrial project carried out by H.Geiger, in which the application of our previously sketched mathematical modeling approach showed unexpected economic effects. In the project a quality control problem in production of tiles had to be solved (cf. [Gei94]). Among others, a problem was to detect breaks in the edges of the upper and lower surface of a tile, deformation of edges, pores in the upper surface (more than about $3mm$ diameter has to be rejected). Uneven distribution of the embedded material must be avoided, certain changes in color are unacceptable. Rough spots on the surface have to be singled out and failures during polishing can even cause breaks in the surface. Time constraints had to be taken into account. All the checks must be done within a 2 seconds time interval per tile. At least 4 camera images ($512 * 512$ pixels per image) are necessary to check all the edges and surfaces. A main requirement is the flexibility of use. A change of the material of the test objects should be possible within less than 1 hour and without changing the software. The system should have a minimal robustness concerning disturbances (like changes of light, position, speed). Typical for such real world problems is the great difficulty to achieve a symbolic, logic and closed form mathematical description of the whole scenario. It turned out in that project that a key problem was that the criteria to be checked were not amenable to easy implementations in numerical algorithms. This is a typical situation where one can successfully apply neural network techniques.

Finally, in the industrial project, the strategy of solution to these practical problems is based on a combination of ANN approaches and classical methods. The network used is designed as a multilayer network with preprocessing facilities, feature extraction and classification neurons at the output. As described in [Gei94], direct exploitation of our joint mathematical modeling approach in ANN simulation with NeuroTools (exploiting the association step and the commutativity of the corresponding diagram) led to considerable reductions in production costs.

6   Concluding Remarks, Prospects

In this note, our interest concentrated on the net structure of an ANN. The networks we consider are (locally) regularly structured in terms of certain configurations which can be interpreted as geometric configurations. This local structuring is homogeneously distributed over the entire net, i.e. the same type of (geometric) configurations can be found locally in each node. Of future interest will be the investigation of the "simplicial structure" of a network in the sense of simplex configurations (cf. [Pfa87]) possibly leading to a corresponding group operation on a net. Besides that, other outcomes of interest include a semantic modeling approach : interpreting a net structure as a general relational structure and working with the category PATH (cf. [Pfa94]). Another idea is to apply categorical constructions like limits, colimits (including products, coproducts) in order to construct large networks by small components (modules); in such processes morphisms play a basic role.

References

[AHS90]   J. Adámek, H. Herrlich, and G.E. Strecker. *Abstract and Concrete Categories*. John Wiley & Sons, 1990.

[And88]   J. André. Endliche nichtkommutative Geometrie. *Annales Univ. Saraviensis. Ser. Math.*, 2(1):1–136, 1988.

[And92]   J. André. Configurational conditions and digraphs. *Journal of Geometry*, 43:22–29, 1992.

[And93]   J. André. On non-commutative geometry. *Annales Univ. Saraviensis. Ser. Math.*, 4(2):93–129, 1993.

[Eck90]   R. Eckmiller. Concerning the emerging role of geometry in neuroinformat-
           ics. In *Parallel Processing in Neural Systems and Computers. R.Eckmiller,
           G.Hartmann, G.Hauske (Eds.), North-Holland,* 1990.

[Gei94]   H. Geiger. Optical quality control with selflearning systems using a com-
           bination of algorithmic and neural network approaches. Proceedings of the
           Second European Congress on Intelligent Techniques and Soft Computing,
           EUFIT'94, Aachen, September 20-23, 1994.

[GP]   H. Geiger and J. Pfalzgraf. Modeling a connectionist network paradigm:
           Geometric and categorical perspectives. In preparation.

[GP95]   H. Geiger and J. Pfalzgraf. Quality control connectionist networks supported
           by a mathematical model. In *Proceedings of the International Conference on
           Engineering Applications of Artificial Neural Networks (EANN'95), 21-23
           August 1995, Helsinki. A.B.Bulsari, S.Kallio (Editors), Finnish AI Society,*
           1995.

[Lan98]   S. Mac Lane. *Categories for the Working Mathematician.* Springer Verlag,
           Graduate Texts in Mathematics **5**, 2nd ed., 1998.

[LS96]   F.W. Lawvere and S.H. Schanuel. *Conceptual Mathematics: A First Intro-
           duction to Categories.* Cambridge University Press, 1996.

[NKK94]   D. Nauck, F. Klawonn, and R. Kruse. *Neuronale Netze und Fuzzy-Systeme.*
           Vieweg Verlag, 1994.

[Pfa85]   J. Pfalzgraf. On a model for noncommutative geometric spaces. *Journal of
           Geometry,* 25:147–163, 1985.

[Pfa87]   J. Pfalzgraf. A note on simplices as geometric configurations. *Archiv der
           Mathematik,* 49:134–140, 1987.

[Pfa94]   J. Pfalzgraf. On a general notion of a hull. In *Automated Practical Rea-
           soning, J.Pfalzgraf and D.Wang (eds.). Texts and Monographs in Symbolic
           Computation, Springer Verlag Wien, New York,* 1994.

[Pfa95]   J. Pfalzgraf. Graph products of groups and group spaces. *J. of Geometry,*
           53:131–147, 1995.

[Pfa98]   J. Pfalzgraf. On a category of geometric spaces and geometries induced by
           group actions. *Ukrainian Jour. Physics,* 43,7:847–856, 1998.

[Pie91]   B.C. Pierce. *Basic Category Theory for Computer Scientists.* The MIT
           Press, Cambridge, Massachusetts, 1991.

[Roj93]   R. Rojas. *Theorie der neuronalen Netze.* Springer Verlag, 1993. Springer-
           Lehrbuch.

A New Artificial Intelligence Paradigm for Computer-Aided Geometric Design

Andres Iglesias and Akemi Gálvez

Department of Applied Mathematics and Computational Sciences, University of
Cantabria, Avda. de los Castros, s/n, E-39005, Santander, Spain
iglesias@unican.es

Abstract. Functional networks is a powerful and recently introduced
Artificial Intelligence paradigm which generalizes the standard neural
networks. In this paper functional networks are used to fit a given set of
data from a tensor product parametric surface. The performance of this
method is illustrated for the case of Bézier surfaces. Firstly, we build the
simplest functional network representing such a surface, and then we use
it to determine the degree and the coefficients of the bivariate polynomial surface that fits the given data better. To this aim, we calculate
the mean and the root mean squared errors for different degrees of the
approximating polynomial surface, which are used as our criterion of a
good fitting. In addition, functional networks provide a procedure to describe parametric tensor product surfaces in terms of families of chosen
basis functions. We remark that this new approach is very general and
can be applied not only to Bézier but also to any other interesting family
of tensor product surfaces.

1 Introduction

1.1 Preliminars

Computer-Aided Geometric Design (CAGD) is devoted to constructing a precise
mathematical description of the shape of a real object, and focuses on the efficient computer representation of its geometry. Its range of applications includes
areas like publicity, animation, multimedia tools, virtual reality, computer vision,
robotics, etc. For an introduction to the field, the reader is referred to [8,10,12].

The main aspect of CAGD is the study of free-form curves and surfaces. They
are essential tools (among others) in the automotive, aircraft and shipbuilding
industries [1]. Roughly speaking, free-form curves and surfaces are parametric
functions governed by a set of points (called *control points*) that more or less
determine the shape of the curve or surface and many of its geometric properties.

Free-form curves and surfaces have been extensively applied to fit data. In
general, we are given a set of data and we look for the curve or surface following
some functional structure and that minimizes the error on the prescribed data. A
number of different methodologies to solve this problem have been described. The
one proposed here takes advantage of a recent Artificial Intelligence paradigm

J.A. Campbell and E. Roanes-Lozano (Eds.): AISC 2000, LNAI 1930, pp. 200–213, 2001.

which generalizes the standard neural networks: the so-called *functional networks* [2].

In this paper we restrict ourselves to the case of Bézier surfaces. Firstly, we obtain the simplest functional network representing such a surface, and then we use it to determine the degree and the coefficients of the biparametric polynomial surface that fits the given data better. However, our proposal is very general and the same scheme can be successfully applied to any other interesting family of tensor product surfaces in CAGD.

The structure of this paper is the following: firstly we introduce some mathematical concepts and definitions. In addition, a set of data to be used later is obtained. Then, functional networks are motivated by introducing an example of a problem which cannot be well described in terms of neural networks. Section 2 describes the functional networks paradigm. Differences between neural and functional networks will also be discussed in this section. Section 3 gives a general methodology to work with these networks. The required steps of the method will be illustrated by its application to the parametric surfaces problem. Some interesting features of the method, like the possibility to determine the degree and the coefficients of the approximating surface that fits the given data better are also shown in this section. Finally, the paper closes with the main conclusions of this work.

1.2   Some Mathematical Definitions

A *Bézier curve of degree m* is given by

$$\mathbf{C}(s) = \sum_{i=0}^{m} \mathbf{P}_i B_i^m(s) \tag{1}$$

where $\{\mathbf{P}_i;\ i = 0, \ldots, m\}$ is a set of $(m + 1)$ two- or three-dimensional points called *control points* and $B_i^m(s)$ are the *Bernstein polynomials of degree m*, defined as

$$B_i^m(s) = \binom{m}{i} s^i (1-s)^{m-i} \qquad \text{where} \qquad \binom{m}{i} = \frac{m!}{i!\,(m-i)!}$$

To make this definition useful practically, we focus on the parameter interval $[0, 1]$ (see [8,10]). Note that in this paper vectors are denoted in bold.

With this notation, a *tensor product Bézier surface of degree m×n* is given by

$$\mathbf{P}(s,t) = \sum_{i=0}^{m} \sum_{j=0}^{n} \mathbf{P}_{ij} B_i^m(s) B_j^n(t) \tag{2}$$

where $\{\mathbf{P}_{ij}|\ i = 0, \ldots, m;\ j = 0, \ldots, n\}$ are also control points and $B_i^m(s)$, $B_j^n(t)$ are the Bernstein polynomials of degrees m and n respectively. Once again, the variables s and t are to be valued onto the square $[0, 1] \times [0, 1]$.

1.3   Obtaining a Set of Data

To describe how functional networks work we need some data. In general, they come from an unknown surface to be obtained. However, since our primary goal is to show the performance of the functional networks in fitting surfaces, we will focus on data given from a parametric Bézier surface. Note that this limitation is motivated by academic purposes only, and the reader will not find any trouble in generalizing our statements to a set of points coming from any unknown surface.

To this aim, we have selected a set of 121 data $\{T_{pq};\ p,q = 1, \ldots, 11\}$ in a regular 11×11 grid from a Bézier surface. This surface has been generated from (2) for the case $m = n = 3$, with 16 control points which are listed in Table 1.

(x, y, z)	(x, y, z)	(x, y, z)	(x, y, z)
$(1, 1, 1)$	$(1, 3, 3)$	$(1, 5, 2)$	$(1, 7, 5)$
$(3, 1, 3)$	$(3, 3, 6)$	$(3, 5, 1)$	$(3, 7, 6)$
$(5, 1, 2)$	$(5, 3, 1)$	$(5, 5, 6)$	$(5, 7, 1)$
$(7, 1, 6)$	$(7, 3, 5)$	$(7, 5, 3)$	$(7, 7, 4)$

Table 1. Control points used to define the parametric tensor product Bézier surface.

The resulting surface is called a *bicubic tensor product Bézier surface* in CAGD. Its final expression is given by:

$$\begin{pmatrix} x(s,t) \\ y(s,t) \\ z(s,t) \end{pmatrix} = \begin{pmatrix} 1 + 6\,s \\ 1 + 6\,t \\ 1 + 6s - 9s^2 + 8s^3 + 6t + 9st - 45s^2t+ \\ 27s^3t - 9t^2 - 45st^2 + 171s^2t^2 - 120s^3t^2+ \\ 7t^3 + 33st^3 - 135s^2t^3 + 99s^3t^3 \end{pmatrix} \quad (3)$$

In order to check the robustness of the proposed method, the third coordinate of the 121 points $\{(x_k, y_k, z_k)\}$ was slightly modified by adding a uniform random variable $U(-0.05, 0.05)$. Such a random variable plays the role of a measure error that usually appears in many realistic situations.

1.4   Motivating Functional Networks

Artificial neural networks have been recognized as a powerful tool for learning and simulating systems in a great variety of fields (see [7] and [9] for a survey of this field). However, not every approximation problem is adequately described in terms of a neural network. The following example illustrates this situation:

Example: *Suppose that we look for the most general family of parametric surfaces* $\mathbf{P}(s,t)$ *such that their isoparametric curves (see [8] and [10] for a description)* $s = s_0$ *and* $t = t_0$ *are linear combinations of the sets of functions:* $\mathbf{f}(s) = \{f_0(s), f_1(s), \ldots, f_m(s)\}$ *and* $\mathbf{f}^*(t) = \{f_0^*(t), f_1^*(t) \ldots, f_n^*(t)\}$ *respectively.*

To be more precise, we look for surfaces $\mathbf{P}(s,t)$ *such that they satisfy the system of functional equations*

$$\mathbf{P}(s,t) \equiv \sum_{j=0}^{n} \boldsymbol{\alpha}_j(s) f_j^*(t) = \sum_{i=0}^{m} \boldsymbol{\beta}_i(t) f_i(s) \qquad (4)$$

where the sets of coefficients $\{\boldsymbol{\alpha}_j(s); j = 0, 1, \ldots, n\}$ *and* $\{\boldsymbol{\beta}_i(t); i = 0, 1, \ldots, m\}$ *can be assumed, without loss of generality, as sets of linearly independent functions. Note that if they are not, we can rewrite equations in (4) in the same form but with linearly independent sets.*

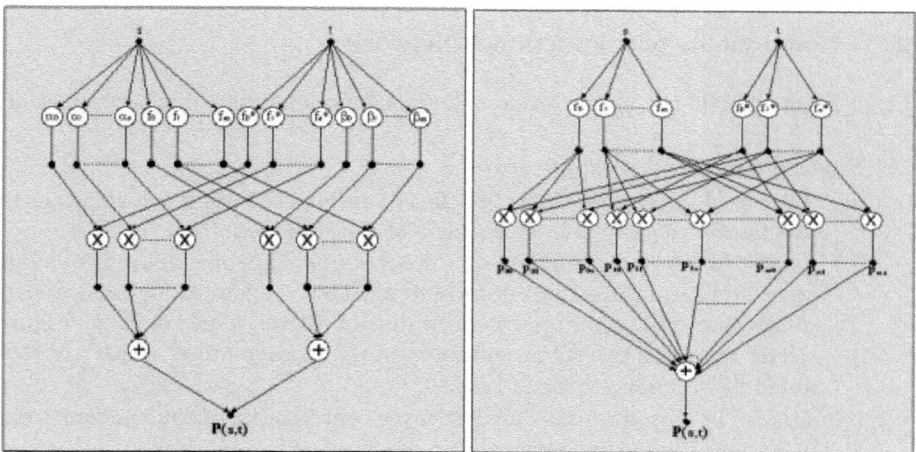

Fig. 1. (left) Graphical representation of a functional network for the parametric surface of eq. (4); (right) Functional network associated with eq. (5). It is equivalent to the functional network on the left.

This problem admits the graphical representation given in Figure 1(left) which, at first sight, looks like a neural network. However, the previous description in terms of neural networks presents the following problems:

- Neural functions in neural networks are identical, whereas neural functions in our example are different. For instance, we may find product and sum operators (indicated in Figure 1 by the symbols '×' and '+' respectively).
- The neuron outputs of neural networks are different; however, in our scheme, some neuron outputs in the example are coincident (this is the case of the outputs in Figure 1(left) associated with the last layer of neurons leading to the value of $\mathbf{P}(s,t)$).

These and other disadvantages suggest that the neural networks paradigm is very restrictive and can be improved in several directions. Recently, a powerful extension of neural networks has been introduced [2] and successfully applied to several problems [3,4], without exhibiting the previous shortcomings. Such an extension is based on the idea of allowing the activation functions of the neurons to be unknown functions from a given family, which will be estimated during the learning process.

Until the present, functional networks has not been extensively applied to CAGD. The first example of application of functional networks to CAGD was given in [5]. However, only implicit and explicit surfaces were considered there.

2   Functional Networks

2.1   Components of a Functional Network

From Figure 1(left) the main components of a functional network become clear:

1. **Several layers of storage units**.
   (a) *A layer of input units*. This first layer contains the input information. In this figure, this input layer consists of the units s and t.
   (b) *A set of intermediate layers of storage units*. They are not neurons but units storing intermediate information. This set is optional and allows connecting more than one neuron output to the same unit. In Figure 1(left) there are two intermediate layers of storage units, which are represented by small circles in black.
   (c) *A layer of output units*. This last layer contains the output information. In Figure 1(left) this output layer reduces to the unit $\mathbf{P}(s,t)$.
2. **One or more layers of neurons or computing units**. A neuron is a computing unit which evaluates a set of input values, coming from the previous layer, of input or intermediate units, and gives a set of output values to the next layer, of intermediate or output units. Neurons are represented by circles with the name of the corresponding neural function inside. For example, in Figure 1(left), we have three layers of neurons. The first one gives outputs of functions with one variable. The second layer exhibits the same function for all its neurons, the product operator. Similarly, the last layer exhibits the sum operator for its two neurons.
3. **A set of directed links**. They connect the input or intermediate layers to its adjacent layer of neurons, and neurons of one layer to its adjacent intermediate layers or to the output layer. Connections are represented by arrows, indicating the information flow direction. We remark here that information flows in only one direction, from the input layer to the output layer.

All these elements together form the *network architecture* or *topology* of the functional network, which defines the functional capabilities of the network. For example, since units are organized in series of layers, the functional network in Figure 1(left) is a *multilayer network*.

2.2 Differences between Functional and Neural Networks

Some of the differences between functional and neural networks were already introduced in Section 1.4. In these paragraphs, we discuss these differences and the advantages of using functional networks instead of standard neural networks.

1. In neural networks each artificial neuron receives an input value from the input layer or the neurons in the previous layer. Then it computes a scalar output $y = f\left(\sum w_{ik}x_k\right)$ from a linear combination of the received inputs $x_1, x_2, \ldots, x_n$ using a set of weights w_{ik} associated with each of the links and a given scalar function f (the *activation function*), which is assumed the same for all neurons. That is, each neuron returns an output $y = f(\sum w_{ik}x_k)$ that only depends on the value $\sum w_{ik}x_k$. Therefore, their neural functions have only one argument. On the contrary, neural functions in functional networks can have several arguments.
2. In neural networks the neural functions are *univariate*: neurons can show different outputs but all of them represent the same values. In functional networks, the neural functions can be *multivariate*.
3. In a given functional network the neural functions can be *different*, while in neural networks they are *identical*.
4. In neural networks there are weights, which must be learned. These weights do not appear in functional networks, where neural functions are learned instead.
5. In neural networks *the neuron outputs are different*, while in functional networks *neuron outputs can be coincident*. As we shall see, this fact leads to a set of functional equations, which have to be solved. These functional equations impose strong constraints leading to a considerable reduction in the degrees of freedom of the neural functions. In most cases this implies that neural functions can be reduced in dimension or expressed as functions of smaller dimensions.

All these features show that the functional networks exhibit more interesting possibilities than standard neural networks. This implies that some problems (e.g. the one introduced in Section 1.4) require functional networks instead of neural networks for their solution. In the next section, we shall take advantage of this fact by solving the problem of the characterization of parametric tensor product surfaces.

3 Working with Functional Networks

In this section, we describe the functional networks methodology, which is organized, for clarity, into eight different steps. These steps are described by their application to the parametric surface example previously introduced in Section 1.

Step 1 (Statement of the problem): Understanding the problem to be solved. This is a crucial step, which has been done in Section 1.

Step 2 (Initial topology): Based on the knowledge of the problem, the topology of the initial functional network is selected. Thus, the system of functional equations (4) leads to the functional network in Figure 1(left). Note that the above equations can be obtained from the network by considering the equality between the two values associated with the links connected to the output unit. We also remark that each of these values can be obtained in terms of the outputs of the preceding units by writing the outputs of the neurons as functions of their inputs, and so on.

Step 3 (Simplification): In this step, the initial functional network is simplified using functional equations. Given a functional network, an interesting problem consists of determining whether or not there exists another functional network giving the same output for any given input. This leads to the concept of equivalent functional networks. Two functional networks are said to be *equivalent* if they have the same input and output units and they give the same output for any given input. The practical importance of this concept is that we can define equivalent classes of functional networks, that is, sets of equivalent functional networks, and then choose the simplest in each class to be used in applications.

Coming back to the example, it seems that the functions $\{\alpha_j(s); j = 0, 1, \ldots, n\}$ and $\{\beta_i(t); i = 0, 1, \ldots, m\}$ have to be learned. However, the functional equations (4) put strong constraints on them. In fact, the general solution of this functional equation is given by the following theorem (see reference [11] for details):

Theorem 1. *The most general family of parametric surfaces* $\mathbf{P}(s, t)$ *such that all their isoparametric curves* $s = s_0$ *and* $t = t_0$ *are linear combinations of the sets of linearly independent functions:* $\mathbf{f}(s) = \{f_0(s), f_1(s), \ldots, f_m(s)\}$ *and* $\mathbf{f}^*(t) = \{f_0^*(t), f_1^*(t) \ldots, f_n^*(t)\}$ *respectively, is of the form*

$$\mathbf{P}(s, t) = \sum_{i=0}^{m} \sum_{j=0}^{n} \mathbf{P}_{ij} f_i(s) f_j^*(t) = \mathbf{f}(s).\mathbf{P}.(\mathbf{f}^*(t))^T \tag{5}$$

where $(.)^T$ *indicates the transpose of a matrix and* $\mathbf{P}_{ij}$ *are elements of an arbitrary matrix* $\mathbf{P}$*; therefore,* $\mathbf{P}(s, t)$ *is a tensor product surface.*

Two important conclusions can be derived from this theorem:

1. No other functional forms for $\mathbf{P}(s, t)$ satisfy equations (4). So, no other neurons can be replaced by neurons β_i, α_j, f_i and f_j^*. Therefore, eq. (5) provides a characterization of the tensor product surfaces, a pressing question in CAGD.
2. The functional structure of the solution is (5). This equation shows that the functional network in Figure 1(right) is equivalent to the functional network in Figure 1(left).

Step 4 (Uniqueness of representation): Here, conditions for the neural functions of the simplified functional network must be obtained. For eq. (5), two cases must be considered:

1. **The $f_i(s)$ and $f_j^*(t)$ functions are given**: Assume that there are two matrices $\mathbf{P} = \{\mathbf{P}_{ij}\}$ and $\mathbf{P}^* = \{\mathbf{P}_{ij}^*\}$ such that

$$\mathbf{P}(s,t) \equiv \sum_{i=0}^{m}\sum_{j=0}^{n} \mathbf{P}_{ij} f_i(s) f_j^*(t) = \sum_{i=0}^{m}\sum_{j=0}^{n} \mathbf{P}_{ij}^* f_i(s) f_j^*(t) \qquad (6)$$

Solving the uniqueness of representation problem consists of solving equation (6). To this aim, we write (6) in the form

$$\sum_{i=0}^{m}\sum_{j=0}^{n} \left(\mathbf{P}_{ij} - \mathbf{P}_{ij}^*\right) f_i(s) f_j^*(t) = 0 \qquad (7)$$

Since the functions in the set $\{f_i(s)\, f_j^*(t) \mid i = 0, 1, \ldots, m \; ; \; j = 0, 1, \ldots, n\}$ are linearly independent because the sets $\{f_i(s) \mid i = 0, 1, \ldots, m\}$ and $\{f_j^*(t) \mid j = 0, 1, \ldots, n\}$ are linearly independent, from (7) we have

$$\mathbf{P}_{ij} = \mathbf{P}_{ij}^* \quad ; \quad i = 0, 1, \ldots, m \; ; \; j = 0, 1, \ldots, n$$

that is, the coefficients $\mathbf{P}_{ij}$ in (5) are unique.

2. **The $f_i(s)$ and $f_j^*(t)$ functions are to be learned**: In this case, assume that there are two sets of functions $\{f_i(s), f_j^*(t)\}$ and $\{\tilde{f}_i(s), \tilde{f}_j^*(t)\}$, and two matrices $\mathbf{P}$ and $\tilde{\mathbf{P}}$ such that

$$\mathbf{P}(s,t) \equiv \sum_{i=0}^{m}\sum_{j=0}^{n} \mathbf{P}_{ij} f_i(s) f_j^*(t) = \sum_{i=0}^{m}\sum_{j=0}^{n} \tilde{\mathbf{P}}_{ij} \tilde{f}_i(s) \tilde{f}_j^*(t) \qquad (8)$$

Then we have

$$\sum_{i=0}^{m}\sum_{j=0}^{n} \mathbf{P}_{ij} f_i(s) f_j^*(t) - \sum_{i=0}^{m}\sum_{j=0}^{n} \tilde{\mathbf{P}}_{ij} \tilde{f}_i(s) \tilde{f}_j^*(t) = 0 \qquad (9)$$

According to Theorem 1 in [6], the solution satisfies

$$
\begin{pmatrix}
\sum\limits_{i=0}^{m} \mathbf{P}_{i0} f_i(s) \\
\sum\limits_{i=0}^{m} \mathbf{P}_{i1} f_i(s) \\
\vdots \\
\sum\limits_{i=0}^{m} \mathbf{P}_{in} f_i(s) \\
\hline
\sum\limits_{i=0}^{m} \tilde{\mathbf{P}}_{i0} \tilde{f}_i(s) \\
\sum\limits_{i=0}^{m} \tilde{\mathbf{P}}_{i1} \tilde{f}_i(s) \\
\vdots \\
\sum\limits_{i=0}^{m} \tilde{\mathbf{P}}_{in} \tilde{f}_i(s)
\end{pmatrix}
= \begin{pmatrix} \mathbf{P}^T \\ \hline \mathbf{B} \end{pmatrix} \mathbf{f}^T(s) \quad ; \quad
\begin{pmatrix}
f_0^*(t) \\
f_1^*(t) \\
\vdots \\
f_n^*(t) \\
\hline
-\tilde{f}_0^*(t) \\
-\tilde{f}_1^*(t) \\
\vdots \\
-\tilde{f}_n^*(t)
\end{pmatrix}
= \begin{pmatrix} \mathbf{I} \\ \hline \mathbf{C} \end{pmatrix} (\mathbf{f}^*(t))^T.
$$

$$(10)$$

with

$$(\mathbf{P} \quad | \quad \mathbf{B}^T) \begin{pmatrix} \mathbf{I} \\ -- \\ \mathbf{C} \end{pmatrix} = \mathbf{0} \Leftrightarrow \mathbf{P} = -\mathbf{B}^T \mathbf{C} \tag{11}$$

From (10) we get

$$\tilde{\mathbf{P}}^T \; \tilde{\mathbf{f}}^T(s) = \mathbf{B}\mathbf{f}^T(s) \qquad ; \qquad (\tilde{\mathbf{f}}^*(t))^T = -\mathbf{C} \; (\mathbf{f}^*(t))^T, \tag{12}$$

Expression (12) gives the relations between both equivalent solutions and the degrees of freedom we have.

However, if we have to learn $\mathbf{f}(s)$ and $\mathbf{f}^*(t)$ we can approximate them as:

$$\mathbf{f}(s) = \phi(s) \; \mathbf{B} \qquad ; \qquad \mathbf{f}^*(t) = \psi(t) \; \mathbf{C}, \tag{13}$$

and we get

$$\mathbf{P}(s,t) = \mathbf{f}(s).\mathbf{P}.(\mathbf{f}^*(t))^T = \phi(s).\mathbf{B}.\mathbf{P}.\mathbf{C}^T.\psi(t)^T = \phi(s).\tilde{\mathbf{P}}.\psi(t)^T, \tag{14}$$

which is equivalent to (5) but with functions $\{\phi(s), \psi(t)\}$ instead of $\{\mathbf{f}(s),$ $\mathbf{f}^*(t)\}$. Thus, this case reduces to the first one.

Step 5 (Data collection): This step has been already done in Section 1.3.

Step 6 (Learning): At this point, the neural functions are estimated (learned), by using some minimization method. In functional networks, this learning process consists of obtaining the neural functions based on a set of data $D = \{(I_i, O_i)|i = 1, \ldots, n\}$ given in the previous step, where I_i and O_i are the i-th inputs and outputs, respectively, and n is the sample size.

Usually, the learning process is based on minimizing the sum of squared errors of the actual and the observed outputs for the given inputs

$$Q = \sum_{i=1}^{n} (O_i - F(I_i))^2, \tag{15}$$

where F is the compound function given the outputs, as a function of the inputs, for the given network topology. One learning alternative consists of approximating each neural function f_i by a linear combination of functions in a given family $\{\phi_{i1}, \ldots, \phi_{im_i}\}$. Thus, the approximated neural function $\hat{f}_i(\mathbf{x})$ becomes

$$\hat{f}_i(\mathbf{x}) = \sum_{j=1}^{m_i} a_{ij}\phi_{ij}(\mathbf{x}), \tag{16}$$

where $\mathbf{x}$ are the inputs associated with the i-th neuron. Note that the above function F includes all the neural functions in the network, and therefore it depends only on the coefficients a_{ij}, which are estimated in the learning process.

In the case of our example, the problem of learning the above functional network reduces to estimate the neuron functions $x(s,t)$, $y(s,t)$ and $z(s,t)$ from a given sequence of triplets $\{(x_k, y_k, z_k), \ k = 1, \ldots, 121\}$ which depend on s and t so that $x(s_k, t_k) = x_k$ and so on. To this aim we build the sum of squared errors function:

$$Q_\alpha = \sum_{k=1}^{121} \left(\alpha_k - \sum_{i=1}^{I} \sum_{j=1}^{J} a_{ij} \phi_i(s_k) \psi_j(t_k) \right)^2 \qquad (17)$$

where, in the present example, we should consider an error function for each variable x, y and z. However, since we have just introduced a measure error into the z coordinate, eq. (17) must be interpreted as an equation for $\alpha = z$ only. The optimum value is obtained when

$$\frac{\partial Q_\alpha}{2\partial a_{rs}} = \sum_{k=1}^{121} \left(\alpha_k - \sum_{i=1}^{I} \sum_{j=1}^{J} a_{ij} \phi_i(s_k) \psi_j(t_k) \right) \phi_r(s_k) \psi_s(t_k) = 0 \qquad (18)$$

$$r = 1, \ldots, I \qquad ; \qquad s = 1, \ldots, J.$$

To fit the 121 data points of our example, we have used monomials in s and t variables for the functions $\{\phi_i(s) = s^i | i = 0, 1, \ldots, I\}$ and $\{\psi_j(t) = t^j | j = 0, 1, \ldots, J\}$ in (17). Of course, every different choice for I and J yields to the corresponding system (18), which must be solved. In particular, as the data points come from a bicubic parametric surface, we have taken values for I and J from 2 to 4. Solving the system (18) for all of these cases, we always obtain the values $1 + 6\,s$ and $1 + 6\,t$ for $x(s, t)$ and $y(s, t)$ respectively (as expected, because they are not affected by any perturbation). But, of course, the corresponding approximation for $z(s, t)$ depends on the I and J values.

Step 7 (Model validation): At this step, a test for quality and/or the cross validation of the model is performed. Checking the obtained error is important to see whether or not the selected family of approximating functions is adequate. A cross validation of the model is also convenient.

To cross validate the model:

1. we have calculated the mean, the maximum and the root mean squared (RMS) errors, for the 121 training data points. The obtained results for the different values of I and J are reported in Table 2. As the reader can appreciate, the best choice corresponds to $I = J = 3$, for which the mean and the RMS errors are 0.0055 and 0.0001 respectively. Since errors are small, the selected approximating third degree bivariate polynomial was considered adequate.
2. we have also used the fitted model to predict a new set of 1681 testing data points, and calculated the mean, the maximum and the root mean squared (RMS) errors, obtaining the results shown in Table 3. Once more, the smallest error is obtained for $I = J = 3$. A comparison between mean and RMS error values for the training and testing data shows that, for this choice, they are comparable. Thus, we can conclude that no overfitting occurs. Note

TRAINING POINTS

	J=2	J=3	J=4
	0.1641	10.8169	0.1399
I=2	0.6731	27.9846	0.5128
	0.0194	1.2192	0.0163
	10.8169	0.0055	0.0072
I=3	27.9846	0.0405	0.0443
	1.2192	0.0001	0.0009
	5.9127	65.8192	0.0079
I=4	32.4473	335.8790	0.0503
	0.8665	9.5675	0.0009

Table 2. Mean, maximum and root mean squared errors of the 121 training points for different values of I and J.

TESTING POINTS

	J=2	J=3	J=4
	0.1443	10.5287	0.1205
I=2	0.6731	27.9846	0.5128
	0.0044	0.3131	0.0036
	1.6696	0.0055	0.0065
I=3	10.748	0.0405	0.04516
	0.0580	0.0001	0.0002
	5.6439	62.3208	0.0068
I=4	32.4474	335.879	0.0515
	0.2171	2.3869	0.0002

Table 3. Mean, maximum and root mean squared errors of the 1681 testing points for different values of I and J.

that a variance for the training data significantly smaller than the variance for the testing data is a clear indication of overfitting. This does not occur here.

As a conclusion, we have obtained $I = J = 3$ as the best choice for fitting the data points. In this case, the approximate bivariate polynomial for $z(s,t)$ is given by:

$$
\begin{aligned}
z(s,t) = & \ 0.995522 + 6.07021 \ s - 9.26584 \ s^2 + 8.24071 \ s^3 + 5.96881 \ t + \\
& 10.0693 \ s \ t - 47.0224 \ s^2 \ t + 27.6718 \ s^3 \ t - 8.86878 \ t^2 - \\
& 48.6516 \ s \ t^2 + 178.817 \ s^2 \ t^2 - 123.711 \ s^3 \ t^2 + 6.89967 \ t^3 + \\
& 35.7064 \ s \ t^3 - 141.16 \ s^2 \ t^3 + 102.271 \ s^3 \ t^3
\end{aligned}
\tag{19}
$$

where 6-digit precision has been used for calculating all the coefficients. Comparison of these results with (3) indicates that we have obtained a good approximation to the true surface. This fact is illustrated in Figure 2, where the fitted surface and the data points are shown. Figure 3(left) shows the errors associated with the data points on the grid for such a surface.

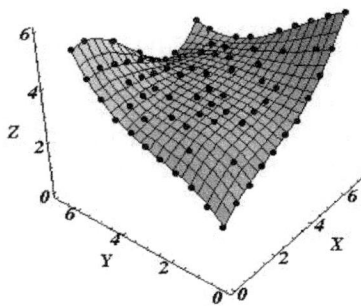

Fig. 2. Fitted surface and used data points.

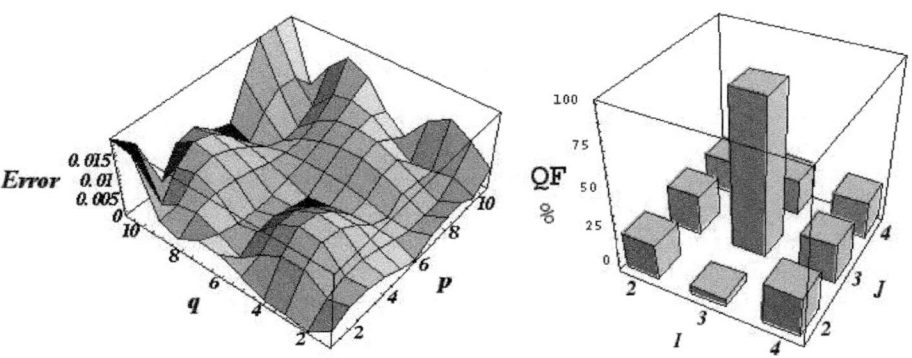

Fig. 3. (left) Error surface (error measured at data points T_{pq} on the grid); (right) Quality function values (see eq. 20) for different choices of I and J.

In spite of the previous results, it could be argued that only the best choice for I and J into the set $\{2, 3, 4\}$ is obtained, and perhaps higher degrees might lead to a better fitting of the data points. The discussion of higher degrees is not included here because of limitations of space. It is enough to say that it is not the case; on the contrary, there is an optimal value for the degrees I and J and smaller or higher values just increase both the mean and the RMS errors. As an illustration, we have calculated the values of a *quality function, QF,* given by:

$$QF = 100 \left(1 - \frac{RMS(TrP) - RMS(TeP)}{RMS(TrP)} \right) \tag{20}$$

where TrP and TeP indicate the training and the testing points, respectively, as a function of the degrees I and J. This function QF returns a number p on the interval $[0, 100]$, which can be interpreted as a measure of the "goodness" of the approximation of the original Bézier surface (given by eq. (3)) by the resulting bivariate polynomial. Thus, the closer to 100 this value is, the more similar these surfaces are.

Figure 3(right) shows the obtained results. Clearly, the choice $I = J = 3$ is indeed associated in practice with a value of 100. This implies that the surface obtained (19) reproduces the original one, with a very small errors, as shown in Figure 3(left).

From Figure 3(right) it becomes clear that this quality function is particularly appropriate for determining the optimal value for I and J. Different values for I or J lead to a significantly larger error. So, this method can be used to approximate a Bézier surface (a parametric tensor product surface, in general) by polynomials, an important issue in CAGD.

Step 8 (Use of the model): Once the model has been satisfactorily validated, it is ready to be used in predicting new points on the surface.

4   Conclusions and Recommendations

In this paper, functional networks, a powerful extension of neural networks, are applied to fit a given set of data from a tensor product surface, a very important geometric entity in CAGD. As an example, we consider here a Bézier surface, which is approximated by using a monomial basis family. The method proposed in this paper is useful not only to get the coefficients of the approximating polynomial surface but also to determine the optimal degree of such a surface, in the sense that it minimizes the error measured at data points. From this point of view, functional networks provide a good tool to describe Bézier surfaces by using the monomial basis. Futhermore, since we are free to choose the basis family for the approximation, our assertions can be generalized to the cases in which the Bézier surface is approximated by any other family of functions, such as trigonometric functions, Hermite polynomials, Laguerre polynomials, etc.

Finally, we remark that the scheme presented here is very general. In fact, we can apply these ideas to approximate any other interesting family of tensor product surfaces in CAGD, such as B-splines. The problem is currently under investigation and the obtained results will be reported elsewhere.

Acknowledgments

The authors would like to acknowledge the CICYT of the Spanish Ministry of Education (project PB98-0640) and the European Fund FEDER (Contract 1FD97-0409) for partial support of this work.

References

1. S. Bu-Qing and L. Ding-Yuan: Computational Geometry, Curve and Surface Modeling. Academic Press, San Diego (1989)
2. E. Castillo: Functional Networks, Neural Processing Letters **7** (1998) 151-159
3. E. Castillo, A. Cobo, J. M. Gutiérrez and R. E. Pruneda: Working with differential, functional and difference equations using functional networks, Applied Mathematical Modeling **23** (1999) 89-107
4. E. Castillo and J. M. Gutiérrez: Nonlinear time series modeling and prediction using functional networks. Extracting information masked by chaos, Physics Letters A, **244** (1998) 71-84
5. E. Castillo, A. Iglesias, J.M. Gutiérrez, E. Alvarez and J. I. Alvaro: Functional Networks. An application to fitting surfaces. In: Fourth International Conference on Information Systems, Analysis and Synthesis, ISAS-98, Orlando, FL (1998) 579-586
6. E. Castillo and A. Iglesias: Some characterizations of families of surfaces using functional equations, ACM Transactions on Graphics **16**(3) (1997) 296-318
7. J.A. Freeman: Simulating Neural Networks with Mathematica, Addison Wesley, Reading, MA, (1994)
8. G. Farin: Curves and Surfaces for Computer Aided Geometric Design, Academic Press, San Diego (1993)
9. J. Hertz, A. Krogh and R.G. Palmer: Introduction to the Theory of Neural Computation, Addison Wesley, Reading, MA (1991)
10. J. Hoschek and D. Lasser: Fundamentals of Computer Aided Geometric Design, A.K. Peters, Wellesley, MA (1993)
11. A. Iglesias and A. Gálvez: A characterization of the tensor-product surfaces (submitted for publication)
12. D.F. Rogers and J. A. Adams: Mathematical Elements for Computer Graphics, Second Edition, McGraw-Hill, New York (1990)

How Symbolic Computation Can Benefit Computer-Aided Geometric Design

Andres Iglesias

Department of Applied Mathematics and Computational Sciences, University of
Cantabria, Avda. de los Castros, s/n, E-39005, Santander, Spain
iglesias@unican.es

Abstract. *Computer-Aided Geometric Design (CAGD)* is one of the
most important fields in Computer Graphics. Usually, CAGD is handled
in traditional programming languages, such as Fortran, Pascal or C. By
contrast, this paper supports the idea that *Symbolic Computation Systems (SCS)* should be used instead. To this aim, the paper shows how
some mathematical expressions for Bézier curves and surfaces can be
easily translated to the *Mathematica* programming language. Then, they
are used to prove symbolically some mathematical properties related to
these geometric entities.

The term *Computer-Aided Geometric Design (CAGD)* was invented by R. E.
Barnhill and R.F. Riesenfeld in 1974 to describe the more mathematical aspects
of Computer-Aided Design (CAD). Since then, CAGD becomes one of the most
important fields in Computer Graphics [1,2]. Usually, CAGD is handled in traditional programming languages, such as Fortran, Pascal or C. However, the
appearance of the *Symbolic Computation Systems (SCS)*, such as *Mathematica*
or *Maple*, opens new and exciting possibilities.

Recently, a *Mathematica* [7] package to deal with Bézier curves and surfaces
(see [5] for a description), one of the most important topics in CAGD, has been
implemented by the author [3,4]. In this paper, this package is used to show how
symbolic computation can be successfully applied to CAGD.

The first task to be done in this process is to represent the geometric entities
by computer. This representation cannot be chosen at random; on the contrary,
it must satisfy some conditions: it should be clear, unambiguous and easy to manipulate. In this sense, one of the most remarkable SCS capabilities is the ability
to represent expressions in a compact, easy and intuitive way. Table 1 shows
how some mathematical expressions can be easily translated to the *Mathematica* programming language. As the reader can appreciate, the powerful symbolic
Mathematica capabilities allow shorter, simpler and more elegant codes, which
simply reproduce the mathematical structure of the equation to implement. In
most cases, these capabilities include pattern recognition and object-oriented
programming features [6].

For instance, the patterns `weights_List` and `weights_?MatrixQ` identify vectors and matrices respectively, the pattern `pts:{{_,_}..}|{{_,_,_}..}` is applied

J.A. Campbell and E. Roanes-Lozano (Eds.): AISC 2000, LNAI 1930, pp. 214–218, 2001.

Mathematical expression	Translation to the *Mathematica* language

$B_i^n(t) = \binom{n}{i} t^i (1-t)^{n-i}$
 	
 Bernstein polynomial
	```Bernstein[i_,n_,t_]:=```
	
	 ```Binomial[n,i] t^i (1-t)^(n-i)```

$\mathbf{B}(t) = \sum_{i=0}^{n} \mathbf{P}_i B_i^n(t)$
 	
 Bézier curve
	```BezierCurve[pts:{{_,_}..}|{{_,_,_}..},t_]:=```
	 ```Module[{n=Length[pts]-1},```
	 ``` Simplify[```
	 ``` Table[Bernstein[i,n,t],{i,0,n}].pts]]```
	 ```]```

$\mathbf{S}(s,t) = \sum_{i=0}^{m} \sum_{j=0}^{n} \mathbf{P}_{ij} B_i^m(s) B_j^n(t)$
 	
 Bézier surface
	```BezierSurface[pts:{{{_,_,_}..}..},{s_,t_}]:=```
	 ```Module[{m=Length[pts]-1,```
	 ``` n=Length[First[pts]]-1,U,V},```
	 ``` {U,V}=MapThread[Table[Bernstein[i,#1,#2],```
	 ``` {i,0,#1}]& {{m,n},{s,t}}];```
	 ``` Plus @@ (U.pts*V) //Simplify]```

$\mathbf{B}(t) = \dfrac{\sum_{i=0}^{n} \mathbf{P}_i w_i B_i^n(t)}{\sum_{i=0}^{n} w_i B_i^n(t)}$
 	
 Rational Bézier curve
	```RationalBezierCurve[pts:{{_,_}..}|```
	 ``` {{_,_,_}..},weights_List,t_]:=```
	 ```Module[{n=Length[pts]-1,lisfun},```
	 ``` If[Length[pts]==Length[weights],```
	 ``` lisfun=Table[Bernstein[i,n,t],```
	 ``` {i,0,n}];```
	 ``` Simplify[(Plus @@```
	 ``` (pts*weights*lisfun))/```
	 ``` (lisfun.weights)```
	 ```],```
	 ``` Message[RationalBezierCurve::badnum]]]```

$\mathbf{S}(s,t) = \dfrac{\sum_{i=0}^{m} \sum_{j=0}^{n} \mathbf{P}_{ij} w_{ij} B_i^m(s) B_j^n(t)}{\sum_{i=0}^{m} \sum_{j=0}^{n} w_{ij} B_i^m(s) B_j^n(t)}$
 	
 Rational Bézier surface
	```RationalBezierSurface[pts:{{{_,_,_}..}..},```
	 ``` weights_?MatrixQ,{s_,t_}]:=```
	 ```Module[{m=Length[pts]-1,```
	 ``` n=Length[First[pts]]-1,U,V},```
	 ``` If[Take[Dimensions[pts],2]==```
	 ``` Take[Dimensions[weights],2],```
	 ``` {U,V}=MapThread[```
	 ``` Table[Bernstein[i,#1,#2],```
	 ``` {i,0,#1}```
	 ```]&,{{m,n},{s,t}}];```
	 ``` Plus @@ (U.(pts*weights)*V)/```
	 ``` (Plus @@ (U.weights*V))```
	 ``` //Simplify,```
	 ``` Message[```
	 ``` RationalBezierSurface::badnum]]]```

Table 1. Examples of translation of some mathematical entities and expressions used in CAGD (left column) to their equivalent symbolic *Mathematica* commands (right column). Vectors are denoted in bold.

to represent an arbitrary number of either two- or three-dimensional control points and the pattern `pts:{{{_,_,_}..}..}` represents a matrix (with an arbitrary number of rows and columns) of three-dimensional points.

Mathematical expression	How to prove it in *Mathematica*
| $$\sum_{i=0}^{n} B_i^n(t) = 1$$ <br><br> Partition of unity | ```
PowerExpand[
    FullSimplify[
        Sum[Bernstein[i,n,t],{i,0,n}]
    ]
]
``` |
| $B_i^n(t) = B_{n-i}^n(1-t)$ <br> Symmetry | ```
rule=Binomial[n_,n_-i_]-> Binomial[n,i]
Bernstein[n-i,n,1-t] /. rule
``` |
| $$\sum_{i=0}^{n} \frac{i}{n} B_i^n(t) = t$$ <br><br> Linear precision for <br> Bernstein polynomials | ```
PowerExpand[
    FullSimplify[
        Sum[(i/n)*Bernstein[i,n,t],{i,0,n}]
    ]
]
``` |
| $$\sum_{i=0}^{n} \left[\left(1 - \frac{i}{n}\right)\mathbf{P} + \frac{i}{n}\mathbf{Q}\right] B_i^n(t)$$ <br> $$= (1-t)\mathbf{P} + t\mathbf{Q}$$ <br><br> Linear precision for <br> Bézier curves | ```
Collect[
    PowerExpand[
        FullSimplify[
            Sum[((1-i/n)*p+(i/n)*q)*Bernstein[i,n,t],
            {i,0,n}]
        ]
    ],
    p]
``` |

Table 2. Examples of some mathematical properties (left column) and how they can be proved by using *Mathematica* (right column).

In addition, all the expressions are manipulated in a symbolic way avoiding the spurious behavior and round-off errors obtained when numerical methods are applied. For example, given a set of three-dimensional control points and their corresponding weights:

```
In[1]:=  pts={{{0,0,0},{2,0,3},{4,0,0}},
              {{0,2,0},{2,2,3},{4,2,2}}};
         weights={{1,4,5},{1,2,3}};
```

the command

```
In[2]:=  RationalBezierSurface[pts,weights,{u,v}]
```

$$\text{Out}[2]:= \left\{ \frac{4v(4-2u+v)}{(1+(6-4\ u)\ v+2\ (-1+u)\ v^2)}, \frac{2(u+2uv)}{1+(6-4\ u)\ v+2\ (-1+u)\ v^2}, \right.$$
$$\left. \frac{6v(4-4v+u(-2+3v))}{1+(6-4\ u)\ v+2\ (-1+u)\ v^2} \right\}$$

returns the mathematical expression of the corresponding Bézier surface. Note that this output corresponds to the symbolic equation of a parametric rational function of degrees $(1, 2)$ in directions (u, v), as the input is given by 2×3 control points.

We can take advantage of the symbolic results to evaluate the curves or surfaces with infinite precision, or determine some of their mathematical properties. Table 2 shows some properties of Bernstein polynomials and Bézier curves and how they can be proved by using *Mathematica*.

Furthermore, the SCS graphical capabilities can also be applied to visualize the geometry of the objects under analysis. Figure 1 shows a Bézier curve (left) and a surface (right) associated with two different sets of control points.

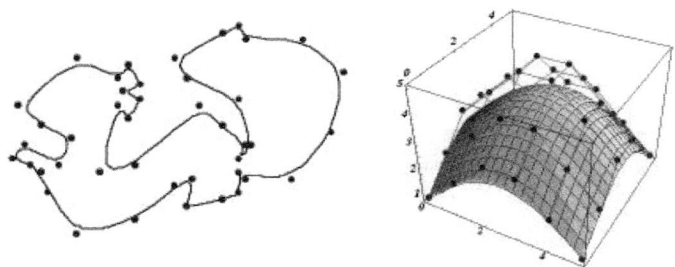

Fig. 1. Examples of: (left) a Bézier curve; (right) a Bézier surface.

The author thanks the CICYT of the Spanish Ministry of Education (project PB98-0640), the European Fund FEDER (Contract 1FD97-0409) and the University of Cantabria for partial support of this work.

References

1. G. Farin: Curves and Surfaces for Computer Aided Geometric Design, Academic Press, San Diego (1993)
2. J. Hoschek and D. Lasser: Fundamentals of Computer Aided Geometric Design, A.K. Peters, Wellesley, MA (1993)
3. A. Iglesias, F. Gutiérrez and A. Gálvez: A Mathematica package for CAGD and Computer Graphics. In: Eurographics Workshop on Computer Graphics and Visualization Education, GVE'99, Coimbra (1999) 51-57
4. A. Iglesias: A package for CAGD: Bézier curves and surfaces, (submitted for publication in: The Mathematica Journal).

5. A. Iglesias and A. Gálvez: A New Artificial Intelligence Paradigm for Computer-Aided Geometric Design, (this volume).
6. R. Maeder: Programming in Mathematica, Second Edition, Addison-Wesley, Redwood City, CA (1991)
7. S. Wolfram: The Mathematica Book, Fourth Edition, Wolfram Media/Cambridge University Press, Champaign, IL (1999)

CDR: A Rewriting Based Tool to Design FPLA Circuits

Zahir Maazouzi, Nirina Andrianarivelo, Wadoud Bousdira, and Jacques Chabin

Laboratoire d'Informatique Fondamentale d'Orléans 45067 Orléans Cedex 02 (Fr.)
{maazouzi, andria, bousdira, chabin}@lifo.univ-orleans.fr

Abstract. A rewriting based method to design circuits on FPLA electronic devices is presented. It is an improvement of our previous work. In comparison with this latter, the number of boolean vectors generated during the design process is reduced. This is done thanks to new forms of rewriting rules denoting new interesting properties on boolean vectors, associated to boolean products. Only boolean products which are implicants of the circuit to design are computed. Thus, this new design process is more efficient than the previous one.

1 Introduction

FPLA (Field Programmable Logic Array) devices are the core of complex circuits as random logic circuits, interface logic, and other applications that require decoding of device inputs. The aim of this work is to design logical circuits for such kind of devices.

It is the first rewriting based method dealing with designing circuits on FPLA chips [1]. It is a correct and complete method, in the sense that, if solutions of the design exist, they will be deduced by our method within a finite period of time. And if no solution exists, then within a finite period of time, a signal of failure is also displayed. Let us note that theoretically, neither simulation nor validation of the computed solutions is needed. Thanks to the formalism of constraint we used, all the properties to be satisfied by the solution are specified in the constraints of the rewrite rules, and they have to be considered in all the steps of the process. Such properties could be conditions on the layout of the logical gates in the device, for example. This approach has been implemented in our prototype CDR (Circuit Design by Rewriting).

New properties of boolean vectors representing boolean products are stated. To the best of our knowledge, we are the first to present all those properties. They will be used to redefine our new system of constrained rewriting rules.

2 Boolean Vectors

We deal with basic notions on boolean algebra [4, 9, 8, 5]. The novelty of this approach consists of handling the boolean functions in the sum of products normal form by a vectorial representation, unlike the classical ones where an

J.A. Campbell and E. Roanes-Lozano (Eds.): AISC 2000, LNAI 1930, pp. 219–222, 2001.
© Springer-Verlag Berlin Heidelberg 2001

algebraic representations is used. With this representation, we prove interesting new properties that will be used to perform the design. It is independent of any design method, which can be used to improve the classical ones.

A boolean vector $\overrightarrow{v} = (v_1, v_2, \ldots, v_n)$ is any n-uple of values all in the set $\{0, 1\}$. Let $\overrightarrow{\mathbb{E}}^n$ be the set of all boolean vectors of size 2^n. For example, $\overrightarrow{\mathbb{E}}^2 = \{(0000), (0001), (0010), (0011), (0100), (0101), (0110), (0111), (1000), (1001), (1010), (1011), (1100), (1101), (1110), (1111)\}$. n is the dimension of a vector of $\overrightarrow{\mathbb{E}}^n$, and 2^n its size. Let us now define the following notions in order to introduce these new properties:

– the split operation: $\forall n \geq 2$

$$Split : \begin{cases} \overrightarrow{\mathbb{E}}^n \to \overrightarrow{\mathbb{E}}^2 \times \overrightarrow{\mathbb{E}}^2 \times \ldots \times \overrightarrow{\mathbb{E}}^2 \\ \overrightarrow{v} \mapsto ((v_1, v_2, v_3, v_4), \ldots, (v_{2^n-3}, v_{2^n-2}, v_{2^n-1}, v_{2^n})) \end{cases}$$

– Let $\overrightarrow{v} = (v_1, \ldots, v_{2^n})$ and $\overrightarrow{u} = (u_1, \ldots, u_{2^n})$ be two vectors in $\overrightarrow{\mathbb{E}}^n$:

$$\overrightarrow{v} \preceq \overrightarrow{u} \text{ iff } \forall i \in [1 \ldots 2^n], \text{ if } u_i = 0 \text{ then } v_i = 0.$$

Let us note that in the literature[8, 5], $\preceq$ is denoted *implication*. When $\overrightarrow{v} \preceq \overrightarrow{u}$, we say that $\overrightarrow{v}$ *implies* $\overrightarrow{u}$, we say also that $\overrightarrow{v}$ is an *implicant* of $\overrightarrow{u}$.

In the same way, we note $\overrightarrow{\mathbb{P}}^n$ all the vectors of $\overrightarrow{\mathbb{E}}^n$ corresponding to products. We call them *product vectors* of dimension n. In order to distinguish between the two notions: vectorial-syntactic, we represent a syntactic form with a pattern p without arrow, and its vectorial value by $\overrightarrow{p}$.

2.1 Recursive Properties on Products

The following theorem is an interesting result on the boolean algebra in the sense that it is independent of any design method. It makes up the basis of our method.

Theorem 1. *Let be* $\overrightarrow{p} \in \overrightarrow{\mathbb{P}}^n$ *such that* $Split(\overrightarrow{p}) = (\overrightarrow{b}_1, \cdots, \overrightarrow{b}_k)$.

1. $\forall i, j \leq k$, *if* $\overrightarrow{b}_i \neq (0000)$ *and* $\overrightarrow{b}_j \neq (0000)$ *then* $\overrightarrow{b}_i = \overrightarrow{b}_j$.
2. $\forall i \leq k$, $\overrightarrow{b}_i \in \overrightarrow{\mathbb{P}}^2$.

3   The Rewriting Approach

We use the paradigm of conditional rewriting[3] to perform the circuit design. The formalism of constraints used here is also the usual one. For more details see [6], and [7]. All the notations used here are the same as in our previous work [2]. A constrained conditional rule is noted $C \wedge L \Rightarrow s = t$, in which C is a list of constraints to be solved by using algorithms in the predefined algebras, or by unification, and L is a list of equations to be solved by rewriting techniques.

When the L part does not exist then we call it an equation. All the inference rules of the maximal-unit-strategy described in [2] are included in our system. The main inference rules we use are *deduction of new rules by Conditional-narrowing, and simplifying conditional rules and inequations*. Let us point out that this inference rule works between an inequation (a pure negative *conditional rule*) and an equation as well. As usual in refutation based methods, we infer in our system until the empty clause is generated. The proof of this empty clause yields the design of the circuit.

4   The Design Specification

The specification of the circuit is divided into three main parts. The axioms represent the inputs, called also equations. All the outputs are represented by a pure negative clause, and finally five conditional rules with the corresponding constraints specify the design of the FPLA and the chip specification. As usual in refutation based methods, CDR infers new clauses until the empty clause is generated. The proof of this empty clause yields the design of the circuit.

All the axioms are ground equations of the form $P\$(out(\overrightarrow{v}), 2) = tt$, where $\overrightarrow{v}$ is a product vector of dimension 2, the second argument of $P\$$.

There are 5 conditional rules specifying the design of the FPLA device. Three of those rules does deduce product vectors, and the remaining ones deduce the output sums. For example the first conditional product rule is as follows:
$$[r \neq 1 \wedge r < n - 2] \wedge P\$(out(\overrightarrow{x}), 2) = tt \wedge P\$(out(t), r) = tt$$
$$\Rightarrow P\$(out(ab(\overrightarrow{x}, t)), r + 2) = tt$$

This rule exploits the theorem 1 in order to deduce recursively product vectors of upper dimension. The operator ab keeps the resulting product vector in a compact way without expanding it in its vectorial representation. It allows to reduce to memory consuming.

The following rule performs a sum between a product and an already deduced sum. This deduction is performed only through two conditions as shown in the constraint part.
$$[\overrightarrow{x} = simp(t, n) \wedge \overrightarrow{x} \npreceq \overrightarrow{y} \wedge x \not< y] \wedge P\$(out(t), n) = tt \wedge S\$(out(\overrightarrow{y})) = tt$$
$$\Rightarrow S\$(out(or(\overrightarrow{x}, \overrightarrow{y}))) = tt$$

The constraint $\overrightarrow{x} \npreceq \overrightarrow{y}$ forbids redundant sums. For example, if x_i's are boolean variables, then the sum between $x_1 x_2 x_3$ and $x_1 x_2 + x_2 x_3'$ are not to be performed, because $x_1 x_2 x_3 \preceq x_1 x_2$. The second constraint $x \not< y$ means that the syntactic representation of x must be syntactically smaller than y with respect to an arbitrary lexicographic order $<$, defined on the algebraic representation of products and which is extended over the sums. The vectors $\overrightarrow{x}$ and $\overrightarrow{y}$ take part of the predefined algebra, thus all the information corresponding to them are wired like integers and booleans. They are in the low hierarchy level [2]. For example, $x_1 x_2 < x_2 x_3 + x_1' x_3$ while $x_3 \not< x_2 x_3 + x_1' x_3$. This constraint allows us to reduce the process in the search space avoiding the deduction of the same sum (i.e. the same semantic value) though syntactically different.

Finally, the outputs is represented by an inequation. All these outputs are the subterms of the variadic term $A\$(\ldots)$. The inequation is as follows:
$$A\$(S\$(out(\overrightarrow{o}_1)), \ \ldots, \ S\$(out(\overrightarrow{o}_k))) = tt \Longrightarrow$$
$$A\$(tt,\ldots,tt) \to tt$$
where $\overrightarrow{o}_1,\ldots,\overrightarrow{o}_k$ are the column vectors of the truth table. Once all the subterms $S\$(\ldots)$ in $A\$(\ldots)$ are reduced to tt, the second standard rewrite rule reduces the negative clause to the nil one, thus we get the refutation.

5   Conclusion

In comparison with our previous work[1], thanks to those new properties combined with a goal-oriented deletion criteria, the number of clauses generated is tremendously reduced. This second version of CDR is more efficient than the first one. All the product vectors are stocked in a particular data structure as the BDDs[?]. The preliminary statistics show that the memory consuming is reduced to 30% with respect to the first version of CDR. Moreover, thanks to the constraints introduced in the product and sum rules, as presented above, only implicants of the outputs are deduced. The number of these deduced products is divided in some cases by half with respect to the previous version. Unfortunately, this saving is performed only for the product vectors, due to the main property stated in theorem 1. We think that additional improvements of our approach are still possible, especially for vectors sum deduction.

References

[1] N. Andrianarivelo, W. Bousdira, J. Chabin, and Z. Maazouzi. Designing FPLA combinational circuits by conditional rewriting. In H. Prade, editor, *John Wiley and Sons*, pages 373–377, Brighton, UK, 1998. 13th European Conference on Artificial Intelligence.
[2] N. Andrianarivelo, W. Bousdira, and J-M. Talbot. On theorem-proving in Horn theories with built-in algebras. In J. Calmet, J.A. Campbell, and J. Pfalzgraf, editors, *Lecture Notes in Computer Science*, volume 1138, pages 320–338, Steyr, Austria, 1996. Third International Conference on Artificial Intelligence and Symbolic Computation.
[3] N. Dershowitz and J-P. Jouannaud. Rewriting Systems. In J. Van Leuven, editor, *Handbook of Theoretical Computer Science*. Elsevier Science Publishers North-Holland, 1990.
[4] P.R. Halmos. *Lectures Notes on Boolean Algebras*. Springer, Berlin, 1974.
[5] Randy H. Katz. *Contemporary logic design*. Benjamin Cummings/Addison Wesley Publishing Company, 1993.
[6] C. Kirchner, H. Kirchner, and M. Rusinowitch. Deduction with Symbolic Constraints. *Revue Française d'Intelligence Artificielle*, 4(3):9–52, 1990. Special issue on Automatic Deduction.
[7] G. Smolka. *Logic Programming over Polymorphically Order-Sorted Types*. PHD Thesis, Universitat Kaiserslautern, FB Informatik, West Germany, 1989.
[8] John F. Wakerly. *Digital Design Principles and Practices*. Prentice Hall International Editions, 1991.
[9] Ingo Wegener. *The Complexity of Boolean Functions*. J. Wiley and Sons, 1987.

Locally Effective Objects and Artificial Intelligence*

Julio Rubio

Departamento de Matemáticas y Computación. Edificio Vives.
Calle Luis de Ulloa s/n. Universidad de La Rioja. 26004 Logroño (La Rioja)
jurubio@dmc.unirioja.es

Abstract. The concept of *locally effective objects* was introduced by Sergeraert in the field of Effective Algebraic Topology, where this tool was used to represent potentially infinite data structures. This notion, borrowed from symbolic computation, was later used to produce, in an innovative way, code implementing the well-known search algorithms in Artificial Intelligence. In this paper, we show how these implementations can be appropriately reinterpreted and specified, by using some recent advances in the algebraic specification setting. As a by-product, the concept of locally effective graphs provides a framework in which the *Production Systems* and the *State Space* programming *metaphors* can be formally integrated.

In his very general proposal to deal with the problem of computability in Algebraic Topology, Sergeraert introduced the notion of *locally effective objects* [7]. This concept allows the programmer to work with *potentially* infinite data structures. This idea was put into practice by Sergeraert and the author to develop a Common Lisp symbolic computation system called EAT (Effective Algebraic Topology). EAT computes homology groups of iterated loop spaces [8] leading to results which had not previously been calculated (either by hand or by computer).

The relevance of locally effective matters was subsequently explored by the author and coworkers, following two different lines. In the first one, algebraic specification tools were used to analyze the mathematical properties of locally effective objects in a Category Theory setting (see [4]). In the second one, the notion of locally effective graphs was used to develop Common Lisp programs which implement, in an innovative and very generic way, the well-known algorithms for searching in state-spaces (see [6]).

This work stems from the confluence of these two lines. The results relating to algebraic specifications obtained in [4] are extended to the case of locally effective graphs. These theoretical results lead to some interesting consequences for interpreting the Common Lisp code developed in [6]. Our approach shows how some high-level mathematical tools (such as Category Theory) produce

---

* Partially supported by DGES, project PB98-1621-C02-01

J.A. Campbell and E. Roanes-Lozano (Eds.): AISC 2000, LNAI 1930, pp. 223–226, 2001.

important (and natural) consequences in the design of very concrete software systems (such as, in our case, generic search procedures in Artificial Intelligence).

The standard perspective in algebraic specifications (that is to say, the *initial semantics* approach) is not convenient for studying locally effective objects. (See [5], for example, for elementary definitions of algebraic specifications.) The suitable context in which locally effective matters can be dealt with is a recently discovered setting known as *hidden specifications* [3]. As the signature *leGraph* for locally effective graphs, we have chosen for the visible part the usual signature for (effective) lists (with main sort *lst*) and for the hidden part (with *grp* as the only hidden sort):

$$
\begin{array}{rclcccl}
vrt - eql & : & grp & vrt & vrt & \rightarrow & bool \\
adj - lst & : & grp & vrt & & \rightarrow & lst \\
vrt - goal & : & grp & vrt & & \rightarrow & bool \\
heuristics & : & grp & vrt & & \rightarrow & nat \\
edge - cost & : & grp & vrt & vrt & \rightarrow & nat
\end{array}
$$

The first two operations encode the graph itself. The relevant information on the *vertices set* is the equality test *vrt-eql*, while the knowledge of the *edges set* is represented by *adj-lst* in the form of adjacency lists. The last three operations store information for the search process: *vrt-goal* is the termination test, *heuristics* estimates the relevance of each vertex and *edge-cost* is intended to evaluate the cost of each edge.

Let us briefly explain why Sergeraert called this kind of data structures *locally effective objects*. According to Sergeraert's terminology [7], an *effective object* is the representation in computer memory of a usual finite datum, such as a list or a graph. For instance, to deal with an *effective graph* we should add to the previous signature a complete set of construction operations (to construct the empty graph, to adjoin a vertex or an edge, and so on). Then, standard algebraic specifications techniques [5], as *initial semantics*, are enough to formally study these data. Even in the absence of constructors, we can manage the same family of data. For instance, in the example, this can be achieved by adjoining an operation $vrts : grp \rightarrow lst$, collecting the vertices set of the graph. The initial semantics of the new signature is void, but we are able to *reconstruct* from the corresponding data structures any of the graphs belonging to the initial object. Thus, they can also be called *effective graphs*. Obviously, infinite graphs cannot be specified by any of the two last signatures. Let us go back to the signature *leGraph*. Obviously, the signature does not allow to construct instances of *leGraph*. However, this signature can be *implemented* (see [4]) and we are able to work with *leGraphs* in a computer. For instance, the (infinite) graph where each natural number has as adjacency list its set of (proper) divisors can be directly implemented. Evidently, such graphs can be used in an algorithmic way (in the example, an algorithm for computing the gcd of two numbers, without using any arithmetical knowledge, can be easily designed), but the kind of information which is accessible from them is very different and much poorer than that of effective objects. Only *local* information is available. If a vertex is given, one can determine whether it is a goal. If two vertices are given, one can ask whether one is a (direct) descendant of the other. But no *global* information is available. For instance, it is undecidable whether the graph is connected or even whether the graph is empty. To understand the meaning of this last claim, let us remark that the example covers an *infinite* graph (thus, it is trivially non-empty), but, in fact, the main characteristic of locally effective objects is that any information about *cardinality* is missing. In general, locally effective objects will be used when the underlying space is infinite but also when it is so huge that no explicit storing would be sensible. This situation appears in a natural way in the field of Symbolic Computation in Algebraic Topology, but also, as it is well-known, in the field of Artificial Intelligence (*combinatorial explosion*).

But before talking about applications, let us go back to the formal analysis of hidden signatures [3] such as *leGraph*. Let us denote by $HAlg(leGraph)$ the hidden category of *leGraph*-algebras (see [3]). Then a functional description of a final object in $HAlg(leGraph)$ is given in the following result.

Theorem 1. *There exists a final object I in $HAlg(leGraph)$ in which:*

$$I(grp) := \{[f_{vrt-eql}, f_{adj-lst}, f_{vrt-goal}, f_{heuristics}, f_{edge-cost}]\},$$

where each element in $I(grp)$ is a tuple of functions so that: $f_{vrt-eql} : D_{vrt} \times D_{vrt} \to D_{bool}, f_{adj-lst} : D_{vrt} \to D_{lst}, f_{vrt-goal} : D_{vrt} \to D_{bool}, f_{heuristics} : D_{vrt} \to \mathbb{N}, f_{edge-cost} : D_{vrt} \times D_{vrt} \to \mathbb{N}$.

The definition of the operations in I is the natural one. For instance,

$$I(vrt - eql)([f_{vrt-eql}, \ldots], v_1, v_2) := f_{vrt-eql}(v_1, v_2).$$

The object described in the last theorem is very close to certain formalisms for modeling object-oriented programming, based on Cardelli's metaphor of "objects as records of functions" (see [1]). This relationship is quite natural because one of the explicit goals of hidden specifications is also to find formal models for object-oriented concepts (see [3]). This result can also be interpreted as a specially easy presentation of a more general result in [3] on the existence of final objects in hidden categories. (The careful reader will notice that the object introduced in the previous theorem is not isomorphic to the object defined in [3]. This is not surprising because, in fact, the object of [3] is not final. Nevertheless, a minor modification of the construction in [3] gives the right object which is, obviously, isomorphic to the functional version presented here.)

From the programming point of view, this final object can be directly implemented, if functional programming is available. This has been the case in the symbolic computation systems for Algebraic Topology [8] which have been developed in Common Lisp. This same direct implementation as a tuple of functions has also been applied in Artificial Intelligence [6]. In [6] the final object of the theorem is stored in a Common Lisp record (struct) similar to:

```
(defstruct legraph vrt-eql adj-lst vrt-goal ...)
```

By using also functional programming, the unique morphism F which exists from one object of $HAlg^D(leGraph)$ to I is computable: this is deduced from the *exponential map* (see [4]). For instance, if J is an object in $HAlg^D(leGraph)$, then given $x \in J(grp)$ the tuple $F(x) = [f_{vrt-eql}, \ldots] \in I(grp)$ is defined by:

$$f_{vrt-eql}(v_1, v_2) := J(vrt - eql)(x, v_1, v_2),$$

and so on. In addition, the object J is *behaviourally equivalent* (see [3], [4]) to the sub-object $F(J)$ of I. Roughly speaking, this implies that the code associated to J can be replaced by $F(J)$ without changing the meaning of the client programs.

These theoretical results casted new light on the work in [6], where we were able to directly *reuse* Forbus-de Kleer's programs [2] without worrying about specific examples or application fields. We can now explain how the program called `problem->legraph` in [6], which transforms a `problem` (that is to say, the data structure used by Forbus-de Kleer to encapsulate the information about state-space problems) into a `legraph` (the Common Lisp `struct` evoked previously) is, in fact, the composite of two more elementary functions: one which deals with a `problem` as an implementation of the signature *leGraph* and the other (universal) function translating this implementation into the final object I of Theorem 1.

Thus, the conclusion of this analysis is that any production (or ruled-based) system *is* a (representation of a) locally effective graph and therefore that, thanks to the final property of I, the generic search procedures of [6] can be applied to it. From this point of view, any production system interpreter or rule-based inference engine (based on pattern-matching, see [2], or on some kind of unification algorithm) can be considered, in particular, as an implementation of the signature *leGraph*, achieving in this way the formal integration of the Production Systems and State-Space programming metaphors.

References

1. Abadi, M., Cardelli, L.: A Theory of Object, Springer, 1996.
2. Forbus, K.D., de Kleer, J.: Building Problem Solvers, MIT Press, 1993.
3. Goguen, J., Malcolm, G.: A Hidden Agenda. Theoretical Computer Science **245** (2000) 55–101.
4. Lambán, L., Pascual, V., Rubio, J.: Specifying implementations, in Proceedings ISSAC'99, ACM Press (1999) 245–251.
5. Loeckx, J., Ehrich, H. D., Wolf, M.: Specification of Abstract Data Types, Wiley-Teubner, 1996.
6. Rubio, J., Bañares, J.A., Muro, P.R.: A functional approach to the implementation of search algorithms (in Spanish), in Proceedings CAEPIA'99, vol. I, Murcia University (1999) 88–97.
7. Rubio, J., Sergeraert, F.: Locally effective objects and Algebraic Topology, in Computational Algebraic Geometry, Birkhäuser (1993) 235–251.
8. Rubio, J., Sergeraert, F., Siret, Y.: An overview of EAT. Symbolic and Algebraic Computation Newsletter **3** (1998) 69–79.

Negotiation Algorithms for Multi-agent Interactions

Marco A. Arranz*

Centre for Agent Research and Development
Manchester Metropolitan University
Chester Street, Manchester M1 5GD, U.K.
M.Arranz@doc.mmu.ac.uk ; timarticus_69@yahoo.com

Abstract. The paper describes a general interaction algorithm for coordinating multi-agent plans. Triggered by a communication and negotiation protocol the coordination framework reconciles situations with negative interferences as well as it handles positive opportunities for mutual benefits. Coordination is understood then as a mechanism to reconcile plans evaluating interactions among agents. The process is a dynamic representation of the environment where the structure of tasks goes from being a set of uncoordinated plans to be a set of coordinated plans.

1 Introduction

Multi-Agent planning is the process of generating a plan among multiple agents where agents' actions and potential interactions are previously specified. In such a plan agents reason about the potential consequences of their actions and about the particular order in which these actions are to be executed. In this way the planner is able to detect and control conflictive interactions induced by incompatible states or by an incompatible resource usage, as well as positive interactions. Coordination algorithms are required to manage agents' behaviour as follows. On one hand it was a way of allocating particular tasks to particular agents sharing a common goal [5], and on the other hand it was a medium to achieve better coordination by aligning behaviour of agents towards different goals, with an explicit division of labour [11]. Tools and methodologies in the design of agent communities have been traditionally based on such a dichotomy. Although both paradigms are closely related both research lines have considered exclusive algorithms *ad hoc*. Different sorts of coordination mechanisms were required in each case depending of pre-specified structural features.

This paper relaxes the strength of such a dichotomy providing a coordination algorithm to manage interaction processes in different sorts of agent societies. In our domain agents have to share the same environment and the same set of physical resources, so they have to plan taking into account the potential set of interactions in such a domain. Triggered by a communication and negotiation protocol the coordination framework we propose reconciles situations with negative interferences as well as

---

* Supported by research grant from the Basque Government, Spain.

J.A. Campbell and E. Roanes-Lozano (Eds.): AISC 2000, LNAI 1930, pp. 227-239, 2001.
© Springer-Verlag Berlin Heidelberg 2001

it handles positive opportunities for mutual benefits. The algorithm itself is flexible enough to keep open the choice of introducing in further studies new ways of interaction. It has as one of its main advantages the capability of being adequate to inherently cooperative domains [6,13] as well as self-interested agent domains [17]. This is an interesting feature which allows us to work in different environments with the same general and suitable framework.

2 Interactions Handling

The study of plan relationships among autonomous agents is a crucial topic in multi-agent systems research. Autonomous agents make plans with the intention *a priori* of avoiding conflicts. Nevertheless they might face situations where the best way of fulfilling their own purposes is to interact with the rest of the agents in the community [1]. For instance, some relationships fit into situations where tasks are too long, or too difficult to be carried out by a single agent [19,20]. In such a case the affected agent can ask for help to share the task[1] assuming some kind benevolence assumption in the community. Some other non-necessarily cooperative relationships relax this of benevolence property to allow situations where the interaction is motivated by mutual benefits [17], or the help relationship is just replaced by a favour relationship [14]. In such a case agents are not motivated by any natural benevolent impulse[2].

Agents could also face situations which might prevent one or both of the plans from being executed as intended. The detection of negative relations is crucial for a successful plan execution. The negative interactions we take as reference are situations where agents need the same resource at the same time or situations where agents have incompatible strategies in achieving their plans [4]. Temporal relations are then associated with actions to ascertain whether a potential resource conflict exists. For instance, access to a non-consumable resource means no conflict if the time periods for utilizing such a resource do not overlap. This is not so if the resource is consumable. Temporal reasoning is also involved in solving conflicts when two actions require exclusive states to exist. For instance, domain-specific heuristics are needed to prevent two agents from occupying the same conference room at the same time for giving different talks. This means that pre-specified axioms are needed to state which actions should not occur simultaneously. Temporal reasoning is important to find out the

---

[1] Task-sharing is a form of cooperation in which agents assist each other by sharing the load in the problem solving. Task-sharing processes have been recently studied also for self-interested agents in contract net protocols [18].

[2] Quantitative arguments are then provided in self-interested domains to evaluate the global benefits of exploiting a favour relationship [2]. Even though doing a favour would mean additional costs for the agent who is required to do the favour, agents should reason about the entire set of plan relationships. Doing a favour might, for instance, imply personal benefits in terms of reducing the costs of dealing with some other current potential conflict. This is usually understood as an implicit utility transfer. Utility transfers forms the basis for agreements in these sorts of domains, and they are used as a negotiation currency to compensate one agent for a disadvantageous agreement.

order in which the conflicts have to be resolved. The solution of a conflict might make the solution of another conflict easier (or impossible); or if a conflict is settled, another conflict might vanish or a new one might arise.

Nevertheless, in this paper we are not specifically focused on defining the set of interactions that agents have to face. Obviously such dependencies will depend on the organizational structure that agents are involved in. For instance, different interactions should be defined for a community of agents where every one works to get the same common goal, and for a community of agents attempting to maximize their own good. The study of agent dependencies is beyond the scope of this paper. Our concern here is about the kind of coordination algorithm -communication protocol and negotiation process- agents need in order to produce coordinated plans assuming a pre-specified set of plan relationships.

3   Coordinating Agents

3.1   A General Coordination Algorithm

Coordination is the process of managing interdependencies between activities. This section focuses on the problem of handling these interdependencies in a domain-independent way.

The general role of any coordination mechanism is to provide constraints for agents' plans. Constraints are understood here as a modification process inside the plans structure, or as a process of making commitments. The knowledge state of a coordinating agent includes its own plan, its knowledge about the rest of the agents plans and the relationships it holds with them. For each agent g its knowledge state is a 3-tuple defined as follows:

$$S_g = <LP_g, EP_g, PR_g> \quad \text{where}$$

- LP_g denotes the set of leaf-actions intended by g's plan.
- EP_g denotes the set of leaf-actions intended by the rest of agents.
- PR_g denotes the set of interactions g maintains with the rest of the agents.

The input state is given by $<Sg_1,...,Sg_n>$ where $\forall g\ EP_g = \varnothing$ and the output state is obtained when $PR_g = \varnothing$.

The coordination process reconciles plans evaluating the possible interactions among agents, hypothesizing the best solution for a coordinated plan, and starting a negotiation process to arrive at mutually-agreed ways of handling every existing relationship. The coordination process then is a dynamic representation of the environment where the structure of tasks goes from being a set of uncoordinated plans to be a set of coordinated plans. In other words, the coordination problem has as input a set of uncoordinated plans and as output a modified set of plans which are finally coordinated.

230   Marco A. Arranz

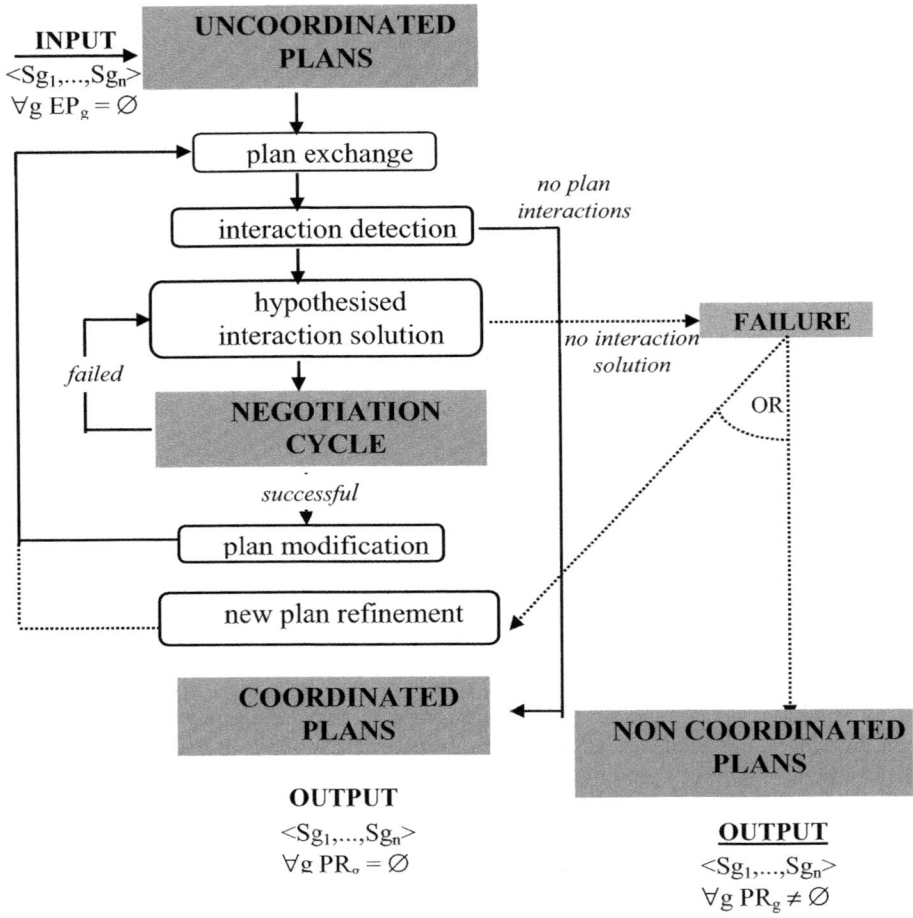

Fig. 1. The coordination algorithm

Figure 1 shows us the steps agents must carry out to complete the whole process. The first step says that, after agents have developed their plans, they pass on the initial information they consider relevant. A plan may be refined or specified by a group of finer grained tasks which describe the agent's goal in more detail[3]. When plans are

---

[3] Having actions at different levels of abstraction is useful to preserve agents' autonomy. It would be desirable that agents exchange only the information they consider relevant. This is only possible if actions are planned and exchanged at different levels of generality. So the structure of a plan can be visualised as a direct graph with a general task as root node. The

refined agents would exchange just the most effective actions, as initial information, preserving the overall plan structure and the history of its refinement.

Exchanging plans in the beginning of the process allows every agent to update its own knowledge base about the rest of the agents' actions. Knowing one's own plan and the actions of the other agents' plans they are able to detect plan relationships. If any agent detects a real or potential plan relationship it hypothesizes the best solution for it and starts a conversation with the agent involved in such a relationship. If the agreement is finally reached agents modify their respective plans according to the deal and they let the community know the new plan changes in order to establish whether or not the new agreement affects someone else. The coordination process ends when it is ensured that there are no more relationships in the community.

3.2   The Negotiation Cycle

Negotiation is a communication process between two or more agents in order to get an agreement. Negotiation varies depending on the kind of organizational structure agents are working in. For instance, negotiation could be to get a common goal (help or favour relationships), to resolve negative interactions (resource and time incompatibilities), or to get mutual benefits (self-interested domains). Many negotiation mechanisms rely on techniques like Argumentation [16], Contract Net Protocols [19,20], Auction-Based Protocols [15], or techniques from economical paradigms such as Utility Theory and Decision Theory [17,21].

Negotiation is going to be understood here as reaching a commitment between agents through a structured message-passing where we specify:

- Who the agents involved in a conversation are and who communicates to whom;
- What the messages exchanged between negotiators are; and
- How this process is conducted and when it takes place.

In the beginning of the whole coordination process agents exchange their plans or the part of their plans they consider transferable. We will not consider lies in our approach. Our *honesty assumption* excludes lies in our message-passing protocol. Nevertheless, some researchers have allowed agents to lie in the negotiation process as a way of knowing whether lies could be a useful resource in societies composed by autonomous self-interested agents [21]. If they detect[4] plan relationships in such information agents then start a negotiation process to resolve them. The protocol is intended to respect agents' autonomy, to allow the flux of planned actions, and to provide the system with several options of getting an agreement.

The negotiation cycle then can be understood as a simple state diagram as it is shown in figure 2. The model is general enough to keep open the choice of introduc-

---

immediate successor nodes from the root node are the most abstract actions in the plan. The leaf nodes represent the most effective actions.

[4] The basic algorithm to detect an interaction is directly obtained from the definition of interaction [14,6,2]. In this way every agent in the community should be able to recognize the set of potential plan interactions the system is dealing with.

ing, in further studies, new ways of interaction. The diagram is composed of a set of states that agents are able to occupy, a set of messages which agents are able to send, and a set of transition rules which establishes the states in which agents are able to send or receive every message.

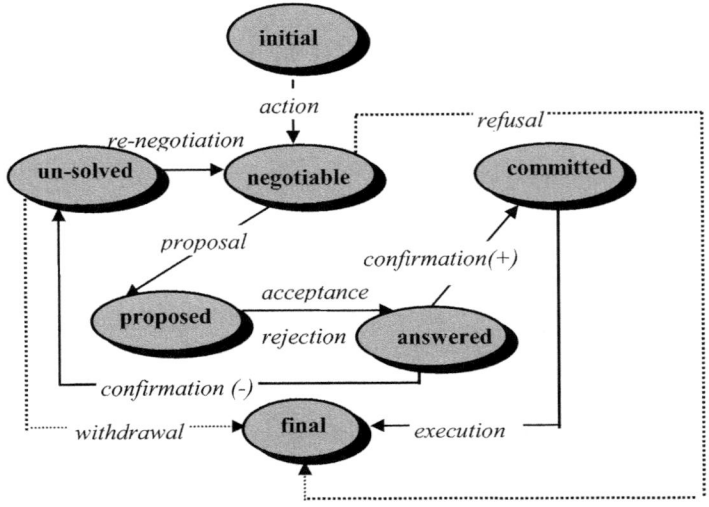

Fig. 2. The Negotiation Cycle

(A) The set of conversation states in which agents are able to take part (*C-States*):
* *initial:* starting state.
* *negotiable:* actions in this state are involved in a plan relationship.
* *proposed:* state in which an agent proposes modifications (solutions) to the concerning action.
* *answered:* state in which an agent answers a proposal.
* *committed:* state in which agents get an agreement (proposal acceptance is confirmed).
* *un-solved:* state in which there is no deal after refusing a proposal.
* *final:* final state in which an action is approved (or rejected) for execution.

(B) The set of message types which agents are able to send (*M-Types*):
* *action:* an agent announces the details of the action involved in a plan relationship.
* *proposal:* the required agent proposes a modification to such an action to the promotor agent.
* *refusal:* the required agent refuses to negotiate about such an action or plan relationship.
* *acceptance/ rejection:* possible reactions to the proposal.
* *confirmation (+/−):* positive or negative confirmation from the required agent. This message closes the initial negotiation or it opens a new re-negotiation phase.

• *re- negotiation*: The agent can start a re-negotiation phase due to any message rejecting a proposal.

• *execution/withdrawal:* the agent announces that the action is going to be executed as previously agreed, or the agent announces why the negotiation processes is going to be dissolved.

(C) The conversation rules which specify the state transitions made through the message types. The communicational behaviour of an agent is specified by a 5-tuple

$< C\text{-}States,\ M\text{-}Types,\ \text{receive, send},\ S_0 >$ where S_0 is the initial state and

• *C-States* are the previously defined ones.
• *M-Types* are the previously defined ones.
• receive: $M\text{-}Types \times$ C-States $\rightarrow$ C-States
• C-States $\rightarrow$ C-States $\times$ $M\text{-}Types$

Every agent is able to open an initial negotiation phase. The negotiation cycle is opened by a message from an agent who has detected a real or potential plan relationship. He requires the attention of the agent he considers relevant for solving the interaction sending a message to him announcing the details of the involved action (duration, required resources, etc.). That would mean that both agents have already started a negotiation process. The agent receiving this message reasons about the best solution for him and, if so, replies with a proposal to the promoter agent. It can also announce that the negotiation process is going to be dissolved. Once the agent that sent the original message receives the agent's reply (if any) he can accept or reject the proposed plan-interaction solution. In the case that the proposal is finally accepted the required agent would confirm the agreement closing the negotiation cycle[5]. In case the proposal is finally rejected the required agent would confirm that no agreement has been reached and he could open a new re-negotiation phase or he could announce that the negotiation process is going to be dissolved.

In this protocol a commitment means that one agent bind itself to a potential deal while waiting for the other agent to either accept or reject its offer. If the other party accepts, both parties are bound to a contract. On the other hand, if the first offer has been rejected agents may exchange more information about their own plans - a new possible plan refinement- in order to see the best way of solving the conflict (re-negotiation proposal).

4 An Example

Lets take an example about two agents having plans for a regular working day in the same research institute. Agents' plans are independent but they have to share the same space and the same set of physical resources so they have to coordinate in order to successfully carry out their planned actions. In figure 3 we show each agent's plan and

---

[5] Notice that such an agreement might not be reached. The negotiation protocol will allow agents to decide whether they want to take part in resolving the conflict or not.

the temporal intervals required for each action involved. The institute provides them with a set of common resources such as, for instance, a printer and a projector. We should also consider the common usage of a conference room used by the institute for every organised event. The so called *room A* is designated for such a purpose.

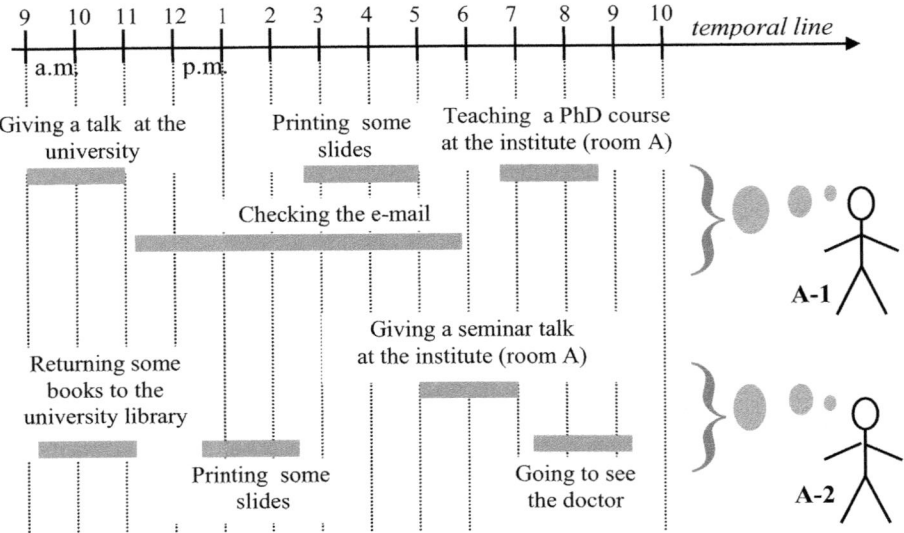

Fig. 3. Two coordinating agents

At the moment of starting the coordination process the initial knowledge states are the following ones:

S_{A-1} = < LP_{A-1}, EP_{A-1}, PR_{A-1} > where
    LP_{A-1} = {*talk-university, printing-slides, checking-e-mail, PhD-course-institute*}
    $EP_{A-1} = \varnothing$
    $PR_{A-1} = \varnothing$
S_{Frida} = < LP_{A-2}, EP_{A-2}, PR_{A-2}> where
    LP_{A-2} = {*returning-books, printing-slides, seminar-talk-institute, going-doctor*}
    $EP_{A-2} = \varnothing$
    $PR_{A-2} = \varnothing$

The agents' knowledge states are moving on as long as the coordination process evolves. Such a process depends on several factors related to the amount of refinements in each action and the particular negotiation protocol selected. In this example we just show how this general coordination algorithm can be used to resolve every plan interaction.

After exchanging mutual information the agents are able to detect every (potential) current interaction. Figure 4 shows a set of four interactions where the first interaction is understood as a favour relationship between agents; the second interaction is just a potential conflict induced by a consumable resource usage; and the last two interactions are real conflicts induced by some time/resources incompatibilities.

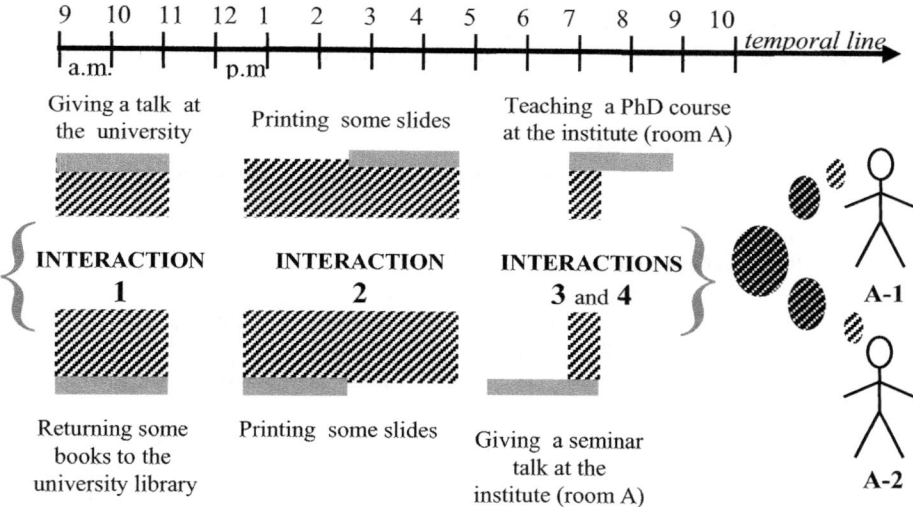

Fig. 4. Set of Agents' Interactions

Agents' knowledge states would then become updated as follows:

S_{A-1} = < LP_{A-1}, EP_{A-1}, PR_{A-1} > where

LP_{A-1} = {*giving-talk-university, printing-slides, checking-e-mail, PhD-course-institute*}

EP_{A-1} = LP_{A-2}

PR_{A-1} = {(*talk- university*$_{A-1}$, *return-books*$_{A-2}$, Interaction-1),
(*printing-slides*$_{A-1}$, *printing-slides*$_{A-2}$, Interaction-2),
(*PhD-course-projector*$_{A-1}$, *seminar-talk-projector*$_{A-2}$, Interaction-3),
(*PhD-course-roomA*$_{A-1}$, *seminar-roomA*$_{A-2}$, Interaction-4)}

S_{A-2} = < LP_{A-2}, EP_{A-2}, PR_{A-2}> where

LP_{A-2} = {*returning-books, printing-slides, seminar-talk-institute, going-doctor*}

EP_{A-2} = LP_{A-1}

PR_{A-2} = { (*return-books*$_{A-2}$, *talk- university*$_{A-1}$, Interaction-1),
(*printing-slides*$_{A-2}$, *printing-slides*$_{A-1}$, Interaction-2),
(*seminar-project.*$_{A-2}$, *PhD-course-projector*$_{A-1}$, interaction-3),
(*seminar-roomA*$_{A-2}$, *PhD-course-roomA*$_{A-1}$, Interaction-4)}

Once conflicts have been detected agents should start resolving them adjusting their knowledge states along the way. It is here where the negotiation protocol is applied.

In order to resolve the first of our conflicts A-2 sends a message to A-1 proposing to it to carry and return its books because A-1's journey is necessary. We assume the university is several miles away from the research institute where they go to work every day. In this way A-2 could delete such an action from its plan if A-1 finally accepts the request. After receiving the message A-1 evaluates the proposal positively because the favour is not going to cost any great additional effort[6]. Thus it sends an acceptance message to A-2. The solution for the favour interaction-1 is finally concluded through a confirmation message to get such an agreement. A-2 will delete the *returning books* action from its plan and A-1 will add the action to its own plan. Then both agents update their respective knowledge states as follows[7]:

S_{A-1} = < LP_{A-1}, EP_{A-1}, PR_{A-1} > where

LP_{A-1} = {*Talk-university, returning-A-2's-books, printing-slides, checking-mail, PhD-course-institute*}

EP_{A-1} = LP_{A-2}

PR_{A-1} = {(*printing-slides*$_{A-1}$, *printing-slides*$_{A-2}$, Interaction-2),
(*PhD-course-projector*$_{A-1}$, *seminar-talk-projector*$_{A-2}$, Interaction-3),
(*PhD-course-roomA*$_{A-1}$, *seminar-roomA*$_{A-2}$, Interaction-4)}

S_{A-2} = < LP_{A-2}, EP_{A-2}, PR_{A-2}> where

LP_{A-2} = {*printing-slides, seminar-talk-institute, going-doctor*}

EP_{A-2} = LP_{A-1}

PR_{A-2} = {(*printing-slides*$_{A-2}$, *printing-slides*$_{A-1}$, Interaction-2),
(*seminar-talk-projector*$_{A-2}$, *PhD-course-projector*$_{A-1}$, interaction-3),
(*seminar-talk-roomA*$_{A-2}$, *PhD-course-roomA*$_{A-1}$, Interaction-4)}

In the next potential conflict agents evaluate whether the printer toner is full enough to print all the agents' printouts. Due to the printer being considered in this case a consumable resource, agents have to make sure that the printer is going to work -is going to have enough ink- during the interval covering both actions. Fortunately they find out that the toner has been recently replaced and there is no need to start a negotiation process. This means that the potential conflicting resource is not going to create any negative interaction. Then

PR_{A-1} = {(*PhD-course-projector*$_{A-1}$, *seminar-talk-projector*$_{A-2}$, interaction-3),
(*PhD-course-roomA*$_{A-1}$, *seminar-talk-roomA*$_{A-2}$, Interaction-4)}

PR_{A-2} = {(*seminar-projector*$_{A-2}$, *PhD-course-projector*$_{A-1}$, Interaction-3),
(*seminar-talk-roomA*$_{A-2}$, *PhD-course-roomA*$_{A-1}$, Interaction-4)}

---

[6] Quantitative criteria are required in positive interactions in order to estimate benefits [14], [17]. Game-theoretic techniques are considered suitable tools for designing self-interested automated negotiation.

[7] After every plan modification agents should inform the other agents in the community in order to see whether the changes affect anyone else. This step is not necessary if the community is composed of two agents. Both of them already know about each other's changes.

The two remaining conflicts are closely related in the sense that the order in which they are handled could help to resolve them more efficiently. On the one hand we see that both actions are planned to be carried out at the same place in overlapping temporal intervals, and on the other hand the same overhead projector is required in both actions. The overlapping time is not too long, and both agents know that if both actions were not overlapped there would be no non-consumable resource-driven conflict. This is why one of the two agents sends a proposal to reduce both action intervals. For instance, A-2 could finish her talk 5 minutes earlier and A-1 could start 5 minutes later his PhD course. This proposal avoids the conflict caused by incompatible actions and removes the conflict related to the shared usage of the same overhead projector. Finally, if the agreement is successfully reached agents'plans would be coordinated so that $PR_{A-1} = \varnothing$ and $PR_{A-2} = \varnothing$.

5 Concluding Remarks and Related Work

Since the early 80s the process of decentralized coordination has been concerned with the particular task of avoiding harmful plan interactions in communities composed of autonomous agents [11]. Nevertheless the study of interaction algorithms for coordinating plans has been mainly related to the intended multi-agent system and its particular organizational structure. For instance, Partial Global Planning (PGP) [9] is one of the most successful multi-agent coordination mechanisms to manage Distributed Problem Solving in cooperative domains such as Distributed Vehicle Monitoring Testbed domains. Nevertheless, PGP does not seem to be suitable enough to coordinate communities of agents with autonomous and independent goals. PGP assumes a common goal to achieve general and coherent results sharing the same set of evaluation criteria. PGP was generalised some years later [8] in order to provide cooperative systems with a set of flexible and domain-independent coordination mechanisms.

On the other hand, game theoretic techniques provide tools to describe rational agency in terms of self-interested purposes [10]. The standard notions of Game Theory are traditionally used in Distributed Artificial Intelligence to see the extent to which self-interested agents are able to cooperate *via* negotiation without any prefixed benevolence assumption. Agents agree to achieve the goal after a negotiation process based on probabilistic reasoning or any other economically rational criteria.

Nowadays universal coordination algorithms are needed to work in different environments with the same general and suitable framework [3,4,7]. The general interaction algorithm we have shown in this paper is flexible enough to be used in cooperative domains as well as in domains where agents work independently for their own goals. The paper describes a mechanism to detect and control conflicting interactions induced by incompatible states or by incompatible resource usage, as well as positive interactions. The study of agent dependencies, however, is beyond the scope of this paper. Our concern is about how the community should behave in order to produce coordinated plans assuming a pre-specified set of plan relationships. Coordination is used then to reconcile plans evaluating interactions among agents, hypothesizing the best solution for a coordinated plan, and starting a negotiation process to resolve every existing conflict. It is a dynamic representation of the environment where the structure of tasks goes from being a set of uncoordinated plans to be a set of coordinated plans.

6   Future Work

Every coordination mechanism is expected to be integrated somehow as part of the agents' architecture or as part of the agents reasoning. In our approach the coordination process is *a priori* independent of the kind of agency we want for our agents. Nevertheless it would be necessary to see the extent to which such an external coordinating approach affects the internal behaviour of every member in the community, and vice versa. In other words, it would be interesting to show how the coordinating module is related to each agent's knowledge base in terms, for instance, of mental attitudes about the world or about the rest of agents.

An interesting domain where we shortly expect to apply the negotiation model we have presented here is the electronic commerce scenario. Agent-based e-commerce is about any kind of electronic commercial transaction -buying and selling products on the net- trying to compare prices and features of products from different vendors in order to get a good (preferably the best) transaction. Particularly interesting are virtual agent-based markets and methods for utilitarian coalition formation among rational information agents [12]. The kinds of interactions that agents should face in this environment are about *how* to determine the terms of the transaction such as prices, quality, delivery, etc. Negotiation would finally take place at that stage. Depending on the particular market we are considering, the complexity of the negotiation process can vary. In some markets prices and other aspects of the transaction are often fixed leaving no room for negotiation. In other markets (e.g. stock, automobile, fine art, local markets, etc.) the negotiation of price or other aspects of the deal depends on the product and the merchant' behaviour. The model we have presented here is intended for extension and application in the near future to manage agent-based e-commerce interactions.

References

1. Alonso, E., How individuals negotiate Societies. In *proc. ICMAS'98*, IEEE Computer Society Press , 1998, pp.: 18-25.
2. Arranz, M.A., Towards a Formal Framework for Coordination in Self-Interested Multi-Agent Domains. In *Proc. of the ICCS'99*, Donostia-San Sebastian, pp.: 75-82, 1999.
3. Arranz, M.A., Interaction Protocols for Agent-Oriented Systems, In Proc. of the *UK-MAS'99*, Bristol, 1999.
4. Arranz, M.A., "Reconciling Plans in Artificial Agent Communities". In Proc. of the *EMCSR 2000*, Austrian Society for Cybernetic Studies, Vienna, pp.: 597-602, 2000.
5. Corkill, D.D. and V. Lesser 83, The Use of Meta-level Control for Coordination in a DPS network. In Proc. of IJCAI-83, Karlsruhe, Germany, 1983, pp. 748-756.
6. Decker, K., Environment Centered Analysis and Design of Coordination Mechanisms. Ph.D. thesis, University of Massachusetts, Amherst, 1995.

7. Decker, K., Coordinating Human and Computer Agents. Conen W. & G. Neumann (eds.) *Coordination Technology for Collaborative Applications-Organizations, Processes and Agents.* LNCS # 1364, pp. 77-98, Springer-Verlag, 1998.
8. Decker, K. and V. Lesser, Generalizing the Partial Global Planning Algorithm. *International Journal of Intelligent and Cooperative Information Systems, 1*: 319-346, 1992.
9. Durfee, E. and V.Lesser, Negotiating task decomposition and allocation using partial global planning. In Gasser, L. & M.N. Huhns (eds.) *Distributed Artificial Intelligence, Vol. II,* Pitman Publishig Ltd., 1989.
10. Ephrati, E and J.S. Rosenschein, Distributed consensus mechanisms for self-interested heterogeneous agents, in *The 1st International Conference on Intelligent and Cooperative Information Systems*, 1993, pp. 71-79.
11. Georgeff, M.P., Communication and Interaction in Multi-Agent Planning. *In Proc. National Conference on Artificial Intelligence*, Washington, DC, 1983, pp.125-129.
12. Klusch, M. (ed.) *Intelligent Information Agents: Agent-Based Information Discovery and Management on the Internet,* Springer-Verlag, Berlin, 1999.
13. Lesser, V., A Retrospective View of FA/C Distributed Problem Solving. In *IEEE Transactions on Systems, Man and Cybernetics 21*: 1347-1362, 1991.
14. von Martial, F., *Coordinating Plans of Autonomous Agents*, Berlin, Springer-Verlag, 1992.
15. Noriega, P. and C. Sierra, Auctions and Multi-Agent Systems. In Klusch, M. (ed.) *Intelligent Information Agents: Agent-Based Information Discovery and Management on the Internet,* Springer-Verlag, Berlin, pp.: 153-175, 1999.
16. Parsons, S., Sierra, C. and N.R. Jennings, Agents that Reason and negotiate by Arguing. *Journal of Logic and Computation 8(3)*: 261-292, 1998.
17. Rosenschein, J.S. and G. Zlotkin, *Rules of Encounter*, Cambridge, MA: MIT Press, 1994.
18. Sandholm, T. and V. Lesser, Issues in Automated Negotiation and Electronic Commerce: Extending the Contract Net Framework, In Lesser, V. (ed.) *1st International Conference on Multi-Agent Systems*, MIT Press, Menlo Park: CA, 1995, pp. 328-335.
19. Smith, R.G., The contract net protocol: high level communication and control in a distributed problem solver. In *IEEE Transactions on Computers 12*, 1980, pp. 1104-113.
20. Smith, R.G. and R. Davies, Frameworks for Cooperation in DPS. In *IEEE Transactions on systems, man and Cybernetics*, vol. SMC-11, n.1, 1981, pp. 61-70.
21. Zlotkin, G. and J.S. Rosenschein, Incomplete Information and Deception in Multi-Agent Negotiation. In *Proceedings of the Twelfth International Joint Conference on Artificial Intelligence*, Sydney, 1991, pp. 225-231.

Some Techniques of Isomorph-Free Search

Thierry Boy de la Tour

LEIBNIZ - IMAG
46, avenue Félix Viallet, 38031 Grenoble Cedex, FRANCE
Thierry.Boy-de-la-Tour@imag.fr

Abstract. The problem of isomorph-free search is analysed in a group theoretic framework, some algorithms are given for building searches in a modular way, and the context in which they can be applied is characterised as generally as possible. These results are applied to the problem of building finite models of monadic first-order sorted logic.

1 Introduction

When searching for objects with specific properties in huge search spaces, it is often impossible to keep the full space within memory, and the searches have to rely on some technique for generating all possible candidates. The efficiency of the search then relies heavily on the precision with which those candidates are generated with respect to the expected property. But the search is necessary because the objects searched for are not to be constructed directly for lack of a dedicated program. The generating mechanism has to retain some generality.

Of course, the objects produced this way can only be computer representations of some possibly more abstract structure. Some structures have direct computer representation, like integers, or strings. Others however do not, in the sense that a unique representation may not easily be found. The typical example is the structure of graphs; though heavily used in computer science and especially in artificial intelligence, the computer representation of graphs is imperfect in the sense that no polynomial algorithm is known to tell whether two representations correspond to the same graph. This problem is known as graph isomorphism (see [3], also for group related topics).

This means that most search procedures are unable to cope with isomorphisms, and generate many isomorphic representations of abstract objects. This is useless as far as the property searched for is *invariant*, i.e. stable on each isomorphism class (but non-invariant properties are mathematically meaningless). This is dreadful when the isomorphism classes are huge. Some computational effort can therefore be profitably spent on isomorph-free generating mechanisms.

But how is this to be done? It is certainly not practical to embed the structure of interest into a fixed one, to generate for example all graphs of a given size if we are only interested in trees. Is there some general way to analyse a structure, and *deduce* from this analysis an isomorph-free search of the structure? The aim of this paper is to provide some ways of doing this.

J.A. Campbell and E. Roanes-Lozano (Eds.): AISC 2000, LNAI 1930, pp. 240–252, 2001.

In the next section we develop the general setting, in which isomorphism classes are given by an equivalence relation. In section 3 we specialise to classes given through operations of groups, which allows the use of group theoretic constructs and algorithms. Section 4 is devoted to the application of these results to the finite models of monadic logic. Conclusion and perspectives are reached at the end of the paper.

2   The General Framework

For a function $f : A \to B$, a partial function $s : A \to A$, $\mathsf{o} \in A$ and an equivalence relation $\cong$ on B, we say that o, s is a $\cong$-*enumeration of* B *through* f iff

$$\begin{cases} \forall x \in B, \exists n \in \mathbb{N} \mid x \cong f(s^n(\mathsf{o})) & \text{(completeness)} \\ \forall n, n' \in \mathbb{N}, \ f(s^n(\mathsf{o})) \cong f(s^{n'}(\mathsf{o})) \Rightarrow n = n' & \text{(frugality)} \end{cases}$$

When $A = B$ and f is the identity, we say that o, s is a $\cong$-*enumeration of* A. We call o the *initial element* of the enumeration, s its *successor function*.

The reason why s is partial is simply to make the last element of the enumeration $l = s^m(\mathsf{o})$ special by letting $s(l)$ be undefined. Then all $s^k(\mathsf{o})$ for $k > m$ are undefined, and the completeness and frugality conditions above are only meaningful for $n, n' \leq m$. This, of course, does not mean that an implementation of s should be non-terminating on l. The undefinedness of $s(l)$ should rather be signalled by returning a special value, or raising an exception, or any other way the programmer may find convenient, and is free to devise.

A trivial example occurs when $\cong$ has only one class on A: then for any $\mathsf{o} \in A$, we have $\mathsf{o}, \emptyset$ as a $\cong$-enumeration of A.

The reason why we need the set A and function f in the definition is simply that the elements of B may not convey sufficient information to compute their $\cong$-successors, and we may need to compute them through f from the information provided in A. Another application is to replace an enumeration of some structure by the enumeration of another more convenient and isomorphic structure, as is made clear in the following simple theorem:

Theorem 1. *If* o, s *is a* $\cong$-*enumeration of* A, $\approx$ *is an equivalence relation on* B *and* $f : A \to B$ *verifies:*

$$\forall x, y \in A, \ f(x) \approx f(y) \Leftrightarrow x \cong y \quad and \quad \forall y \in B, \exists x \in A \mid f(x) \approx y$$

then o, s *is a* $\approx$-*enumeration of* B *through* f.

Proof. We first prove completeness: for $y \in B$, there exists a $x \in A$ such that $f(x) \approx y$. But by completeness of the $\cong$-enumeration, we have $\exists n \in \mathbb{N}$ such that $x \cong s^n(\mathsf{o})$, and then $y \approx f(x) \approx f(s^n(\mathsf{o}))$. We now prove frugality: $\forall n, n' \in \mathbb{N}$, suppose $f(s^n(\mathsf{o})) \approx f(s^{n'}(\mathsf{o}))$, then we have $s^n(\mathsf{o}) \cong s^{n'}(\mathsf{o})$, and by frugality of the $\cong$-enumeration we get $n = n'$.

This theorem is not very surprising due to the strong property of f; it means that we have been able to fully characterise the equivalence classes of B as equivalence classes of A. It is not in itself helpful for decomposing a difficult search problem into simpler ones. The following theorem is more useful in this respect, since it is often the case that some invariant function like, e.g. the number of edges, of vertices, the degree of a graph, etc., can make the search easier if only one value of the invariant function is considered at a time. Given sets A and B with equivalence relations $\cong$ and $\approx$, a function $\iota : A \to B$ is *invariant* if $\forall x, y \in A$, $x \cong y \Rightarrow \iota(x) \approx \iota(y)$.

Theorem 2. *If o, b is a $\approx$-enumeration of B through $g : B' \to B$, $\cong$ is an equivalence relation on A and $\iota : A \to B$ is invariant; if $\forall i \in B$, we have a $\cong$-enumeration o_i, a_i of $A_i = \{x \in A \mid \iota(x) \approx i\}$ through $f_i : A'_i \to A_i$; then we get a $\cong$-enumeration θ, s of A through $h : A' \times B' \to A$ with: $A' = \bigcup_{k \in \mathbb{N}} A'_{g(b^k(o))}$, $\theta = \langle o_{g(o)}, o \rangle$ and:*

$$\forall \langle y, j \rangle \in A' \times B', h(\langle y, j \rangle) = f_{g(j)}(y)$$
$$s(\langle y, j \rangle) = \langle a_{g(j)}(y), j \rangle \text{ if defined, } \langle o_{g(b(j))}, b(j) \rangle \text{ otherwise.}$$

Proof. It should be clear that any $s^n(\theta)$ is a tuple $\langle a_{g(j)}^m(o_{g(j)}), j \rangle$ with $j = b^k(o)$, and conversely that any such tuple can be obtained exactly once as a $s^n(\theta)$.

We first prove completeness: $\forall x \in A$, let $i = \iota(x)$, then $\exists k \in \mathbb{N} \mid i \approx g(b^k(o))$. We have $x \in A_i = A_{g(b^k(o))}$, so let $j = b^k(o)$, we have $\exists m \in \mathbb{N} \mid x \cong f_{g(j)}(a_{g(j)}^m(o_{g(j)}))$, hence from the preceding remark we have $\exists n \in \mathbb{N} \mid s^n(\theta) = \langle a_{g(j)}^m(o_{g(j)}), j \rangle$. We therefore have $h(s^n(\theta)) = f_{g(j)}(a_{g(j)}^m(o_{g(j)})) \cong x$.

Now suppose that $h(s^n(\theta)) \cong h(s^{n'}(\theta))$, which means by the remark above that for some $k, m, k', m' \in \mathbb{N}$ we have $f_i(a_i^m(o_i)) \cong f_{i'}(a_{i'}^{m'}(o_{i'}))$ with $i = g(b^k(o))$ and $i' = g(b^{k'}(o))$. By the property of invariance of ι we get:

$$i \approx \iota(f_i(a_i^m(o_i))) \approx \iota(f_{i'}(a_{i'}^{m'}(o_{i'}))) \approx i'$$

(since $f_i(a_i^m(o_i)) \in A_i$) which implies that $k = k'$ by frugality of o, b, so that $i = i'$, and then that $m = m'$ by frugality of $f_i = f_{i'}$. By the remark above we get $n = n'$.

3   Enumerations Modulo Group Orbits

In the sequel we will have to consider many different relations on sets of rather complex objects, such as functions or some species of graphs, and relate them in some readable way. We will do this through groups and group operations. An *operation* of a group G on a set A is a function $op : A \times G \to A$ such that $\forall x \in A, \forall \sigma, \pi \in G$; we have $op(op(x, \sigma), \pi) = op(x, \sigma\pi)$. The reader should note that we have left implicit the name of the product in $\sigma\pi$; this is usually harmless since only one group product is linked to a symbol G. Similarly, there is usually only one operation linked to a symbol A, and the standard notation x^σ

for $op(x, \sigma)$ consequently leaves op implicit. x^σ does not have the same meaning whether $x \in A$ or $x \in B$. The same holds for some other notations, such as $G_x = \{\sigma \in G \mid x^\sigma = x\}$, the *stabiliser* of x in G (a subgroup of G; this is noted $G_x{<}G$).

Given a group G and its operation on A, the corresponding equivalence relation we will consider on A is the *G-orbit relation* $x \sim_G y \Leftrightarrow \exists \sigma \in G \mid x^\sigma = y$. Its equivalence classes are called *G-orbits* of A. In what follows, G-enumeration stands for $\sim_G$-enumeration.

In case there is no explicit operation on A and we still consider G-enumerations of A, this means that we consider the trivial operation of G on A, defined as $x^\sigma = x$, and that we refer to a $=$-enumeration.

3.1   Enumeration of Standard Products

One usual way of representing complex objects is to see them as tuples of simpler objects, like e.g. a labelled graph consists of a graph and a labelling function. Hence, given enumerations for A and B, we would like to provide an enumeration of $A \times B$: intuition may suggest that the mere product of the enumerations of A and B is sufficient, but this is wrong. Consider as above $o \in A = B$ and G such that $o, \emptyset$ is a G-enumeration of A (i.e. G is *transitive* on A), then $\langle o, o \rangle, \emptyset$ is *not* a G-enumeration of $A \times A$, since obviously the G-orbit of $\langle o, o \rangle$ is limited to the diagonal of $A \times A$. Even if $A \neq B$ the coordinates may not be independent objects, and we have to suitably refine the equivalence relation on B, which is done by means of stabilisers.

Theorem 3. *Considering a group G with operations on A and B, and the standard extension of these operations on $A \times B$ (i.e. $\langle x, y \rangle^\sigma = \langle x^\sigma, y^\sigma \rangle$), if o, a is a G-enumeration of A through $f : A' \to A$ and $\forall H{<}G$, o_H, b_H is a H-enumeration of B through $g : B' \to B$, let $o^\times = \langle o, o_{G_{f(o)}} \rangle$ and let $s : A' \times B' \to A' \times B'$ the partial function defined by:*

$$s(\langle x, y \rangle) = \langle x, b_H(y) \rangle \text{ if defined, } \langle x', o_{H'} \rangle \text{ otherwise,}$$
$$\text{where } x' = a(x), H = G_{f(x)}, H' = G_{f(x')}$$

then $o^\times, s$ is a G-enumeration of $A \times B$ through $\langle f, g \rangle$.

Proof. First, it should be clear that every $s^n(o^\times)$ is of the form $\langle a^m(o), b_H^k(o_H) \rangle$ with $H = G_{f(a^m(o))}$, that m, k is uniquely determined by n, and reciprocally that every such tuple (as long as it is defined) is equal to some $s^n(o^\times)$.

Let $\langle x, y \rangle \in A \times B$. By completeness of o, a we have $\exists m \in \mathbb{N}, \exists \sigma \in G$ such that $x^\sigma = f(a^m(o))$. Let $x' = f(a^m(o))$ and $H = G_{x'}$. We apply the completeness of o_H, b_H to y^σ: $\exists k \in \mathbb{N}, \exists \pi \in H \mid y^{\sigma\pi} = g(b_H^k(o_H))$. We therefore have

$$\langle x, y \rangle^{\sigma\pi} = \langle x'^\pi, y^{\sigma\pi} \rangle = \langle f, g \rangle(\langle a^m(o), b_H^k(o_H) \rangle)$$
$$= \langle f, g \rangle(s^n(o^\times)) \text{ for some } n \in \mathbb{N}$$

which proves completeness since $\sigma\pi \in G$.

Now suppose $\exists \sigma \in G \mid \langle f, g \rangle (s^n(\mathsf{o}^\times))^\sigma = \langle f, g \rangle (s^{n'}(\mathsf{o}^\times))$, and that $s^n(\mathsf{o}^\times) = \langle a^m(\mathsf{o}), b_H^k(\mathsf{o}_H) \rangle$ with $H = G_{f(a^m(\mathsf{o}))}$, and similarly $s^{n'}(\mathsf{o}^\times) = \langle a^{m'}(\mathsf{o}), b_{H'}^{k'}(\mathsf{o}_{H'}) \rangle$ with $H' = G_{f(a^{m'}(\mathsf{o}))}$. By equating coordinates we get

$$f(a^m(\mathsf{o}))^\sigma = f(a^{m'}(\mathsf{o})) \text{ and } g(b_H^k(\mathsf{o}_H))^\sigma = g(b_{H'}^{k'}(\mathsf{o}_{H'})).$$

From the first we get $m = m'$, thus $H = H'$ and $\sigma \in H$. Hence from the second we get $k = k'$, and this proves $n = n'$.

It should be noted that this theorem encompasses the case where the coordinates are independent, since then the operation of $G_{f(x)}$ on B is exactly the operation of G on B. The successor function for $A \times B$ then exactly corresponds to the intuitive enumeration of the Cartesian product of the set of canonical elements obtained by the enumerations of each coordinate. The wise programmer would however implement a special successor function for this independent product in order to avoid the useless computation of $G_{f(x)}$.

3.2   Refinement to a Subgroup

In the previous theorem we have used a family of enumerations of B, one for each $H < G$. We will now prove that it is possible to compute each of these from the enumeration relative to G, by use of double cosets. For any H, K subgroups of G and any $\sigma \in G$, we call *double coset* of σ the set $H\sigma K = \{\rho\sigma\pi \mid \rho \in H, \pi \in K\}$. It is well-known that for fixed H, K, the double cosets form a partition of G. In other words, the relation $\approx_H^K$ defined by $\forall \sigma, \sigma' \in G$, $\sigma \approx_H^K \sigma' \Leftrightarrow H\sigma K = H\sigma' K$ is an equivalence relation.

Theorem 4. *If o, a is a G-enumeration of A through $f : A' \to A$, H is a subgroup of G, and $\forall x \in A'$, α_x, d_x is a $\approx_{G_{f(x)}}^H$ -enumeration of G, let $\theta = \langle \mathsf{o}, \alpha_\mathsf{o} \rangle$, let $g : A' \times G \to A$ the function defined by $g(\langle x, \sigma \rangle) = f(x)^\sigma$, and*

$$s(\langle x, \sigma \rangle) = \langle x, d_x(\sigma) \rangle \text{ if defined, } \langle a(x), \alpha_{a(x)} \rangle \text{ otherwise;}$$

then θ, s is a H-enumeration of A through g.

Proof. As above, we have a 1-1 correspondence between the n's and the m, k's such that $s^n(\theta) = \langle x, d_x^k(\alpha_x) \rangle$ with $x = a^m(\mathsf{o})$. We now prove completeness.

$\forall x \in A, \exists m \in \mathbb{N}, \exists \sigma \in G \mid x^\sigma = f(a^m(\mathsf{o}))$ by completeness of o, a. Let $x' = a^m(\mathsf{o}) \in A'$, by completeness of $\alpha_{x'}, d_{x'}$ applied to σ^{-1}, we have $\exists k \in \mathbb{N}$ such that $d_{x'}^k(\alpha_{x'}) \in G_{f(x')}\sigma^{-1}H$, thus $\exists \rho \in G_{f(x')}, \exists \pi \in H \mid d_{x'}^k(\alpha_{x'}) = \rho\sigma^{-1}\pi$. By the remark above we have $\exists n \in \mathbb{N} \mid s^n(\theta) = \langle x', \rho\sigma^{-1}\pi \rangle$, thus $g(s^n(\theta)) = f(x')^{\sigma^{-1}\pi} = x^\pi$, which proves completeness.

$\forall n, n' \in \mathbb{N}$, suppose $\exists \pi \in H \mid g(s^{n'}(\theta))^\pi = g(s^n(\theta))$. Let m, k such that $s^n(\theta) = \langle x, \sigma \rangle$ with $x = a^m(\mathsf{o}), \sigma = d_x^k(\alpha_x)$, and m', k' such that $s^{n'}(\theta) = \langle x', \sigma' \rangle$ with $x' = a^{m'}(\mathsf{o}), \sigma' = d_{x'}^{k'}(\alpha_{x'})$. We then have $f(x')^{\sigma'\pi} = f(x)^\sigma$, i.e. $f(x')^{\sigma'\pi\sigma^{-1}} = f(x)$. Since $\sigma'\pi\sigma^{-1} \in G$, by frugality of o, a we get $m = m'$, and then $x = x'$. Hence $\sigma'\pi\sigma^{-1} \in G_{f(x)}$, and $\sigma' \in G_{f(x)}\sigma H$. By frugality of α_x, d_x this implies $k = k'$, and we get $n = n'$.

3.3   Permuting Coordinates

The standard operation on a Cartesian product is not always convenient, and the fair representation of a complex structure as a tuple may require the possibility to allow some permutation of coordinates. For example, forests of k trees of size n can be represented as k-tuples of trees, but the ordering of coordinates imposed by the Cartesian product is too restrictive to encompass forest isomorphisms.

More generally, given a group G operating on A and $H < S_n$ (where S_n is the symmetric group on $\{1..n\}$), we will consider (in some special cases) the independent operation of G on the coordinates of A^n, mixed with the operation of H on the coordinate indices. This operation is actually linked with the wreath product $G \wr H$, but since defining $G \wr H$ and then its operation on A^n (which is *not* the standard extension of an operation on A) would be tedious, we only define the equivalence relation on A^n induced by them: $\langle x_1, \ldots, x_n \rangle \sim_{G \wr H} \langle y_1, \ldots, y_n \rangle \Leftrightarrow \exists \sigma_1, \ldots, \sigma_n \in G, \exists \pi \in H \mid \forall i \in \{1..n\}, x_i^{\sigma_i} = y_{i\pi}$.

We have a trivial case with $H = I$. The operation of $G \wr I$ on A^n is however similar to the standard operation of $G \times \ldots \times G$ on A^n (the Cartesian product of groups is naturally defined with the componentwise product, e.g. $\langle \sigma, \pi \rangle \langle \sigma', \pi' \rangle = \langle \sigma\sigma', \pi\pi' \rangle$), for which theorem 3 provides an enumeration. Of course, the remark following the proof of theorem 3 applies, since in the Cartesian product of groups coordinates are independent.

This trivial case can be generalised in the following way: suppose there exists a partition $X_1, \ldots, X_k$ of $\{1..n\}$ and $K_1, \ldots, K_k$ where each K_i is a permutation group on X_i, such that H is isomorphic to $K_1 \times \ldots \times K_k$ (this can be easily tested on a generator set of H). Then the operation of $G \wr H$ on A^n is similar to the operation of $(G \wr K_1) \times \ldots \times (G \wr K_k)$ on $A^{|X_1|} \times \ldots \times A^{|X_k|}$, and we can compose the $G \wr K_i$-enumerations through theorem 3.

Of course, not every H can be decomposed in this way, and it seems very difficult to devise a general $G \wr H$-enumeration of A^n. Below, we will solve only two special cases: the symmetric product with $H = S_n$ and the cyclic product with $H = C_n$, which is the group generated by the permutation $(1\ 2 \ldots n)$.

3.4   The Symmetric Product

Theorem 5. *If* o, a *is a G-enumeration of A through* $f : A' \to A$, *and* $n \in \mathbb{N}$, *let* $g : A'^n \to A^n$ *be the standard componentwise extension of f,* $\mathsf{o}^\times = \langle \mathsf{o}, \ldots, \mathsf{o} \rangle \in A'^n$ *and* $s(\langle x_1, \ldots, x_n \rangle) = \langle x_1, \ldots, x_{k-1}, x_k', \ldots, x_k' \rangle$ *where* $k = \max\{i \geq 1 \mid a(x_i)$ *is defined*$\}$ *and* $x_k' = a(x_k)$. *Then* $\mathsf{o}^\times, s$ *is a $G \wr S_n$-enumeration of A^n through g.*

Proof. It is clear than $\forall m \in \mathbb{N}, s^m(\mathsf{o}^\times)$ is of the form $\langle a^{m_1}(\mathsf{o}), \ldots, a^{m_n}(\mathsf{o}) \rangle$ with $m_1 \leq \cdots \leq m_n$, and conversely that every such tuple (as long as it is defined) is obtained as a $s^m(\mathsf{o}^\times)$.

$\forall \langle x_1, \ldots, x_n \rangle \in A^n$, by completeness of o, a we have $\exists m_1, \ldots, m_n \in \mathbb{N}$, $\exists \sigma_1, \ldots, \sigma_n \in G \mid \forall i, x_i^{\sigma_i} = f(a^{m_i}(\mathsf{o}))$. By sorting the list $m_1, \ldots, m_n$ we get a permutation $\pi \in S_n$ such that $m_{1\pi} \leq \cdots \leq m_{n\pi}$, and then by the previous remark a m such that $s^m(o^\times) = \langle a^{m_{1\pi}}(o), \ldots, a^{m_{n\pi}}(o) \rangle$, and therefore $\langle x_1, \ldots, x_n \rangle \sim_{G \wr S_n} g(s^m(o^\times))$.

Now suppose $g(s^m(o^\times)) \sim_{G \wr S_n} g(s^{m'}(o^\times))$. For some $m_1 \leq \cdots \leq m_n$ and $m'_1 \leq \cdots \leq m'_n$ we then have $\langle f(a^{m_1}(o)), \ldots, f(a^{m_n}(o)) \rangle \sim_{G \wr S_n} \langle f(a^{m'_1}(o)), \ldots, f(a^{m'_n}(o)) \rangle$, which means $\exists \sigma_1, \ldots, \sigma_n \in G, \exists \pi \in S_n$ such that $\forall i, f(a^{m_i}(o))^{\sigma_i} = f(a^{m'_{i\pi}}(o))$, and by frugality of o, a we get $m_i = m'_{i\pi}$. Therefore we have $m'_{1\pi} \leq \cdots \leq m'_{n\pi}$ while $m'_1 \leq \cdots \leq m'_n$, which means that $m'_i = m'_{i\pi} = m_i$ and by the remark above we get $m = m'$.

3.5   The Cyclic Product

The case $H = C_n$ is much more difficult than the previous one, even though we will come up with a surprisingly simple algorithm for the successor function. In order to keep proofs readable (well, sort of...) we need to develop this algorithm and the corresponding proofs on strings on a finite alphabet V, on which we suppose an =-enumeration o, a. The strict linear ordering $\prec$ of V is defined as $a^i(o) \prec a^j(o) \Leftrightarrow i < j$. The link with our general framework will be through the implicit isomorphism between A^n and the strings of length n.

For $x \in V^n$, we note $|x| = n$ the *length* of x. Then $\forall i \in \{1..n\}$, x_i denotes the ith letter in x, so that $x = x_1 \ldots x_n$. The empty string is noted ε. For $j \in \{1..n\}$, $x_{i,j}$ denotes $x_i \ldots x_j$ if $i \leq j$, and ε otherwise. $x_{1,i}$ is a *prefix* of x. For $y \in V^n$, we note $x \sqcap y$ the greatest (w.r.t. length) common prefix of x and y. The strict lexicographic ordering is defined[1] as $x \sqsubset y \Leftrightarrow |x \sqcap y| < |x|$ and $x_{1+|x \sqcap y|} \prec y_{1+|x \sqcap y|}$. We give without proofs the three following properties:

$$\forall x, y \in V^n, \forall m \in \mathbb{N}, \ x \sqsubseteq y \Rightarrow x^m \sqsubseteq y^m \tag{1}$$

$$\forall x, x' \in V^n, \forall y, y' \in V^m, \ xy \sqsubseteq x'y' \Rightarrow x \sqsubseteq x' \tag{2}$$

$$\forall x, x' \in V^n, \forall y, y' \in V^m, \ x \sqsubset x' \Rightarrow xy \sqsubset x'y' \text{ and } yx \sqsubset yx' \tag{3}$$

A less obvious property that we will need is:

Lemma 1. $\forall x, x', y, y' \in V^n$, if $x \sqsubseteq y, y \sqsubset y'$ and $|x \sqcap x'| > |y \sqcap y'|$ then $x' \sqsubset y'$.

Proof. let $k = |y \sqcap y'| + 1$, we have $y_{1,k} \sqsubset y'_{1,k}$, but $x'_{1,k} = x_{1,k} \sqsubseteq y_{1,k}$ by (2), and therefore $x'_{1,k} \sqsubset y'_{1,k}$, and $x' \sqsubset y'$ by (3).

In the reasoning below we will need a rather unusual operator on strings: for $x \in V^k$ and $n \in \mathbb{N}$, $\overrightarrow{x}^n$ is the string of length n defined by $\forall i, (\overrightarrow{x}^n)_i = x_{(i-1 \bmod k)+1}$. Hence x is simply repeated as much as necessary to get a string of length n.

For $x \in V^n$, let $\kappa(x) = \max\{i \geq 1 \mid a(x_i) \text{ is defined}\}$, which may obviously be undefined (in which case x is maximal w.r.t $\sqsubseteq$). This is the index at which x should be increased in order to get its successor in the lexicographic order. Now,

---

[1] Yes, we will only compare strings of the same length

let $\nu(x) = \overrightarrow{x_{1,k-1}a(x_k)}^{\,n}$, where $k = \kappa(x)$. It is clear that $x \sqsubseteq \nu(x)$, and that the smaller k is, the bigger the interval between x and $\nu(x)$ is likely to be.

We say that x has the *smallest prefix property*, or s.p.p. in short, iff $\forall i, j \geq 1$, if $i + j \leq |x| + 1$ then $x_{1,j} \sqsubseteq x_{i,i+j-1}$. This property is preserved by ν:

Lemma 2. *if $x \in V^n$ has the s.p.p. then so does $\nu(x)$ if defined.*

Proof. let $k = \kappa(x), z = \nu(x), i, j \geq 1 \mid i + j \leq |z| + 1$, and $r = (i - 1 \bmod k) + 1$; we have $z_{i,i+j-1} = z_{r,r+j-1}$ by definition of ν. Hence we have to prove that $z_{1,j} \sqsubseteq z_{r,r+j-1}$. If $r = 1$ this is obvious, so suppose $r > 1$ and consider two cases.

If $r + j - 1 < k$ then $z_{r,r+j-1} = x_{r,r+j-1} \sqsupseteq x_{1,j}$ since x has the s.p.p. But here $j < k$, thus $z_{1,j} = x_{1,j}$.

If $r + j - 1 \geq k$ then $x_{r,r+j-1} \sqsubseteq z_{r,r+j-1}$, and $|x_{r,r+j-1} \sqcap z_{r,r+j-1}| = k - r \leq j - 1$, and we also have $k - r < k - 1$, so that $|x_{1,j} \sqcap z_{1,j}| = \min(j, k - 1) > |x_{r,r+j-1} \sqcap z_{r,r+j-1}|$. We still have $x_{1,j} \sqsubseteq x_{r,r+j-1}$, hence by lemma 1 we get $z_{1,j} \sqsubseteq z_{r,r+j-1}$.

In order to turn ν into a successor function among the strings that have the s.p.p. we still have to prove that they will all be reached. This is where the use of a linear ordering on strings shows its convenience.

Lemma 3. *If $x \sqsubset y \sqsubset \nu(x)$ then y does not have the s.p.p.*

Proof. let $z = \nu(x), n = |x|$ and $k = \kappa(x)$. By definition of κ we have $x_{1,k} \sqsubset y_{1,k}$, and from $y \sqsubseteq z$ by (2) we get $y_{1,k} \sqsubseteq z_{1,k}$. Hence by definition of ν we have $y_{1,k} = z_{1,k}$. When we let $j = |y \sqcap z| + 1$, we obviously have $y_j \prec z_j$. The division of $j - 1$ by k yields $j - 1 = qk + r$, and we have $y_{qk+1,j} \sqsubset z_{qk+1,j}$. But $z = \overrightarrow{z_{1,k}}^{\,n} = \overrightarrow{y_{1,k}}^{\,n}$, thus $z_{qk+1,j} = y_{1,r+1}$.

We say that $x \in V^n$ is *minimal* iff $\forall i \leq n$, $x \sqsubseteq x_{i,n}x_{1,i-1}$. This really means that x is the smallest in the set of strings that can be obtained from x by shifting round its letters, i.e. in its C_n-orbit. It is obvious that all minimal strings have the s.p.p.: let j such that $i + j \leq n + 1$; then from $x \sqsubseteq x_{i,n}x_{1,i-1}$ we get $x_{1,j} \sqsubseteq x_{i,i+j-1}$ by (2). The reverse however is not true, and among the powers of ν from o^n (the smallest minimal string of length n), we still have to test for the minimal ones. From the definition of minimality this seems to require the actual use of the ordering relation $\prec$ on V. But remember that we have *not* defined any orders on the enumerated sets, and that we are only allowed successor functions and initial elements. Fortunately, this test can be replaced by a much simpler and more efficient one.

Lemma 4. *if x has the s.p.p. then $\nu(x)$ is minimal iff $|x| \bmod \kappa(x) = 0$.*

Proof. let $n = |x|, k = \kappa(x), z = \nu(x)$. We first prove the if part; let $i \leq n$, if $i = 1$ then $z \sqsubseteq z_{i,n}z_{1,i-1}$ is trivial, so we take $i > 1$. We have $k \leq n$ and n is a multiple of k. First suppose $k = n$; then $z_{1,n-i+1} = x_{1,n-i+1} \sqsubseteq x_{i,n}$ since x has s.p.p.; we have $x_{i,n} \sqsubseteq z_{i,n}$ by definition of ν, and therefore $z \sqsubseteq z_{i,n}z_{1,i-1}$ by (3).

Now supposing $n = mk$ with $m \geq 2$, we have $z = (z_{1,k})^m$. If we let $r = i - 1 \bmod k$, we have $z_{i,n} z_{1,i-1} = z_{r+1,n} z_{1,r} = (z_{r+1,r+k})^m$ (since we have $r + k < 2k \leq n$). By lemma 2, z has the s.p.p. so that $z_{1,k} \sqsubseteq z_{r+1,r+k}$, thus by (1) we get $z \sqsubseteq z_{i,n} z_{1,i-1}$.

We now prove the converse of the only if part, i.e. the division of n by k now yields $n = qk + r$ with $0 < r < k$. Let $i = qk + 1$, we have $z_{i,n} = z_{1,r}$ by definition of ν. We also have $z_{1,k-r} = x_{1,k-r} \sqsubseteq x_{r+1,k}$ since x has s.p.p. and $x_{r+1,k} \sqsubseteq z_{r+1,k}$ by definition of ν. Therefore, $z_{i,n} z_{1,k-r} = z_{1,r} z_{1,k-r} \sqsubseteq z_{1,r} z_{r+1,k}$ by (3), thus $z_{i,n} z_{1,i-1} \sqsubseteq z$ still by (3), which proves that z is not minimal.

We may now define a successor function s for minimal strings, in a recursive way:

$$s(x) = \text{if } |x| \bmod \kappa(x) = 0 \text{ then } \nu(x) \text{ else } s(\nu(x)).$$

Theorem 6. $\{s^m(o^n) \mid m \in \mathbb{N}\}$ *is the set of minimal strings of* V^n.

Proof. If $x \in V^n$ is minimal, then it has the s.p.p., and so do the $\nu^k(x)$ by lemma 2, and they are the only elements greater than x having the s.p.p. by lemma 3. Hence by lemma 4 $s(x)$ is minimal and no y is minimal if $x \sqsubset y \sqsubset s(x)$. Let l be the last element in the enumeration o, a; since o^n and l^n are minimal, then $\exists m \in \mathbb{N} \mid s^m(o^n) = l^n$. There is no element of V^n greater than l^n, and it is clear that $s(l^n)$ is undefined, hence all minimal elements of V^n are in the set $\{s^m(o^n) \mid m \in \mathbb{N}\}$, and only them.

Termination of s can not be questioned from what precedes, but we may still raise some suspicion about the complexity of the number of recursive calls, so let us relieve them at once.

Lemma 5. *if x has s.p.p. and $\kappa(x) < |x|$ then $\kappa(\nu(x)) > \kappa(x)$*

Proof. let $k = \kappa(x)$ and $z = \nu(x)$, since x has the s.p.p. we have $x_1 \preceq x_k$, and by definition of κ we know that $a(x_k)$ is defined. Since $k < |x|$ we have $z_{k+1} = x_1$, therefore $a(z_{k+1})$ is defined, and $\kappa(z) \geq k + 1$.

Hence the number of recursive calls is bounded by $|x|$. It is easy to see that the complexity of s is $O(n^2 k' + nk)$ where n is the length of the input string x (each letter counted as 1), k the maximal time for computing $a(x_i)$ *if defined*, and k' the time taken by a on the last element l. k and k' are independent of n. This bound can be reached; the reader may check that on $x = ol^{n-1}ol^n$ of length $2n+1$ prime, which is clearly minimal, the computation of $s(x) = \nu^{n+1}(x) = ol^n a(o)^{n+1}$ performs $n + 1$ computations of $a(o)$ and $\frac{n(n+1)}{2}$ computations of $a(l)$.

Theorem 7. *If o, a is a G-enumeration of A through $f : A' \to A$, and $n \in \mathbb{N}$, let g be the componentwise extension of f, $o^n = \langle o, \ldots, o \rangle \in A'^n$, and s defined as above, then o^n, s is a $G \wr C_n$-enumeration of A^n through g.*

Proof. Every defined $s^m(o^n)$ is obviously of the form $\langle a^{m_1}(o), \ldots, a^{m_n}(o) \rangle$, and can be considered as a string on $V = \{a^k(o) \mid k \in \mathbb{N}\}$, so that the framework used above applies.

Letting $x \in A^n$, we have $\exists m_1, \ldots, m_n \in \mathbb{N}, \exists \sigma_1, \ldots, \sigma_n$ such that $\forall i$, $x_i^{\sigma_i} = f(a^{m_i}(o))$. Let $\pi \in C_n$, and $m_i' = m_{i^\pi}$ such that $a^{m_1'}(o) \ldots a^{m_n'}(o)$ is minimal (i.e. the smallest w.r.t $\sqsubseteq$ of the strings that can be obtained this way). By theorem 6, $\exists k \in \mathbb{N} \mid s^k(o^n) = a^{m_1'}(o) \ldots a^{m_n'}(o)$, hence $g(s^k(o^n)) = \langle f(a^{m_1'}(o)) \ldots f(a^{m_n'}(o)) \rangle \sim_{G \wr C_n} x$, which proves completeness.

Suppose $g(s^m(o^n)) \sim_{G \wr C_n} g(s^{m'}(o^n))$, which by the remark above translates to $\langle f(a^{m_1}(o)), \ldots, f(a^{m_n}(o)) \rangle \sim_{G \wr C_n} \langle f(a^{m_1'}(o)), \ldots, f(a^{m_n'}(o)) \rangle$, and then to $\exists \sigma_1, \ldots, \sigma_n \in G, \exists \pi \in C_n$ such that $\forall i$, $f(a^{m_i}(o))^{\sigma_i} = f(a^{m_{i^\pi}'}(o))$. By frugality of o, a we get $\forall i$, $m_i = m_{i^\pi}'$. Hence $s^m(o^n)$ and $s^{m'}(o^n)$ are both in the same C_n-orbit, and since they are minimal by theorem 6, they must be equal. Since $m < m' \Leftrightarrow s^m(o^n) \sqsubset s^{m'}(o^n)$, we must have $m = m'$.

4 Application to Finite Monadic Sorted Algebras

Given a finite set $\mathcal{S}$ of *sorts*, $\forall s, t \in \mathcal{S}$, we will consider the *monadic types* s (for constants of type s), $s \rightarrow o$ (for predicates) and $s \rightarrow t$ (for functions); a *monadic signature* Σ is a tuple of monadic types. A finite Σ-*algebra* $\mathcal{A}$ is given by a family of finite carrier sets $(\mathcal{A}_s)_{s \in \mathcal{S}}$ such that $\forall s, t \in \mathcal{S}$, $\mathcal{A}_s \neq \emptyset$ and $s \neq t \Rightarrow \mathcal{A}_s \cap \mathcal{A}_t = \emptyset$, and by a tuple A such that $\forall i$, if $\Sigma_i = s$ then $A_i \in \mathcal{A}_s$, if $\Sigma_i = s \rightarrow o$ then $A_i \subseteq \mathcal{A}_s$, and if $\Sigma_i = s \rightarrow t$ then A_i is a function from $\mathcal{A}_s$ to $\mathcal{A}_t$.

Two Σ-algebras $\mathcal{A}$ and $\mathcal{B}$ are *isomorphic* iff there is a function σ such that $\forall s \in \mathcal{S}$, σ is bijective from $\mathcal{A}_s$ to $\mathcal{B}_s$, and $\forall i$, if $\Sigma_i = s$ then $\sigma(A_i) = B_i$, if $\Sigma_i = s \rightarrow o$ then $\sigma(A_i) = \{\sigma(x) \mid x \in A_i\} = B_i$, and if $\Sigma_i = s \rightarrow t$ then $\sigma(A_i) = \{\langle \sigma(x), \sigma(y) \rangle \mid \langle x, y \rangle \in A_i\} = B_i$.

Since the set of finite Σ-algebras is infinite, we will only consider the isomorphism relation within the finite set $\mathfrak{A}$ of Σ-algebras for a fixed family of carrier sets $(\mathcal{A}_s)_{s \in \mathcal{S}}$. Then the σ above are permutations of $\biguplus_{s \in \mathcal{S}} \mathcal{A}_s$, and more precisely are the elements of the product $\mathcal{G}$ of the Sym $\mathcal{A}_s$. Then the formulas for $\sigma(A_i)$ above can easily be proved to define operations of $\mathcal{G}$ on the corresponding sets (depending on Σ_i) and can therefore be noted A_i^σ. The sets operated upon will be noted: $\mathcal{C}_s$ for the set of constants of type s, i.e. $\mathcal{C}_s = \mathcal{A}_s$, $\mathcal{P}_s$ the power set of $\mathcal{A}_s$, and $\mathcal{F}_s^t = \mathcal{A}_t^{\mathcal{A}_s}$. Remark that if A_i is a function f, we have $\forall x, f^\sigma(x^\sigma) = f(x)^\sigma$.

Then $\mathfrak{A}$ is a Cartesian product of sets $\mathcal{C}_s, \mathcal{P}_s, \mathcal{F}_s^t$, and $\forall \mathcal{A}, \mathcal{B} \in \mathfrak{A}$, $\mathcal{A}$ is isomorphic to $\mathcal{B}$ iff $\exists \sigma \in \mathcal{G} \mid \forall i, A_i^\sigma = B_i$, i.e. $A^\sigma = B$ under the standard extension of the previous operations to tuples. Therefore, theorems 3 and 4 provide a $\mathcal{G}$-enumeration of $\mathfrak{A}$ from $\mathcal{G}$-enumerations of the sets $\mathcal{C}_s, \mathcal{P}_s, \mathcal{F}_s^t$.

Finding a $\mathcal{G}$-enumeration of $\mathcal{C}_s$ is trivial, and any $x \in \mathcal{C}_s$ may serve as initial element since its $\mathcal{G}$-orbit is $\mathcal{C}_s$, and then $x, \emptyset$ is such an enumeration. The case of predicates is scarcely less trivial, since any two sets of same cardinality can be put in 1-1 correspondence; let $n = |\mathcal{A}_s|$, o, a be any =-enumeration of $\mathbb{N}_n = \{0..n\}$, and a function $f : \mathbb{N}_n \rightarrow \mathcal{P}_s$ such that $\forall i \in \mathbb{N}_n$, $|f(i)| = i$, then by theorem 1, o, a is a $\mathcal{G}$-enumeration of $\mathcal{P}_s$ through f.

4.1   Enumeration of $\mathcal{F}_s^t$ with $s \neq t$

For any two $k, n \in \mathbb{N}$ such that $0 < k \leq n$, a k-*partition* of n is a tuple $p \in \mathbb{N}^k$ such that $0 < p_1 \leq \ldots \leq p_k$ and $\sum_{i=1}^k p_i = n$; we note $\mathrm{Part}_k\, n$ the set of k-partitions of n. For example, $\mathrm{Part}_3\, 6 = \{\langle 1, 1, 4\rangle, \langle 1, 2, 3\rangle, \langle 2, 2, 2\rangle\}$.

With any $f \in \mathcal{F}_s^t$ we associate the cardinality $\iota(f)$ of its image $f(\mathcal{A}_s) \subseteq \mathcal{A}_t$. We have $1 \leq \iota(f) \leq \min n, n'$, where n is the cardinality of $\mathcal{A}_s$ and n' the cardinality of $\mathcal{A}_t$, and it is easy to see that $\forall \sigma \in \mathcal{G}$, $\iota(f^\sigma) = \iota(f)$, i.e. that ι is invariant. Since it is trivial to provide a =-enumeration of integers from 1 to $\min n, n'$, in order to obtain a $\mathcal{G}$-enumeration of $\mathcal{F}_s^t$ we need only, by theorem 2, provide a $\mathcal{G}$-enumeration of $F_k = \{f \in \mathcal{F}_s^t \mid \iota(f) = k\}$ for $1 \leq k \leq \min n, n'$.

For any $f \in F_k$ and any $y \in f(\mathcal{A}_s)$, the cardinality $c_f(y)$ of $f^{-1}(y)$ is a non-zero integer, and we have $\sum_{y \in f(\mathcal{A}_s)} c_f(y) = n$, so that by sorting the $c_f(y)$'s we may associate with f a partition $\mathrm{p}(f) \in \mathrm{Part}_k\, n$. But $\forall \sigma \in \mathcal{G}$, we have $(f^\sigma)^{-1}(y^\sigma) = f^{-1}(y)^\sigma$, so that $c_{f^\sigma}(y^\sigma) = c_f(y)$, and the integers obtained by c_{f^σ} and c_f are the same even though they are not obtained in the same order, which is removed by the sorting: $\mathrm{p}(f^\sigma) = \mathrm{p}(f)$.

The converse also holds; if $\mathrm{p}(f) = \mathrm{p}(g)$, then there is a $\pi \in \mathrm{Sym}\, \mathcal{A}_t$ such that $\forall y \in \mathcal{A}_t$, $c_f(y) = c_g(y^\pi)$, and then $\exists \sigma_y \in \mathrm{Sym}\, f^{-1}(y) \mid f^{-1}(y)^{\sigma_y} = g^{-1}(y^\pi)$. Let $\sigma = \pi \prod_{y \in \mathcal{A}_t} \sigma_y \in \mathcal{G}$, then $\forall x \in \mathcal{A}_s$, let $y = f(x)$, then $x^{\sigma_y} \in g^{-1}(y^\pi)$, and then $g(x^{\sigma_y}) = y^\pi = f(x)^\pi$. Since the $f^{-1}(y)$ and $\mathcal{A}_t$ are disjoint, this translates to $g(x^\sigma) = f(x)^\sigma$, which means that $g = f^\sigma$.

Hence by taking a function $\mathrm{q} : \mathrm{Part}_k\, n \rightarrow F_k$ inverse to p, we are in a position to apply theorem 1, so that any =-enumeration of $\mathrm{Part}_k\, n$, which need not be described here, is a $\mathcal{G}$-enumeration of F_k through q.

4.2   Enumeration of $\mathcal{F}_s^s$

The analysis above fails for $\mathcal{F}_s^s$ since the domain and image of these functions are no longer disjoint, and therefore cannot be considered independently. For $f \in \mathcal{F}_s^s \subseteq \mathcal{A}_s \times \mathcal{A}_s$, we can see f as a directed graph on $\mathcal{A}_s$, though a special one since each vertex has exactly one edge coming out of it. The structure of these *function graphs* is easy to analyse.

Let $x_0 \in \mathcal{A}_s$, define $x_i = f^i(x_0)$. Since $\mathcal{A}_s$ is finite, $\exists i \leq j \mid f(x_j) = x_i$, i.e. there is a cycle in f. Let f' be the graph obtained from f by removing the edge from x_j to x_i; this operation does not disconnect any vertex from x_0, but the connex component of x_0 in f' has one more vertex than edges, and is therefore a tree. Hence the connex component of x_0 in f consists of trees connected to the cycle $x_i, \ldots, x_j$.

Let n be the cardinality of $\mathcal{A}_s$. The number of connex components of f may vary from 1 to n (for the identity function), and is clearly an invariant, so that theorem 2 applies, and we are reduced to enumerating functions with a fixed number k of connex components. If $n_1, \ldots, n_k$ are their sizes, we have $\sum_{i=1}^{k} n_i = n$, so that with each f we may associate an element of $\text{Part}_k n$, and this is an invariant. Then we are once again able to restrict the enumeration to function graphs corresponding to a fixed partition $p \in \text{Part}_k n$.

We call $\mathcal{V}_i$ the set of the connected function graphs of size i. Consider the case where $k = 2$ and $p_1 \neq p_2$. Then we have to find a $\mathcal{G}$-enumeration of $\mathcal{V}_{p_1} \times \mathcal{V}_{p_2}$, and $\langle f, g \rangle$ is isomorphic to $\langle f', g' \rangle$ iff $\exists \sigma \in \mathcal{G} \mid \langle f^\sigma, g^\sigma \rangle = \langle f', g' \rangle$, (since f and g are disjoint), hence we may apply theorem 3, taking into account the independence of coordinates. If however $p_1 = p_2$, then the coordinates may be swapped, and we are in a position to perform a symmetric product of $(\mathcal{V}_{p_1})^2$, according to theorem 5. This is easily generalised to any k, by grouping the identical p_i's. Our problem is therefore reduced to finding a $\mathcal{G}$-enumeration of $\mathcal{V}_i$.

A trivial invariant of elements of $\mathcal{V}_i$ is the length c of their cycle. Another invariant is the partition of the sizes of the trees grafted to the cycles, although this time the order in which these sizes are distributed is relevant. For $p \in \text{Part}_c i$ it is sufficient to consider all possible permutations of p's coordinates that are minimal (w.r.t. lexicographic ordering) in their C_c-orbit. When the tuples p thus obtained have a period $d < c$, i.e. d is the smallest integer such that $p^{(12\ldots c)^d} = p$, then we are in position to perform a d-fold cyclic product of $\frac{c}{d}$-tuple of trees. All that is needed after this is the enumeration of trees of a given size, which is very similar to the enumeration of connex components, though in a recursive way.

5   Conclusion and Future Work

This analysis has led to the implementation in OCAML-2 of a system for isomorph-free generation of these monadic algebras. The system, called BiGFooT, cannot be presented here for lack of space. Let us mention one application though: by considering two sorts e and v and considering two functions from e to v, we can generate the directed multigraphs with fixed number of edges and vertices. Labels can be added by way of predicates, on v and e as well.

Many things remain to be done in this line of work. It is first necessary to extend the work done in section 3 at least to dyadic logic. Dyadic functions and relations are as complex as graphs (see [1]), and we will certainly need some work in the line of [2]. We would also like to extend our theoretical framework to show how automorphism groups can be efficiently computed directly by the successor functions, as it is done in BiGFooT. And this system has to be completed: pruning techniques in the line of [4] should be included, etc.

References

1. Thierry Boy de la Tour. On the complexity of finite sorted algebras. In Ricardo Caferra and Gernot Salzer, editors, *Automated Deduction in Classical and Non-Classical Logics*, Lecture Notes in Artificial Intelligence, LNCS 1761, pages 95–108. Springer Verlag, 1999.

2. Leslie Ann Henderson. Efficient algorithms for listing unlabelled graphs. Technical Report CSR-7-90, University of Edinburgh, Department of Computer Science, 1990.
3. C. Hoffmann. *Group-theoretic algorithms and graph isomorphism.* Lecture Notes in Computer Science 136. Springer Verlag, 1981.
4. Nicolas Peltier. A new method for automated finite model building exploiting failures and symmetries. *Journal of Logic and Computation,* 4(8):511–543, 1998.

Author Index

Lecture Notes in Artificial Intelligence (LNAI)

Lecture Notes in Computer Science